11,519

COMMONWEALTH OF AUSTRALIA

DEPARTMENT OF NATIONAL DEVELOPMENT

BUREAU OF MINERAL RESOURCES, GEOLOGY AND GEOPHYSICS

BULLETIN No. 122

DEVONIAN AND CARBONIFEROUS BRACHIOPODS FROM THE BONAPARTE GULF BASIN, NORTHWESTERN AUSTRALIA

BY

JOHN ROBERTS

Vol. 1: TEXT

Issued under the Authority of the Hon. R. W. Swartz, M.B.E., E.D.

Minister for National Development

1971

A

COMMONWEALTH OF AUSTRALIA

DEPARTMENT OF NATIONAL DEVELOPMENT

MINISTER: THE HON. R. W. SWARTZ, M.B.E., E.D.

SECRETARY: L. F. BOTT, D.S.C.

BUREAU OF MINERAL RESOURCES, GEOLOGY AND GEOPHYSICS

DIRECTOR: N. H. FISHER

THIS BULLETIN WAS PREPARED IN THE GEOLOGICAL BRANCH

ASSISTANT DIRECTOR: J. N. CASEY

ISBN 0 642 98380 1

Published by the Bureau of Mineral Resources, Geology and Geophysics
Canberra, A.C.T.

Printed by Mercury-Walch Pty Ltd, 5-7 Bowen Road, Moonah, Tasmania.

CONTENTS

ILLUSTRATIONS

FIGURES

SUMMARY

Articulate brachiopods from a nearly continuous late Devonian and Carboniferous sequence (Frasnian to early Namurian) in the Bonaparte Gulf Basin are described systematically. The 91 species of brachiopods described here, with the exception of the meekellidid and spiriferidinid fauna of the Septimus Limestone described by Thomas (1970), represent nearly the entire brachiopod fauna. Of the species described, 51 are undescribed previously, 24 are described, and 16 are specifically indeterminate. Twelve new generic names are proposed, including *Dorsoscyphus* (Family Schuchertellidae, Subfamily Dorsoscyphinae nov.), *Schistochonetes* (Subfamily Rugosochonetinae), *Lomatiphora* (Family Productellidae, Subfamily Lomatiphorinae nov.), *Spinocarinifera, Acanthocosta,* and *Spinauris* (all Subfamily Overtoniinae), *Septemirostellum* and *Grammorhynchus* (both Superfamily Rhynchonellacea, Family Trigonirhynchiidae), *Ningbingella* (Family ?Pugnacidae), *Cardiothyris* (Subfamily Athyridinae), *Austrochoristites* (Family Brachythyrididae, Subfamily Choristitidinae), and *Litothyris* (Subfamily Brachythyridinae). Palaeontological studies reveal complex mantle canal systems in the chonetidinid *Delepinea uttingi* Thomas, the productellidid *Lomatiphora aquila* gen. et sp. nov., and the productininid *Productina margaritacea* (Phillips).

Thirteen zones and one informal faunal unit based on brachiopods are recognized in the Frasnian to early Namurian sequence. Most Carboniferous zones supersede and the others are modifications of brachiopod 'zonal assemblages' established by Thomas (1970). One of the Devonian zones, the *Crurithyris apena* Zone, was first recognized in the Canning Basin, Western Australia, by Veevers (1959a). The brachiopod zones are dated from internal evidence as well as from conodonts (Druce, 1969) and foraminifers (Mamet & Belford, 1968), and because they are identified in varied rock-types are extensively used in determining correlations within the basin (Veevers & Roberts, 1968). Some zones, particularly those from the Frasnian and Tournaisian, are recognized in the Canning and Carnarvon Basins in Western Australia, and hence provide a basis for correlation. It is noteworthy, however, that the Famennian brachiopod faunas of reef-complexes in the Bonaparte Gulf and Canning Basins are almost unrelated. In the Carboniferous, the brachiopod faunas of Western and eastern Australia have only a few cosmopolitan species in common, and distinct faunal provinces characterized by relatively large endemic populations are recognized on each side of the Australian continent. In terms of overseas faunal affinities, the Frasnian fauna is related to a world-wide 'platform' fauna, which is best known in Europe, USSR, and North America, and the Famennian fauna is closest to assemblages from North America and the USSR. The richer Carboniferous faunas contain several early Carboniferous cosmopolitan species and in the Tournaisian have affinities with faunas in Europe, USSR, and North America. In the Visean and early Namurian there are remarkably close relationships with faunas from Europe, but relatively few with faunas from continental North America.

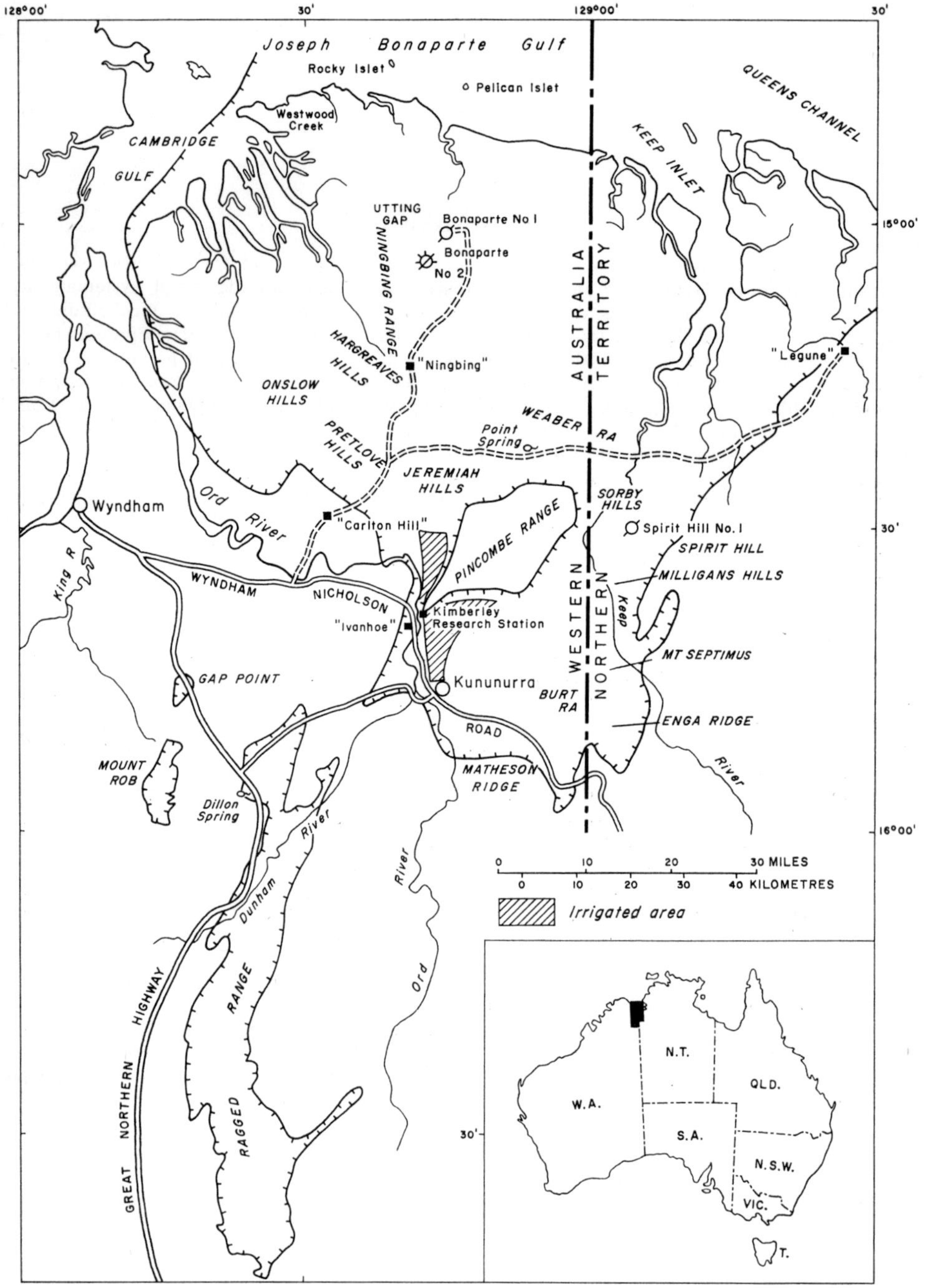

Figure 1. Locality map of the southwestern part of the Bonaparte Gulf Basin.

INTRODUCTION

This Bulletin describes the brachiopod fauna of the Devonian and Carboniferous (Frasnian to early Namurian) platform sediments of the Bonaparte Gulf Basin in northwestern Australia (Fig. 1). It is part of a comprehensive study of the Bonaparte Gulf Basin begun by the Bureau of Mineral Resources in 1963, and follows an earlier study by Thomas (1970), started in the mid-1950's, on brachiopods from the Carnarvon, Canning, and Bonaparte Gulf Basins. Thomas's material from the Bonaparte Gulf Basin, with the exception of that from the Septimus Limestone at Mount Septimus, came largely from samples collected in reconnaissance surveys. Other earlier work on brachiopods from the Bonaparte Gulf Basin includes that of Veevers (1959b) on rhynchonellids and a productacean, and Thomas (1965) on *Delepinea*. A comprehensive description of the Devonian and Carboniferous geology of the basin is given by Veevers & Roberts (1968).

All the specimens described in this Bulletin are housed in the Commonwealth Palaeontological Collection, Bureau of Mineral Resources, Canberra, Australia; type and figured specimens bear the prefix CPC. Locality information is provided in Appendices 1 and 2.

The classification and terminology in the brachiopod systematics mainly follows that in the Treatise on Invertebrate Paleontology, Part H, Brachiopoda (1965). One major exception is the use of the term ventral adminicula (Browne, 1953b), which in the Treatise is defined as the ventral part of the dental plates. In the present usage, dental plates refer to the ridges of calcareous tissue extending along the inner margins of the delthyrium and subtending the teeth, and ventral adminicula are buttressing plates extending from the dental plates to the floor of the pedicle valve.

ACKNOWLEDGMENTS

This study is part of a comprehensive geological and palaeontological investigation of the Bonaparte Gulf Basin. Description of the brachiopod faunas began at the Bureau of Mineral Resources, and continued at the U.S. National Museum, Washington, D.C., and the Department of Geology, University of Illinois, Urbana, Illinois. I wish to thank Dr P. M. Kier, Chairman of the Department of Palaeobiology, U.S. National Museum, and Dr F. A. Donath, Head of the Department of Geology, University of Illinois, for the use of facilities in their departments. At these institutions I received generous assistance and advice from Dr G. A. Cooper (U.S.N.M.) and Dr J. L. Carter (U.I.).

Comparative material was kindly loaned by Professor D. Hill, University of Queensland; Dr P. J. Coleman, University of Western Australia; Dr H. Brunton, British Museum (Nat. Hist.), London; Dr W. H. C. Ramsbottom, Institute of Geological Sciences, Leeds; Dr G. A. Cooper, U.S. National Museum, Washington, D.C.; Dr Mackenzie Gordon Jr., U.S. Geological Survey, Washington, D.C.; Dr C. MacClintock, Peabody Museum, Yale University, New Haven, Connecticut; Dr S. Gould, Museum of Comparative Zoology, Harvard University, Cambridge, Massachusetts; Dr J. L. Carter, University of Illinois; Dr T. E. Bolton, Geological Survey of Canada, Ottawa, Ontario; and Dr M. Gaetani, University of Milan.

I acknowledge helpful discussions with Drs J. J. Veevers and D. J. Belford and Messrs P. J. Jones and E. C. Druce, Bureau of Mineral Resources; Dr K. S. W. Campbell, Australian National University, Canberra; Dr G. A. Thomas, University of Melbourne; Drs G. A. Cooper and R. S. Boardman, U.S. National Museum, Washington, D.C.; Drs Mackenzie Gordon Jr., J. T. Dutro, R. E. Grant, and W. J. Sando, U.S. Geological Survey, Washington, D.C.; Drs J. L. Carter and D. B. Blake, University of Illinois, Urbana, Illinois; Dr C. Collinson, Illinois State Geological Survey, Urbana, Illinois; Dr D. J. McLaren, Institute of Sedimentary and Petroleum Geology, Calgary, Alberta, Canada; and

Dr P. Sartenaer, Institut Royal des Sciences Naturelles, Bruxelles, Belgium. Dr H. Brunton, British Museum (Nat. Hist.), London, Dr P. Copper, Department of Geology, Laurentian University, Sudbury, Ontario, Canada, and Mr R. McKellar, formerly of the Geological Survey of Queensland, provided valuable information in correspondence.

Photographs of some specimens were taken by Messrs J. Zwartko and M. Doyle. Some specimens were etched and prepared by Messrs R. Brown and M. Doyle at the Bureau of Mineral Resources, Canberra.

My work in the United States from September 1967 to June 1969 was carried out during the tenure of a Harkness Fellowship of the Commonwealth Fund of New York.

STRATIGRAPHY

The Bonaparte Gulf Basin in northwestern Australia contains sedimentary rocks ranging in age from early Cambrian to early Cretaceous. The onshore part of the basin occupies an area of about 7000 square miles south of the Joseph Bonaparte Gulf (Figs 1, 2), and is bordered by the Precambrian Kimberley Block to the west, and the Precambrian Halls Creek Mobile Zone to the east and southeast (Traves, 1955, fig. 33; Veevers, 1967, fig. 1). It contains a minimum aggregate of 24,000 feet of Cambrian, Lower Ordovician, Upper Devonian, Carboniferous,

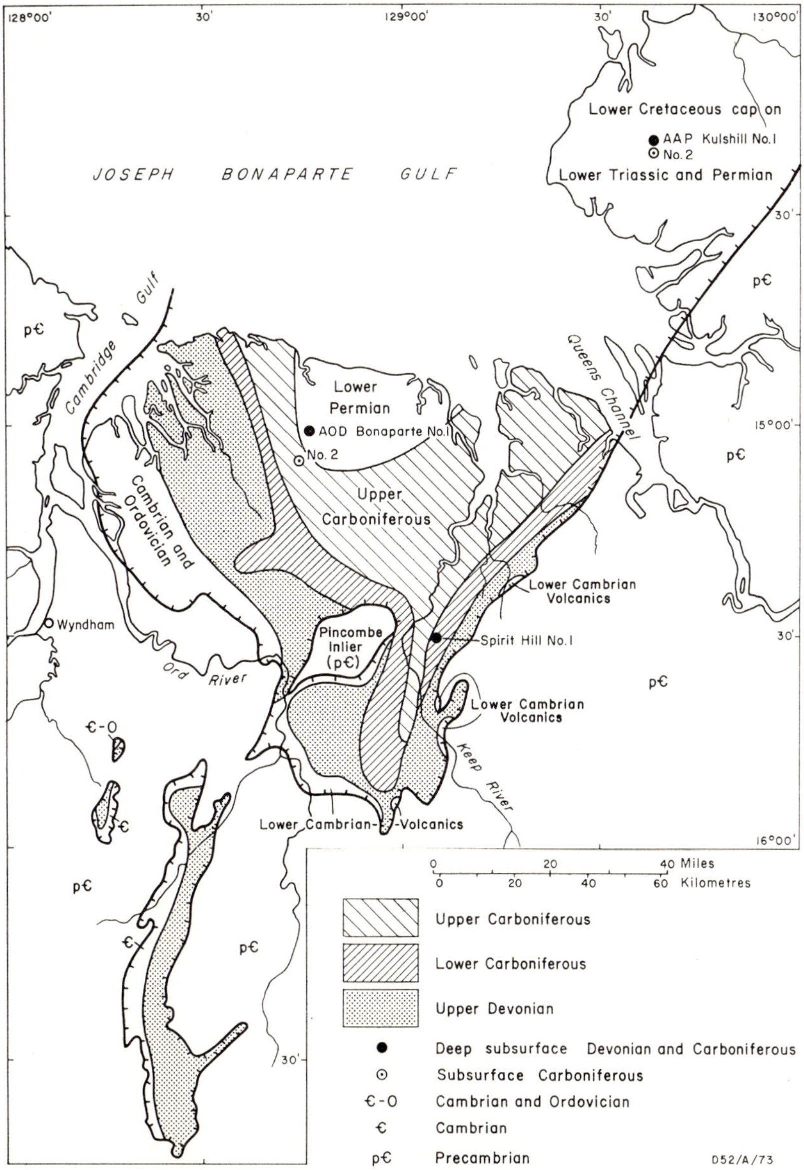

Figure 2. Solid geological map of the Bonaparte Gulf Basin.

Permian, Lower Triassic, and Lower Cretaceous sediments, all mainly marine, resting on Precambrian sedimentary and crystalline rocks. Geophysical surveys in the Timor Sea show that the basin extends over a large area offshore, and Veevers (1967) suggests, from the results of an aerial magnetic survey, that the Bonaparte Gulf and Canning Basins were connected around the northwestern margin of the Kimberley Block during the Phanerozoic.

The Devonian and Carboniferous stratigraphy of the basin is described in detail by Veevers & Roberts (1967; 1968), and Roberts & Veevers (in press), and in summary by Guillaume (1966) and Brady, Jauncey, & Stein (1966), and will be briefly summarized here.

Devonian

Devonian sedimentation began in the early Frasnian or possibly the late Givetian in three provinces within the basin; the platform conglomerate province, the platform carbonate province, and the basinal shale province (Veevers & Roberts, 1967, fig. 2). In general terms, the platform conglomerate province corresponds with the southeastern platform and the platform carbonate province with the northwestern platform recognizable in the Carboniferous (*see* Fig. 3).

The Cockatoo Formation, 5000 feet of Frasnian shallow-water quartz sandstone, conglomerate, limestone, and dolomite, was deposited on the platform. In the south it consists of a thick conglomerate wedge (Ragged Range Conglomerate Member), which passes northwards into two thick tabular bodies of quartz sandstone (Kellys Knob Sandstone Member and Cecil Sandstone Member) which are separated by fossiliferous median members (Fig. 3); the tabular sandstone bodies extend over both the platform conglomerate and platform carbonate provinces. The middle members in the platform conglomerate province comprise the Kununurra Member, a fossiliferous glauconitic quartz sandstone which becomes dolomitic towards the north, and the Abney Member, a fossiliferous red-brown quartz sandstone. In the platform carbonate province the middle member becomes increasingly rich in carbonates towards the north; the Hargreaves Member, consisting of interbedded sandstone, dolomite, and marl, passes northwards into the Westwood Member, which contains back-reef limestone and interbedded lagoonal limestone and sandstone (Veevers, 1968). Overlying the Cecil Sandstone Member is the late Frasnian to possibly early Famennian Jeremiah Member, a sandy dolomite which links the Cockatoo Formation with the overlying Ningbing Limestone.

The Ningbing Limestone is a reef-complex which extends throughout most of the Famennian and into the Tournaisian. It consists of 1000 feet of well-exposed back-reef, a narrow reef, and poorly exposed inter-reef and fore-reef. Behind the reef-complex 1100 feet of interbedded impure limestone and sandstone (Buttons Beds) were deposited in offshore and inshore lagoonal areas (Veevers, 1968, fig. 1b).

In the basinal province, at least 2300 feet of dark siltstone and shale constituting the lower part of the Bonaparte Beds was deposited in deep water during the Famennian and probably during the Frasnian.

Carboniferous

Platform and basinal regions persisted throughout the early Carboniferous. Because of the uniformity of sedimentation and the lack of thick conglomerate wedges the platform areas are termed the southeastern and northwestern platforms (Fig. 3). Sedimentation on the northwestern platform continued probably without interruption from the Devonian into the Carboniferous. On the southeastern platform, following an hiatus in the late Famennian, the Burt Range Formation was deposited behind a reef in a back-reef and interfingering lagoonal environment (Veevers, 1968); the formation consists of 1000 feet of grey calcarenite (reefal sediments) which passes upwards into 500 feet of interbedded limestone, calcare-

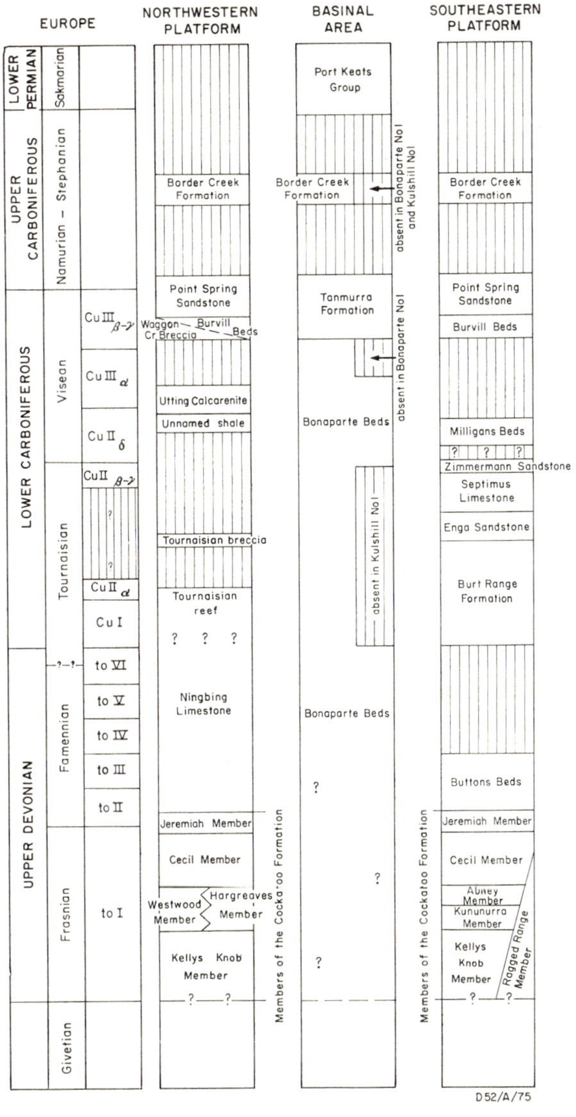

Figure 3. Correlation chart of the Devonian and Carboniferous formations in the Bonaparte Gulf Basin. The possible hiatus in the German Carboniferous zones is based on the work on conodonts by Rhodes et al. (1969) and Druce (1969). Platform and basinal areas of deposition are recognised in the Upper Devonian and Lower Carboniferous only.

B

ous sandstone, and quartz sandstone. The Carboniferous reef-complex envisaged by Veevers (1968) was probably continuous with the Tournaisian reef in the Ningbing Limestone on the northwestern platform, and remained in existence until the middle Tournaisian (Fig. 3). The Burt Range Formation is succeeded by the Enga Sandstone, 530 feet of white quartz sandstone; the Septimus Limestone, 590 feet of crinoidal calcarenite; and the Zimmermann Sandstone, 460 feet of brown to white quartz sandstone.

The northwestern platform was faulted, uplifted, and eroded during the middle Tournaisian, destroying the reef-complex. Isolated Carboniferous outcrops on the northwestern platform consist of an unnamed Tournaisian dolomite breccia, 100 feet thick, deposited on the shore of the Waggon Creek valley after the erosion of the platform (Veevers & Roberts, 1966), and an estimated 400 feet of Visean crinoidal calcarenite, the Utting Calcarenite. Veevers (1968), using a computer classification of limestones, distinguishes the calcarenite in the Utting Calcarenite and the Septimus Limestone from the reefal limestones in the Frasnian, Famennian, and early Tournaisian.

In the basinal area, deposition of dark shale and siltstone began in the Devonian and persisted until the late Visean, with a possible hiatus in the early Tournaisian. The combined Devonian and Carboniferous sequence, which has not been pierced, is almost 9000 feet thick in Bonaparte No. 1 Well. A widespread transgression in the early Visean deposited shale and siltstone over both platforms. On the southeastern platform more than 800 feet of Visean shale and siltstone is termed the Milligans Beds; the shale on the northwestern platform is known from shot-holes only, and is unnamed. A regression in the late Visean probably moved the basinal area seawards, and shoreline and paralic sediments were subsequently deposited over the old basinal areas.

Shoreline sediments, termed the Burvill Beds, consist of sandstone, shale, and interbedded sandy limestone, and have a maximum thickness of 280 feet. They are probably equivalent to the Waggon Creek breccia (Veevers & Roberts, 1966) deposited on the shore of the Waggon Creek valley; the breccia consists of a basal pebbly sandstone 20 feet thick, and an overlying breccia 165 feet thick containing tabular blocks of dolomite up to 30 feet across and subangular to rounded fragments of quartzite, set in a dolomitic matrix. Both the Burvill Beds and the Waggon Creek Breccia are overlain by the Point Spring Sandstone, 890 feet of quartz sandstone. Offshore, the Burvill Beds and the Point Spring Sandstone are represented by approximately 1000 feet of sandstone and interbedded limestone of the Tanmurra Formation; this formation is known in the subsurface only.

After uplift and erosion of the onshore part of the basin in the late Carboniferous, the Border Creek Formation, a fluvial deposit of quartz sandstone, conglomerate, and siltstone, was deposited disconformably over many of the marine early Carboniferous units.

Position of the Devonian-Carboniferous Boundary

The position of the Devonian-Carboniferous boundary, accepted by the 2nd Congress of Carboniferous Stratigraphy and Geology (1937) as the base of the *Gattendorfia* Zone (CuI) in Germany, is shown by Mamet (1967) to lie within the Tournaisian, near the base of the Tn2A, in Belgium and northern France. In

Great Britain, Rhodes et al. (1969) show that the boundary between the Avonian and the Old Red Sandstone of the Avon Gorge area is within the lower part of the *Gattendorfia* Zone, whereas in the northcrop of the Carboniferous Limestone in the South Wales Coalfield the boundary between the Old Red Sandstone and the Lower Limestone Shales coincides with the toVI-CuI boundary. The period boundary at the base of the *Gattendorfia* Zone is accepted by many workers, but others prefer to take it at the base of the Tn1A, proposed by the 1st Congress of Carboniferous Stratigraphy and Geology (1928). These discrepancies have led to considerable confusion in the ages and correlations of the late Devonian and early Carboniferous.

In this paper, the Devonian-Carboniferous boundary is drawn in correlation charts (Figs 3, 4, 7, 9) at the base of the *Gattendorfia* Zone of the German section; this conforms with the boundary used in other papers on the Bonaparte Gulf Basin (Veevers & Roberts, 1968; Druce, 1969).

BRACHIOPOD ZONES

Brachiopod zones in the late Devonian and Carboniferous sequence in the Bonaparte Gulf Basin (Fig. 4) are based on a nominate species and an accompanying assemblage of species, following Opel's original concept (Berry, 1968). In terms of the British Stratigraphical Code (George et al., 1967) they are concurrent-range zones: 'a body of strata characterized by the overlapping ranges of specified fossil groups, from one or more of which it takes its name'. The zones, in most cases, are recognized in measured sections in which the stratigraphic ranges of the brachiopod species are plotted against the thickness of sediment. In the Tournaisian sequence on the southeastern platform a composite section is established to give the most complete sequence (Figs 5, 6). The total stratigraphic ranges of species are determined by projecting into the composite section information from other sections; traced beds and beds of known ages, determined from conodonts and ostracods, are used as datum lines. A conventional biostratigraphical approach cannot be made throughout reef-complexes because the back-reef and lagoon are the only parts that contain continuous sequences of bedded sediments; in the reef, fore-reef, and inter-reef, stratigraphic ranges of species are established mainly from conodont ages, determined by Druce (1969), in terms of the German ammonoid zones. The ranges of brachiopods in the more important sections and faunal lists from other localities are given in Appendix 1. Brachiopod localities in the text are the original field locality numbers. Appendix 2, small-scale maps from Veevers & Roberts (1968), gives the locations of the more important sections and spot samples.

A total of thirteen zones and one informal faunal unit is recognized in the late Devonian and Carboniferous rocks of the Bonaparte Gulf Basin. One of the zones, the *Crurithyris apena* Zone, was originally designated by Veevers (1959a) from the Frasnian of the Canning Basin, Western Australia. The Carboniferous brachiopod zonal assemblages established in the Bonaparte Gulf Basin by Thomas (1970) are extensively revised. Thomas's zonal assemblages 'A. Tournaisian brachiopods from part of the Burt Range Formation and correlated outcrops', and 'B. The Tournaisian *Unispirifer fluctuosus* zonal assemblage', are replaced by a sequence of zones recognized in the richly fossiliferous Tournaisian sequence on the southeastern platform; his assemblage 'C. The late Tournaisian to possibly early Visean *Spirifer spiritus* zonal assemblage' is emended to extend from the uppermost beds of the Septimus Limestone into the Zimmermann Sandstone; and his assemblage 'F. The late Visean to possibly Namurian *Anthracospirifer milliganensis* zonal assemblage' extended from the Burvill Beds into the lower part of the Point Spring Sandstone.

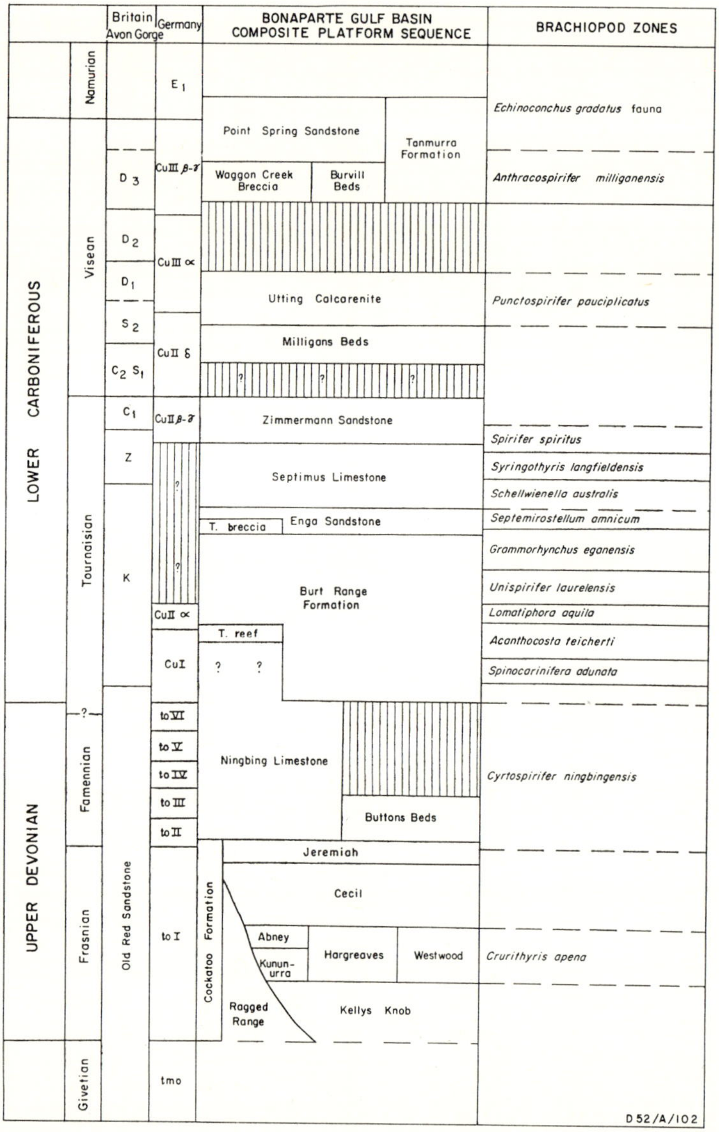

Figure 4. Late Devonian and early Carboniferous brachiopod zones in the Bonaparte Gulf Basin. The possible hiatus between the $CuII_{\alpha}$ and $CuII_{\beta\text{-}\gamma}$ zones and the correlation between the British and German Carboniferous zones is based on work on conodonts by Rhodes et al. (1969) and Druce (1969).

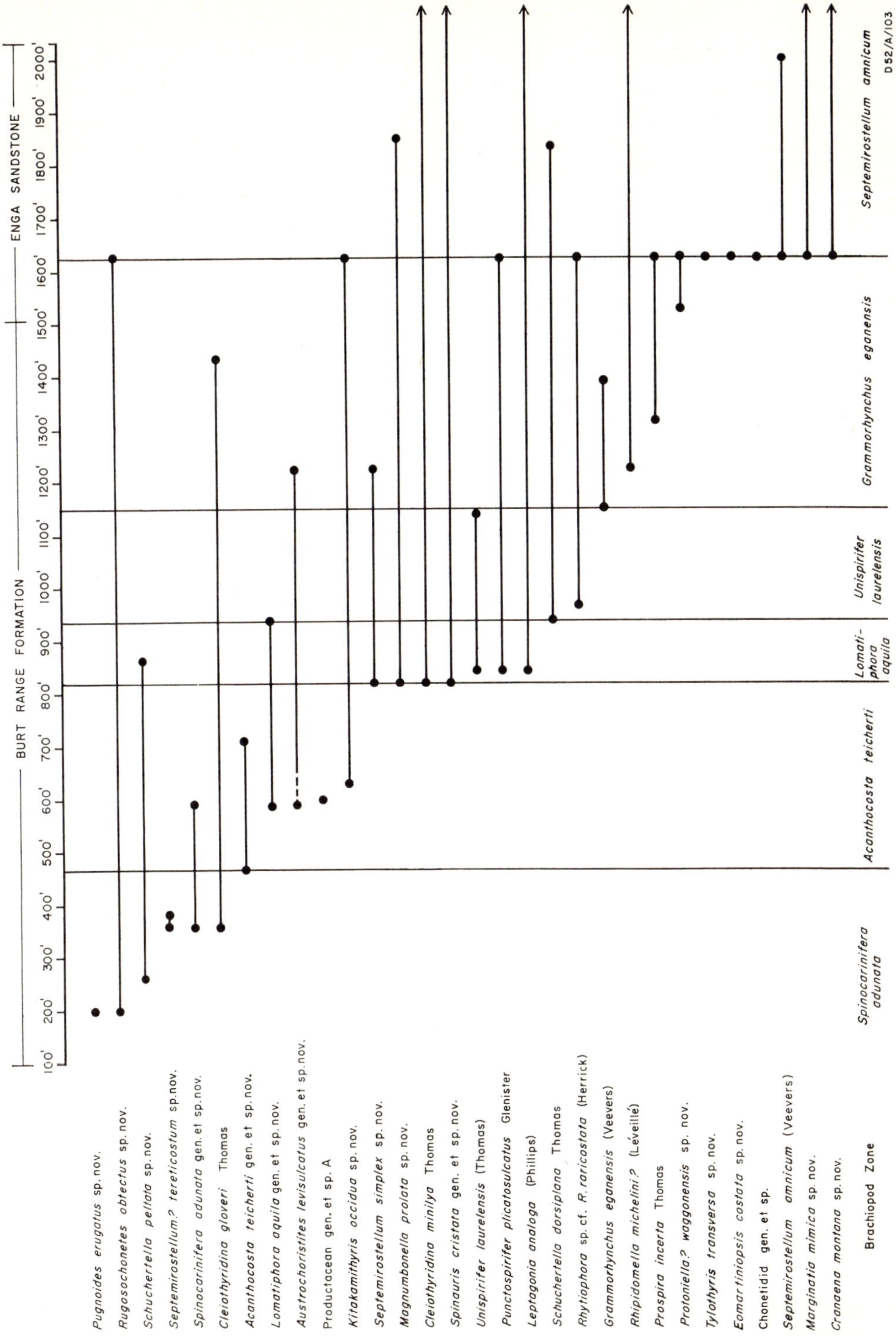

Figure 5. Stratigraphic ranges of brachiopod species in the Burt Range Formation, Burt Range area.

Crurithyris apena ZONE

The *Crurithyris apena* Zone was designated by Veevers (1959a) in the Canning Basin, Western Australia. The zone is restricted to the range of *Crurithyris apena* Veevers, and contains also the longer-ranging *Hypothyridina margarita* Veevers, and *Desquamatia (Synatrypa) kimberleyensis* (Coleman). Veevers suggests that the *apena Zone* at Bugle Gap probably lies between the *Amphipora ramosa* Zone and the 'Upper Manticoceras' Zone of Teichert (1943; 1949), and hence is early Frasnian in age (Veevers, 1959a, table 1, p. 15). This age is confirmed by later work on conodonts and ammonoids by Glenister & Klapper (1966), who demonstrate that rocks at locality K149, Bugle Gap, unconformably overlying reef limestone containing the *apena* Zone, are Frasnian to Iα-toIδ in age from conodonts, and to Iβ-toIδ in age from ammonoids.

In the Bonaparte Gulf Basin, *Crurithyris apena* is recorded from one level in the Kununurra Member, and from near the base of the Westwood Member, of the Cockatoo Formation. Fossiliferous beds in the Westwood Member are approximately 2000 feet thick, and the upper part, in which there are no diagnostic brachiopods, may be younger than *apena* Zone. The Hargreaves Member of the Cockatoo Formation is correlated on fossil evidence with the Kununurra Member and the lower part of the Westwood Member (Roberts et al., 1967). Because of thick unfossiliferous sandstone (Cecil and Kellys Knob Members) above and below these members, the *apena* Zone in the Bonaparte Gulf Basin is arbitrarily

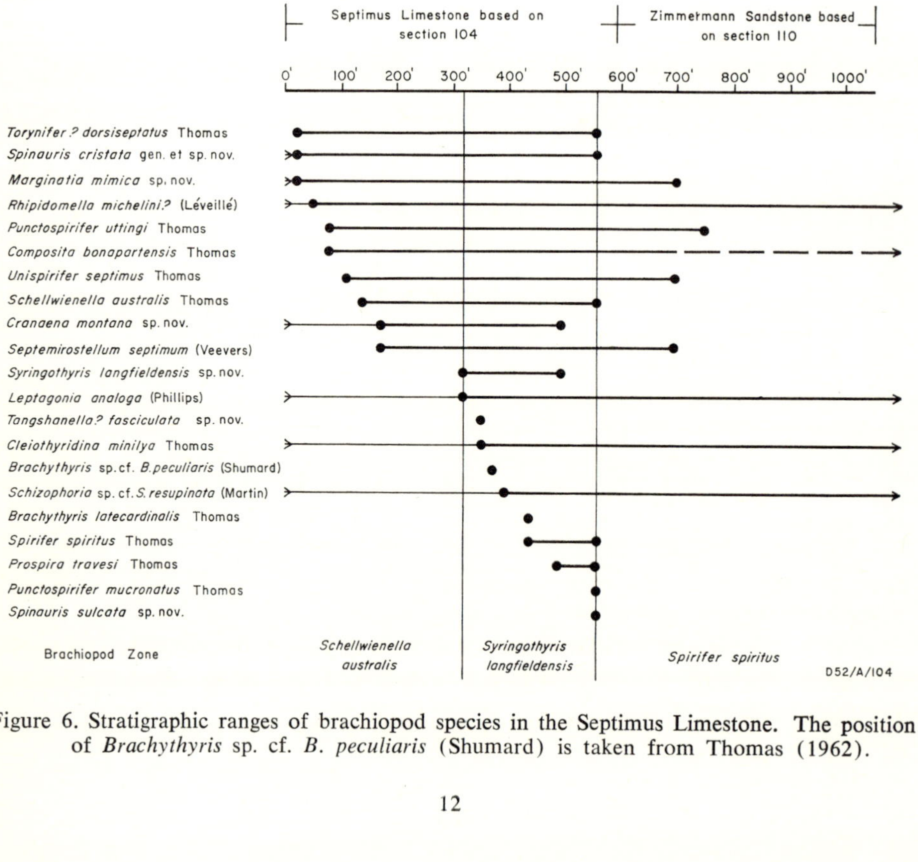

Figure 6. Stratigraphic ranges of brachiopod species in the Septimus Limestone. The position of *Brachythyris* sp. cf. *B. peculiaris* (Shumard) is taken from Thomas (1962).

extended throughout the Kununurra, Hargreaves, and Westwood Members; the Abney Member, overlying the Kununurra Member in the south, may lie within the zone.

Brachiopod species in the *apena* Zone in the Bonaparte Gulf Basin include *Schuchertella* sp. cf. *S. gratillica* Veevers, *Globosochonetes? mathesonensis* sp. nov., *Steinhagella* sp., *Calvinaria?* sp., *Crurithyris apena* Veevers, and *Cyrtospirifer* sp. from the Kununurra Member; *Productella westwoodensis* sp. nov., *Tropidoleptus?* sp., *Spinatrypa prideri larga* subsp. nov., *Tenticospirifer columnaris* sp. nov., *Cyrtospirifer* sp., and an indeterminate rhynchonellid from the Hargreaves Member; and *Retichonetes arenarius* sp. nov., *Productella westwoodensis* sp. nov., *Steinhagella* sp., *Spinatrypa prideri larga* subsp. nov., *Desquamatia (Synatrypa) kimberleyensis* (Coleman), *Tenticospirifer columnaris* sp. nov., *Crurithyris apena* Veevers, *Cyrtospirifer* sp., and the same indeterminate rhynchonellid from the Westwood Member.

Conodonts in the *apena* Zone in the Bonaparte Gulf Basin point to a late Frasnian age (Roberts et al., 1967) rather than the possibly earlier Frasnian age suggested by *Crurithyris apena,* and hence the local age of the zone cannot be determined precisely. Druce (*in* Roberts et al., 1967) admits that the conodont evidence is tenuous because of current disagreement on the specific determinations of icriodod conodonts. The remaining brachiopods suggest a general Frasnian age. *Tenticospirifer columnaris* sp. nov. is morphologically close to *T. cyrtinaformis* (Hall & Whitefield) from the Lime Creek Formation of Iowa (Fenton & Fenton, 1924), and to *T. lictor* Nalivkin from the Frasnian Semiluki Beds of the Russian Platform (Nalivkin, 1930; Lyashenko, 1959). *Desquematia (Synatrypa) kimberleyensis* (Coleman) resembles *D. zonataeformis* Alekseeva, 1962, from the uppermost Givetian and lowermost Frasnian of the Kuznetsk Basin, Siberia, and *D. ciliipes* (Crickmay) from the Frasnian Grumbler Formation of the Northwest Territories of Canada (Crickmay, 1957; 1967). *Spinatrypa prideri larga* subsp. nov. has similar features to *S. planosulcata* (Webster) from the Cerro Gordo Member of the Lime Creek Formation of Iowa (Fenton & Fenton, 1924), and to a species from the Frasnian of Boulonnais, France, currently being described by P. Copper (pers. comm.). It is noteworthy that the genus *Steinhagella* Goldring, represented by *Steinhagella* sp. in the Bonaparte Gulf Basin and *S. numida* Veevers in the Canning Basin, is recorded earlier in Australia (Frasnian toI) than in Europe (Famennian toV-toVI, Goldring, 1957). Jones (1968) confirms a Frasnian age from ostracods.

Cyrtospirifer ningbingensis Zone

The *Cyrtospirifer ningbingensis* Zone is recognized throughout the Famennian part of the Ningbing Limestone, a reef complex, and in the early Famennian Buttons Beds; it may be present in the underlying Jeremiah Member of the Cockatoo Formation. Druce (*in* Roberts et al., 1967) contends on the basis of conodonts that the toIIα and toIV Zones of the Famennian are absent in the Ningbing Limestone, and that there are limited outcrops only of toV-toVI age.

Cyrtospirifer ningbingensis sp. nov. ranges throughout the zone, and is accompanied, in rocks of toII-toIII age, by *Schuchertella* sp. cf. *S. gratillica* Veevers, *Sentosia subquadrata* sp. nov., *Mesoplica? jeremiahensis* sp. nov., *Nayunnella turgida* sp. nov., *Rugaltarostrum australe* sp. nov., *Ptychomaletoechia lucida*

(Veevers), and *Litothyris alticostata* gen. et sp. nov. All these species are found in the back-reef of the Ningbing Limestone, but *Cyrtospirifer ningbingensis* is present in the reef, fore-reef, and inter-reef. *Nayunnella turgida* is recorded also in the fore-reef, and *Sentosia subquadrata* in the reef. In late Famennian (toV-toVI) back-reef sediments, *Cyrtospirifer ningbingensis* is accompanied by *Schuchertella* sp., probably *Ptychomaletoechia lucida* (Veevers), *Sentosia subquadrata* sp. nov., and *Litothyris alticostata* gen. et sp. nov. This late Famennian association is taken mainly from section 433, which may be toV or younger between 380 feet and 1130 feet.

The Buttons Beds, offshore lagoonal sediments equivalent in age to the older part of the Ningbing Limestone (toII-toIII), contain *Cyrtospirifer ningbingensis* together with *Leioproductus buttonensis* sp. nov., Productacean gen. et sp. B, *Cyrtospirifer depressus* sp. nov., and *Meristella?* sp.

Sartenaer (1967) suggests that many rhynchonellids are useful in both local and intercontinental correlations in the late Devonian. *Ptychomaletoechia lucida* (Veevers) is recorded from the Fairfield, Napier, Geikie, and Pillara Formations, and the Bugle Gap Limestone in the Canning Basin, and probably has a range of toII-toVI, the same as the entire *ningbingensis* Zone. Veevers (1959a) has already indicated a close morphological relationship between *lucida* and *'Camarotoechia' sobrina* Stainbrook from the Famennian Percha Shale of New Mexico, USA. *Nayunnella turgida* sp. nov. and *Rugaltarostrum australe* sp. nov. are not closely related to overseas species, but according to Sartenaer (1967) both genera are restricted to the Famennian. Species of *Leioproductus* Stainbrook, 1947, are likewise Fammenian in age (Muir-Wood & Cooper, 1960). *Cyrtospirifer depressus* sp. nov., from the Buttons Beds in the Burt Range area, is morphologically close to two Soviet species, *C. kemerovensis* (Besnosova) from the lower Tournaisian Abishevsky Beds of the Kuznetsk Basin, and *C. ziganensis* Krestovnikov & Karpyshev from the Etroeungt of the Urals. Although the ages of the Soviet species are almost certainly older than indicated by Besnosova (1959) and Krestovnikov & Karpyshev (1948), these relationships suggest a younger age than that indicated by ostracods (Jones, 1968). However, because Jones's correlations within the Buttons Beds are based on a large number of ostracod species, his interpretation is used in plotting the stratigraphic position of the Buttons Beds on the southeast platform (Fig. 3).

Spinocarinifera adunata ZONE

The Tournaisian *adunata* Zone is recognized in the interval 190 feet to 470 feet above the base of the composite section of the Burt Range Formation (Fig. 5). The zone is characterized by *Spinocarinifera adunata* gen. et sp. nov., the nominate species, *Pugnoides erugatus* sp. nov., *Septemirostellum? tereticostum* sp. nov., and *Schuchertella peltata* sp. nov. Long-ranging species appearing for the first time in the zone include *Rugosochonetes obtectus* sp. nov., and *Cleiothyridina gloveri* Thomas. *Spinocarinifera adunata* ranges into the succeeding *teicherti* Zone.

An early Tournaisian age for the zone is indicated by the morphological similarity of *Spinocarinifera adunata* with *S. niger* (Gosselet). *S. niger* is described by Dehee (1929) from the Etroeungt and lower Tournaisian of France, and by Paeckelmann (1931) from the Tournaisian of Germany, and is recorded by

Demanet (1958) from the Tnla, Tnlb, and Tn2a of Belgium. In addition, *S. adunata* is close to *S. inflata* (Sokolskaya) from the early Tournaisian of the Kuznetsk Basin, Siberia (Sarycheva et al., 1963), and possibly to *S? arcuata* (Hall) from the Kinderhook of Missouri (Weller, 1914). Conodonts identified by Druce (1969, fig. 7) indicate a correlation between the *adunata* Zone and the lower half of the CuI Zone of Germany, and the lowermost part of the K Zone in Britain.

Acanthocosta teicherti ZONE

The *teicherti* Zone extends throughout the interval 470 feet to 820 feet of the composite section of the Burt Range Formation. The nominate species, *Acanthocosta teicherti,* ranges throughout the lower two-thirds of the zone. Other diagnostic species include *Spinocarinifera adunata* gen. et sp. nov., *Schuchertella peltata* sp. nov., *Lomatiphora aquila* gen. et sp. nov., and Productacean gen. et sp. A. The longer-ranging species *Kitakamithyris occidua* sp. nov. first appears near the middle of the zone.

A fauna in the Tournaisian reef portion of the Ningbing Limestone (locality 7/1) contains three species characteristic of the *teicherti* Zone: *Acanthocosta teicherti,* Productacean gen. et sp. A, and *Schuchertella peltata. Kitakamithyris occidua,* which is in the reef, first appears in the *teicherti* Zone, and ranges to the base of the *amnicum* Zone. Conodonts from the reef also indicate a correlation with the *teicherti* Zone in the Burt Range Formation; Druce (pers. comm.) gives a correlation with the interval 475 feet to 950 feet, probably 650 feet, above the base of the Burt Range Formation. Apart from the long-ranging species *Leptagonia analoga* (Phillips) and *Schizophoria* sp. cf. *S. resupinata* (Martin), the remainder of the brachiopod fauna is endemic to the reef. It is noteworthy that *Leptagonia analoga* first appears in the southeastern platform in the *aquila* Zone, at 840 feet above the base of the Burt Range Formation, and *Schizophoria* sp. cf. *S. resupinata* first appears at 320 feet above the base of the Septimus Limestone in the *langfieldensis* Zone. A number of species in the reef fauna have strong affinities with forms in the Tournaisian of the northern hemisphere, particularly those in the Chappel Limestone of Texas, described by Carter (1967). *Spirifer otwayi* sp. nov. morphologically resembles *S. chappelensis* Carter from the Chappel Limestone; it is close also to *S. tersiensis* Rotai from the early Tournaisian Taidon and lower Netersinsky Beds of the Kuznetsk Basin, Siberia (Sarycheva et al., 1963). *Brachythyris planulata* sp. nov. is closely comparable to specimens of *B. chouteauensis* (Weller) from the Chappel Limestone (Carter, 1967). *Rhytiophora calhounensis* (Miller), with which I have closely compared a species from the reef, comes from the Chouteau Limestone of Missouri (Moore, 1928) and the Chappel Limestone of Texas. *Crassumbo? jonesi* sp. nov. is possibly related to *Crassumbo inornatus* Carter and *C. turgidus* Carter from the Chappel Limestone. The remaining brachiopod species in the reef are *Ningbingella flexuosa* gen. et sp. nov., *Cardiothyris bisulcata* gen. et sp. nov., and *Ovatia* sp. A.

The late Kinderhookian age suggested by the brachiopod fauna is slightly younger than that of the conodonts, the latest Kinderhookian being equivalent, on the basis of conodonts, to the overlying *aquila* Zone *(see* Druce, 1969, fig. 7). Druce's work on conodonts indicates that, in terms of the British Carboniferous, the *teicherti* Zone is equivalent to the middle part of the K Zone.

Lomatiphora aquila Zone

The *aquila* Zone extends from 820 feet to 935 feet above the base of the composite section of the Burt Range Formation. *Lomatiphora aquila* gen. et sp. nov., the nominate species, is accompanied by *Schuchertella peltata* sp. nov., at its final appearance, *Septemirostellum simplex* sp. nov., *Austrochoristites levisulcatus* gen. et sp. nov., and *Unispirifer laurelensis* (Thomas). Longer-ranging species which first appear at or near the base of the *aquila* Zone include *Spinauris cristata* gen. et sp. nov., *Magnumbonella prolata* gen. et sp. nov., *Cleiothyridina minilya* Thomas, and *Punctospirifer plicatosulcatus* Glenister. The sudden influx of new species between 820 feet and 840 feet above the base of the formation may not have been as abrupt as shown in Figure 5 because of poor outcrop between localities 100/24 and 100/25, and the sandy lithology between localities 101/12 and 101/12C (Veevers & Roberts, 1968, fig. 41).

The *aquila* Zone had probably the shortest duration of all the brachiopod zones in the Bonaparte Gulf Basin. The nominate species is unrelated to any previously described productellidid, and the other characteristic species range into the overlying *laurelensis* Zone. Of these, *Unispirifer laurelensis* (Thomas) is morphologically close to *U. platynotus* (Weller) from the Kinderhook at Burlington, Iowa (Weller, 1914); to *'Fusella' ussiensis* (Tolmachoff) from the early Tournaisian Taidon and Netersinsky Beds of the Kuznetsk Basin, Siberia (Besnosova, 1959); and possibly to specimens referred to *Spirifer* aff. *clathratus* M'Coy by Vaughan (1905) from the Z_1 Zone near Bristol, England. Thomas (1970) also compared *laurelensis* with *U. ussiensis* (Tolmachoff), and in addition mentioned similarities to a small species referred to *Spirifer tornacensis* de Koninck by Sarycheva & Sokolskaya (1952) from the C_1tsch (late Tournaisian) of the Moscow Basin, and *U. minnewankensis* (Shimer) described by Brown (1952) from the Upper Tournaisian part of the Rundle Formation, Alberta, Canada. *Septemirostellum simplex* sp. nov. is close to *S. mitcheldeanensis* (Vaughan) from the Tournaisian K, Z, and basal C Zones of the Avon Gorge section in Great Britain; species identified as *mitcheldeanensis* are recorded by Demanet (1958) from the Tn1a, Tn1b, and Tn2a Zones of Belgium. Conodonts in the *aquila* Zone indicate a correlation with the upper part of the CuII Zone of Germany, the upper part of the Chouteau Limestone in the upper Mississippi Valley area of USA, and an interval within the upper part of the K Zone in Great Britain (Druce, 1969, fig. 7).

Unispirifer laurelensis Zone

The *laurelensis* Zone, named after *Unispirifer laurelensis* (Thomas), extends from 935 feet to 1150 feet above the base of the Burt Range Formation. In addition to the nominate species, the zone is characterized by *Austrochoristites levisulcatus* gen. et sp. nov., *Septemirostellum simplex* sp. nov., and the first occurrences of *Schuchertella dorsiplana* Thomas, and *Rhytiophora* sp. cf. *R. raricostata* (Herrick).

The affinities of *Unispirifer laurelensis* and *Septemirostellum simplex* have already been dealt with in the discussion on the preceding *aquila* Zone. *Rhytiophora raricostata* (Herrick), with which a species ranging from the *laurelensis* Zone to the base of the *amnicum* Zone is closely compared, is recorded from the Kinderhookian Logan Formation of Ohio (Hyde, 1953). The *laurelensis* Zone can be determined to be equivalent to an uppermost part of the K Zone in Britain (*see* Druce, 1969, fig. 7).

Grammorhynchus eganensis Zone

The *eganensis* Zone extends from 1150 feet to 1625 feet above the base of the composite section, throughout nearly all of the upper part of the Burt Range Formation, and into the base of the Enga Sandstone. As well as *Grammorhynchus eganensis* (Veevers), the nominate species, the zone has as characteristic species *Prospira incerta* Thomas and *Protoniella? waggonensis* sp. nov. *Cleiothyridina gloveri* Thomas, *Austrochoristites levisulcatus* gen. et sp. nov., and *Septemirostellum simplex* sp. nov. make their final appearances within the zone, and there is the first appearance of the long-ranging cosmopolitan species *Rhipidomella michelini?* (Léveillé).

In the vicinity of Legune Homestead, in the northeastern part of the basin, *Prospira incerta* may range down as far as the *aquila* Zone. *P. incerta* is morphologically close to *P. prima* Maxwell from the Tournaisian of Queensland (McKellar, 1967, fig. 7); the fauna containing *prima,* from the Bancroft Formation in the Cannindah Creek area, belongs to the *Schellwienella* cf. *burlingtonensis* Zone of eastern Australia (Roberts, 1965; Banks et al., in press). The nominate species, *Grammorhynchus eganensis,* is present in the Tournaisian Laurel Formation in the Canning Basin; further work is required to determine its precise stratigraphic range within that formation. Conodont evidence, summarized by Druce (1969, fig. 7), indicates a correlation between the *eganensis* Zone and an upper part of the K Zone in Great Britain; according to conodonts there are no equivalents in Germany or USA because of breaks in the sequence.

Septemirostellum amnicum Zone

The *amnicum* Zone contains *Septemirostellum amnicum* (Veevers) as nominate species, and is characterized by *Tylothyris transversa* sp. nov., *Eomartiniopsis costata* sp. nov., and Chonetidid gen. et sp. The latter three species are found in shale at locality 102/4 at the northern end of Enga Ridge, and as yet are unknown elsewhere, possibly because the shale is replaced southwards by coarser clastic sediments. Species making their first appearance in the *amnicum* Zone are *Marginatia mimica* sp. nov., and *Cranaena montana* sp. nov. Final appearances include *Rugosochonetes obtectus* sp. nov., *Kitakamithyris occidua* sp. nov., *Punctospirifer plicatosulcatus* Glenister, *Prospira incerta* Thomas, and *Protoniella? waggonesis* sp. nov. at the base of the zone, and *Magnumbonella prolata* sp. nov. and *Schuchertella dorsiplana* Thomas within the zone. The abrupt termination of the ranges of many brachiopod species at the base of the zone and the abundance of bivalves in the higher fossiliferous beds reflect the change in lithology from interbedded calcarenite, shale, and sandstone near the base to pure quartz sandstone in the upper two-thirds of the Enga Sandstone (Veevers & Roberts, 1968, fig. 42).

The *amnicum* Zone extends from 110 feet above the base of the Enga Sandstone, or 1625 feet above the base of the composite section, to the youngest fossiliferous beds at 485 feet above the base of the sandstone. Because there is no outcrop between the northeastern end of Enga Ridge and the overlying Septimus Limestone at the foot of Mount Septimus (Appendix 2, Fig. 10) the boundary between the *amnicum* Zone and the overlying *australis* Zone is arbitrary.

Septemirostellum amnicum (Veevers) is closest morphologically to *S? acutirugatum* (de Koninck) from the Tournaisian of Belgium (de Koninck, 1887) and the late Tournaisian (C_1tsch) of the Moscow Basin (Sarycheva & Sokolskaya, 1952); in Belgium, Demanet (1958) records *acutirugatum* from the Tn1a, Tn1b, Tn2a, Tn2c, and Tn3b. In Western Australia, *amnicum* is present at 140 feet and between 469 feet and 560 feet above the base of the Moogooree Limestone in the Carnarvon Basin (Thomas, 1970). The affinities of *Prospira incerta* Thomas are given in the discussion on the underlying *eganensis* Zone. *Eomartiniopsis costata* sp. nov. has closest affinities with *E. girtyi* (Branson) from the Chouteau Limestone of Missouri (Branson, 1938) and the Chappel Limestone of Texas (Carter, 1967). *Tylothyris transversa* sp. nov. in some respects resembles *T. laminosa* (M'Coy) recorded from the Tournaisian of Belgium (Tn2c and Tn3b *in* Demanet, 1958), and the Z and C Zones of Great Britain (North, 1920); Brunton (pers. comm.) has found possible *laminosa* specimens as high as the lower D Zone. *Marginatia mimica* sp. nov., which ranges from the *amnicum* to near the top of the *spiritus* Zones, is very close morphologically to *M. deruptoides* Sarycheva, and is externally similar to specimens referred to *M. burlingtonensis* (Hall) by Sarycheva et al. (1963); both species are from the Tournaisian of the Kuznetsk Basin, Siberia. *Cranaena montana* sp. nov., which ranges from the *amnicum* to the *langfieldensis* Zones, resembles *C. globosa* Weller from the Burlington Limestone of Missouri (Weller, 1914). Conodont evidence (Druce, 1969, fig. 7) shows that the *amnicum* Zone is equivalent to an uppermost part of the K Zone in Great Britain.

Schellwienella australis Zone

The *australis* Zone extends throughout the lowermost 320 feet of the Septimus Limestone. *Schellwienella australis* Thomas, the nominate species, ranges into the overlying *langfieldensis* and *spiritus* (emended) Zones; Thomas (1970) records *S. australis* from 450 feet to 500 feet, and *S.* sp. cf. *S. australis* Thomas from 550 to 600 feet (Thomas's measurements) above the base of the Septimus Limestone. He provisionally separated the specimens in the higher beds from *australis* because of their smaller size, but admitted that they may be juveniles. Present work indicates that *australis* first appears 135 feet above the base and ranges to the uppermost beds of the Septimus Limestone. In addition to the nominate species, the *australis* Zone is characterized by the first appearances of *Composita bonapartensis* Thomas, *Puntospirifer uttingi* Thomas, *Unispirifer septimus* Thomas, *Septemirostellum septimum* (Veevers), and *Torynifer? dorsiseptatus* Thomas. Thomas's (1962a) assemblage (d) from about 100 feet above the base of the Septimus Limestone lies within the *australis* Zone.

The affinities of the greater part of the brachiopod fauna from the Septimus Limestone are given by Thomas (1970), and where necessary they will be summarized here. Thomas assigns an age of late Tournaisian to early Visean to the entire Septimus Limestone fauna. A slightly earlier age is suggested by conodonts and foraminifers. Druce (1969, fig. 7) uses conodonts to correlate the Septimus Limestone with the uppermost K Zone and lower one-third of the Z Zone of the Avon Gorge section in Great Britain; and Mamet & Belford (1968), on the basis of a sparse foraminifer assemblage, suggest a middle Tournaisian age (Mamet's foraminifer zone 7).

Most brachiopods in the *australis* Zone range throughout the Septimus Limestone. Except for *Leptagonia analoga* (Phillips), *Schizophoria* sp. cf. *S. resupinata* (Martin), *Rhipidomella michelini?* (Léveillé), *Cleiothyridina minilya* Thomas, and *Composita bonapartensis* Thomas, all of which extend well into the Visean, the fauna has Tournaisian affinities. Thomas (1970) compared *Schellwienella australis* Thomas with Belgian specimens of *S. crenistria* (Phillips), and *S. burlingtonensis* Weller from the Burlington Limestone of the Mississippi Valley area, USA, and the late Tournaisian of the Moscow Basin. *Unispirifer septimus* Thomas resembles *Spirifer pentagonus* Sokolskaya and *S. ventricosus* Sokolskaya from the Tournaisian of the Moscow Basin, and *S. mediocris* Tolmachoff and *S. similis* Tolmachoff from the Tournaisian Taidon and Netersinsky horizons in the Kuznetsk Basin, Siberia (Thomas, 1970).

Syringothyris langfieldensis Zone

The *langfieldensis* Zone is characterized by *Syringothyris langfieldensis* sp. nov., *Tangshanella? fasciculata* sp. nov., *Spirifer spiritus* Thomas, *Brachythyris latecardinalis* Thomas, *Brachythyris* sp. cf. *B. peculiaris* (Shumard), and *Prospira travesi* Thomas. In the Septimus Limestone, the *langfieldensis* Zone extends from 320 feet to 560 feet, and its lower part is richly fossiliferous.

Brachiopods from the zone do not indicate a precise age, and the zone is dated by conodonts as lower Z Zone in the sense of the Avon Gorge section in Great Britain (Druce, 1969, fig. 7). *Syringothyris langfieldensis* sp. nov. is close to *S. elongata* North from the Z and C Zones of Ireland, and the D_2 Zone of Yorkshire (North, 1920); it also resembles a Devonian form referred to *S. chemungensis* Cushing by Hyde (1953). *Tangshanella? fasciculata* sp. nov. bears an 'advanced' fasciculate ornament similar to that of younger spiriferaceans, but it is a distinctive species and cannot be compared closely with other taxa. According to Thomas (1970), both *Brachythyris latecardinalis* and *Prospira travesi* also have no obvious similarities with other species. *Brachythyris peculiaris* (Shumard) is recorded from the Chouteau Limestone of the Mississippi Valley area (Weller, 1914), and hence suggests an older age than the conodonts. *Spirifer spiritus* is compared by Thomas (1970) with *S. attenuatus* Sowerby from the late Tournaisian and early Visean of Britain, Belgium, and USSR.

Spirifer spiritus Zone (emended)

Thomas's (1970) late Tournaisian to possibly early Visean *Spirifer spiritus* zonal assemblage, from 450 feet to 600 feet above the base of the Septimus Limestone (Thomas's measurements), is now termed the *Spirifer spiritus* Zone, and emended to extend from 560 feet above the base of the Septimus Limestone into the overlying Zimmermann Sandstone. The top of the zone is arbitrary; the youngest fossiliferous horizon in the Zimmermann Sandstone is at 155 feet above the base of the formation. Nonmarine conglomerate, sandstone, and siltstone of the Border Creek Formation overlie disconformably the Zimmermann Sandstone in the Burt Range area.

The *spiritus* Zone emend. is characterized by *Spirifer spiritus* Thomas, *Punctospirifer mucronatus* Thomas, and *Spinauris sulcata* sp. nov. At the base of the zone there is the final appearance of *Spirifer spiritus* Thomas, *Prospira travesi* Thomas, *Spinauris cristata* gen. et sp. nov., *Schellwienella australis* Thomas, and *Torynifer?*

dorsiseptatus Thomas. The abrupt end of the ranges of these species is probably influenced by the sudden change in lithology from calcarenite of the Septimus Limestone to quartz sandstone of the Zimmermann Sandstone, and possibly by the disconformity between the two formations (Veevers & Roberts, 1968, p. 85). The extent of this hiatus cannot be determined from fossil evidence. Species having their last appearance within the zone include *Marginatia mimica* sp. nov., *Puncto-spirifer uttingi* Thomas, *Unispirifer septimus* Thomas, and *Septemirostellum septimum* (Veevers).

The affinities of *Spirifer spiritus* to the late Tournaisian and early Visean *S. attenuatus* Sowerby have already been mentioned on page 19 and by Thomas (1970). Thomas also noted some resemblances between *Punctospirifer mucronatus* and *P. transversus* (McChesney) from the Chester of North America. Conodonts from the base of the *spiritus* Zone suggest an age of middle Z Zone in the sense of the Avon Gorge section in Great Britain (Druce, 1969).

Punctospirifer pauciplicatus ZONE

Because of a possible hiatus, and the few brachiopods in the Milligans Beds, the *pauciplicatus* Zone is separated from the older *spiritus* Zone by a break equivalent to most of the German IIδ Zone (Fig. 4). Only long-ranging species extend into the zone, and the fauna, which is Visean in age, has a completely different aspect from that of the Tournaisian zones. The *pauciplicatus* Zone is recognized throughout the Utting Calcarenite, and is characterized by *Punctospirifer pauciplicatus* sp. nov., *Dorsoscyphus spinulosus* gen. et sp. nov., *Productina margaritacea* (Phillips), *Serratocrista* sp., and *Schistochonetes abruptus* gen. et sp. nov. Species which first appear in the zone and range into the younger *Anthracospirifer milliganensis* Zone include *Dictyoclostus? funiferus* sp. nov., *Stegacanthia strigis* sp. nov., *Delepinea uttingi* Thomas, *Megachonetes zimmermanni* (Paeckelmann), *Podtsheremia? humilicostata* sp. nov., and *P? thomasi* sp. nov. Present work shows that two distinctive species, *Rugosochonetes macgregori* sp. nov. and *R. ustulatus* sp. nov., are confined to the lowermost part of the Utting Calcarenite in section 107 (Appendix 1, Fig. 18). Further etching of the silicified fauna may lead to the recognition of another brachiopod zone at the base of section 107.

The brachiopods in the *pauciplicatus* Zone are of Visean aspect. *Productina margaritacea* (Phillips) is characteristic of the D Zone in the British Isles (Muir-Wood, 1948; Brunton, 1966a), and the late Visean in the Donetz Basin in USSR (Rotai, 1931); in Belgium it ranges from the latest Tournaisian (Tn3c) to the late Visean (V3b) (Demanet, 1958); and in Germany, Paeckelmann (1931) records it from the late Visean Oberen Kohlenkalk. *Serratocrista* Brunton, 1968, has previously been recorded from the lower part of the D Zone in Ireland (Brunton, 1968). Also suggesting an early Visean age is *Delepinea uttingi* Thomas, which is comparable with *D. destinezi* (Vaughan) from the C_2S_1 Zone of Britain and the early Visean of Belgium, and with *D. carinata* (Garwood) from the C_2 Zone of England and the early Visean of Belgium (Thomas, 1965). *Megachonetes zimmermanni* (Paeckelmann) is common in the Visean of Poland, Belgium, and France (Paeckelmann, 1930), in the C_1mkh (Visean) of the Russian Platform (Sokolskaya, 1950), and possibly from the Carboniferous Limestone of Yorkshire, England (Paeckelmann, 1930); Sarycheva et al. (1963) record *zimmermanni* from the Tournaisian of the Kuznetsk Basin, Siberia.

Demanet (1958) did not list *zimmermanni* in Belgian faunas, and may have included it in *M. papilionacea* (Phillips). Both *Podtsheremia? humilicostata* sp. nov. and *P? thomasi* sp. nov. have affinities with the *Spirifer duplicostatus* Phillips group of species from the Visean of Belgium and the British Isles (Davidson, 1858). *Anthracospirifer shoshonensis* (Branson & Greger, 1918) from the early Meramecian (lower or middle Salem) is the closest North American species to both *humilicostata* and *thomasi. Dorsoscyphus spinulosus* gen. et sp. nov. and *Schistochonetes abruptus* gen. et sp. nov. are known from the Bonaparte Gulf Basin only. The two species of *Rugosochonetes* are remarkably close to species from the Visean of Great Britain. *R. macgregori* bears a close resemblance to *R. silleesi* Brunton from the low D Zone of Ireland, and to the fine-ribbed holotype group of *R. celticus* Muir-Wood from North Wales (Brunton, 1968); and *R. ustulatus* is morphologically close to *R. transversalis* Brunton, also from the low D Zone of Ireland (Brunton, 1968).

Conodonts in the *pauciplicatus* Zone indicate a slightly younger age (early Visean) than that suggested by some of the brachiopods (Druce, 1969), but foraminifers identified by Mamet & Belford (1968) belong to Mamet's foraminifer zones 14 or 15, equivalent to the V3a and V3b (late Visean) Zones of Belgium.

Anthracospirifer milliganensis Zone (emended)

The *milliganensis* Zone was designated by Thomas (1970) as the late Visean to possibly Namurian *Anthracospirifer milliganensis* zonal assemblage. Thomas recognized the assemblage in the Burvill Beds in the Weaber Range and the Milligans Hills. The assemblage is now given zonal status, and emended to extend from the base of the Burvill Beds to about 100 feet above the base of the overlying Point Spring Sandstone. The *milliganensis* Zone has *Anthracospirifer milliganensis* Thomas as the nominate species, and is characterized also by *Syringothyris fontanalis* sp. nov., *Schellwienella weaberensis* Thomas, *Rugosochonetes* sp. cf. *R. kennedyensis* Maxwell, *Globosochonetes burvillensis* gen. et sp. nov., and *Cleiothyridina?* sp. The faunal content of the zone varies from locality to locality (Appendix 1, Table 1), probably because of differences in environment accentuated by proximity to the shore, and the difficulty in collecting brachiopods from hard ferruginized calcareous gravel, the most common fossiliferous rock-type in the Burvill Beds. A number of species, for example, *Globosochonetes burvillensis* and *Pleuropugnoides* sp., are known from one locality only, and correlations within the basin are based mainly on the association of *Schellwienella weaberensis, Dictyoclostus? funiferus,* and *Rugosochonetes* sp. cf. *R. kennedyensis;* the nominate species, *Anthracospirifer milliganensis,* is known so far from Milligans Hills and Bonaparte No. 1 Well.

Thomas (1970) determined the age of the zone as late Visean to possibly Namurian on the affinities of *Anthracospirifer milliganensis* and *Schellwienella weaberensis.* He compared *milliganensis* with the *Anthracospirifer bisulcatus* (Sowerby) group of species in the late Visean and Namurian of Britain and Belgium, with *Spirifer parabisulcatus* Semichatova from the late Visean of the Moscow Basin, and with *Spirifer nox* Bell, from the upper Windsor of Nova Scotia. *Schellwienella weaberensis* is compared with *S. ornata* Demanet from the late Visean of Belgium, and with *S. reprinki* Sokolskaya from the middle Visean of the Moscow Basin (Thomas, 1970).

Conodonts from the Burvill Beds indicate an age of not older than D_2, in terms of the British coral-brachiopod zones, and possibly as young as Namurian (Druce, 1969). Foraminifers from the same beds are in Mamet's lower zone 16, equivalent to the uppermost V3b and lower V3c (late Visean) of Belgium (Mamet & Belford, 1968).

Echinoconchus gradatus FAUNA

The youngest Carboniferous brachiopod fauna in the Bonaparte Gulf Basin is found in outcrop at locality 458/12, 600 feet above the base of the Point Spring Sandstone. The fauna is missing, in some instances by erosion, or has not been found, in other sections measured through the Point Spring Sandstone along the southern scarp of the Weaber Range (Veevers & Roberts, 1968, fig. 59). The fauna, which is poorly preserved, consists of *Echinoconchus gradatus* Campbell, *Ovatia* sp. B, *Punctospirifer* sp., and *Brachythyris* sp. cf. *B. peculiaris* (Shumard). Of these, both *Echinoconchus gradatus* and *Ovatia* sp. B are recorded from the *milliganensis* Zone in the Burvill Beds, and *Punctospirifer* sp. from the Waggon Creek Breccia and Core 4 of Bonaparte No. 1 Well. Other probable occurrences of the *gradatus* fauna are in the subsurface Tanmurra Formation in Core 2 (677 feet to 690 feet) in Bonaparte No. 1 Well, and in Core 23 (6697 feet to 6718 feet) in Kulshill No. 1 Well. Core 2 of Bonaparte No. 1 Well contains *Echinoconchus gradatus,* and the same species is recorded with *Composita* sp. and *Schellwienella* sp. in Core 23 of Kulshill No. 1 Well.

The age of the *gradatus* fauna in the Point Spring Sandstone is inferred as Namurian because of its stratigraphic position above the late Visean to Namurian *milliganensis* Zone. Support for a Namurian age is provided by foraminifers of Mamet's zone 17 (early Namurian) in the upper part of the Tanmurra Formation (Mamet & Belford, 1968, fig. 2). Using foraminifers in Kulshill No. 1 Well, Caye & Le Fevre (appendix VI (d) *in* Duchemin & Creevey, 1966) date the interval 6060 feet to 6100 feet as late Visean to early Namurian, and the 7760-foot level as Visean. In eastern Australia, *Echinoconchus gradatus* Campbell is recorded from the Visean *Orthotetes australis* and *Delepinea aspinosa* Zones (Roberts, 1965; Banks et al., 1969).

CORRELATION WITHIN THE BONAPARTE GULF BASIN

In the Bonaparte Gulf Basin, in contrast to most microfossils, brachiopods are found in a wide variety of marine sediments, and brachiopod zones play an important role in determining correlations within the basin. The brachiopods, along with most of the other fossil groups, are more common in the platform sequences of sandstone, limestone, and calcarenite than in the deeper-water basinal shale and siltstone. Brachiopod zones have been of most value in determining stratigraphic relationships in the widespread Tournaisian sediments, particularly in correlating sandy nearshore sediments in the northeastern part of the basin with the standard platform sequences in the south. Evidence from brachiopods, conodonts, foraminifers, and spores was used in the preparation of a correlation chart of the important sections in the Bonaparte Gulf Basin given by Veevers & Roberts (1968, fig. 40) and reproduced in Figure 7. The following discussion will substantiate, and in some cases modify, those correlations based on brachiopods.

DEVONIAN

One zone only, the *Crurithyris apena* Zone, is recognized within the Frasnian Cockatoo Formation. Brachiopods from the zone provide a correlation between the Kununurra Member and possibly the Abney Member in the south and the Hargreaves and Westwood Members in the north (Roberts et al., 1967). *Crurithyris apena* Veevers and *Steinhagella* sp. are present in the Kununurra and

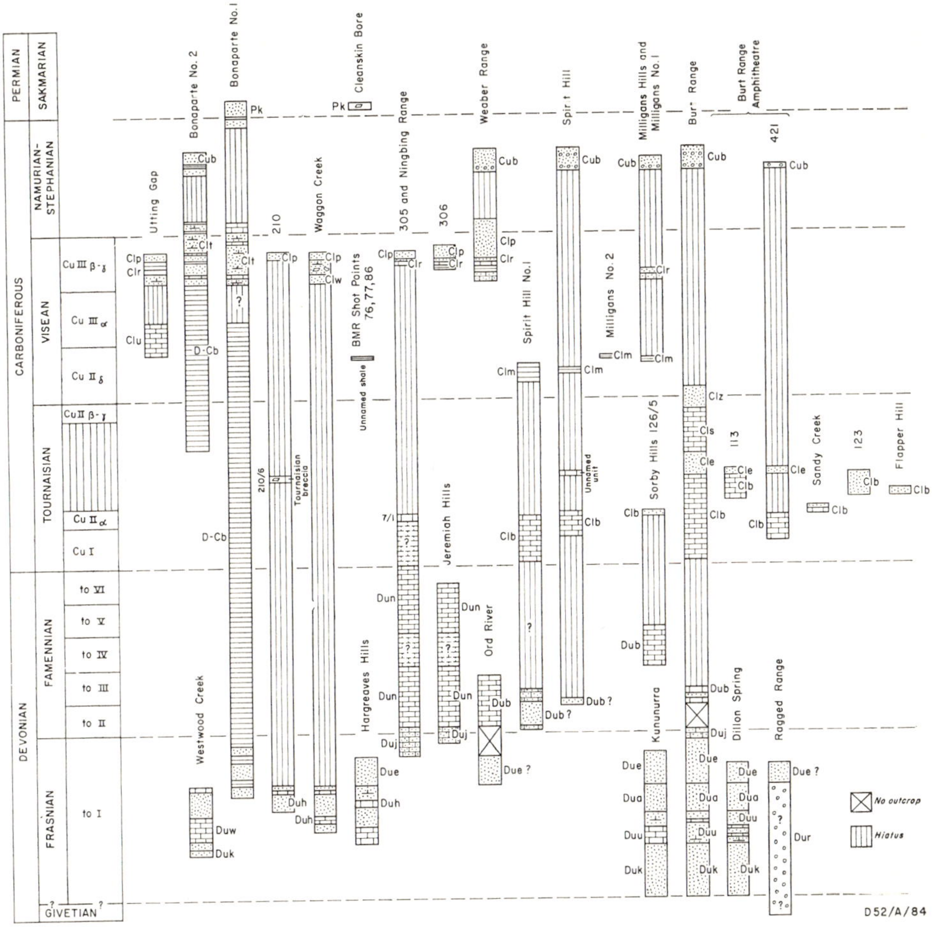

Figure 7. Correlation of important Devonian and Carboniferous sections in the Bonaparte Gulf Basin.

Westwood Members, and *Productella westwoodensis* sp. nov., *Spinatrypa prideri larga* subsp. nov., *Tenticospirifer columnaris* sp. nov., and an indeterminate rhynchonellid in the Hargreaves and the Westwood Members. These correlations are supported by conodonts identified by E. C. Druce (Roberts et al., 1967).

A reef-complex in the Famennian hampers a conventional biostratigraphical study because continuous sequences of bedded sediments are present in the back-reef and associated lagoonal sediments only. One zone, the *Cyrtospirifer ningbin-*

gensis Zone, is recognized throughout the Famennian portion of the Ningbing Limestone and in the early Famennian Buttons Beds. The Buttons Beds are offshore lagoonal sediments deposited behind the back-reef (Veevers, 1968) to the south and southeast of the main reef-complex.

CARBONIFEROUS

The *Spinocarinifera adunata* Zone, the oldest Tournaisian zone, is recognized in the westernmost outcrops of the Burt Range Formation on the plain to the west of Mount Septimus and Enga Ridge. It has not been identified elsewhere in the basin.

The succeeding *Acanthocosta teicherti* Zone is identified in the Tournaisian part of the Ningbing Limestone (locality 7/1) by the presence of *Acanthocosta teicherti* gen. et sp. nov., *Kitakamithyris occidua* sp. nov., *Schuchertella peltata* sp. nov., and Productacean gen. et sp. A. Elsewhere, the zone is identified in the Burt Range Formation in section 421, at locality 421/5, 120 feet above the base of the section at the northern end of the Burt Range Amphitheatre (Appendix 1, Fig. 12).

One of the most widespread zones in the Tournaisian platform sequence, the *Lomatiphora aquila* Zone, is recognized throughout the Burt Range area, and along the eastern margin of the basin to near Legune Homestead. The zone is identified in the Burt Range Formation in section 421 at the northern end of the Burt Range Amphitheatre; at locality 421/17, 465 feet above the base of the section, the fauna consists of *Lomatiphora aquila* gen. et sp. nov. and *Punctospirifer plicatosulcatus* Glenister (Appendix 1, Fig. 12). In measured sections this is stratigraphically the first appearance of *plicatosulcatus*; small specimens of *Punctospirifer* are known from cuttings in Ivanhoe water bore No. 2 (Appendix 2, Fig. 10) from probably an even lower stratigraphic level in the Burt Range Formation. At Sorby Hills at the northern end of the Pincombe Range, and at Spirit Hill, faunas from the Burt Range Formation containing *Lomatiphora aquila* are less diverse than those in the Burt Range area. *Lomatiphora aquila* and the accompanying species range down into the *teicherti* Zone; however, because *Acanthocosta teicherti* gen. et sp. nov. is absent, the faunas are referred to the *aquila* Zone. The fauna from locality 126/5 in the Sorby Hills contains *Lomatiphora aquila, Cleiothyridina gloveri* Thomas, and two rhynchonelloids. At Spirit Hill, the zone is in dolomitic limestone at locality 117/1, Spirit Hill Cave; the fauna from this locality includes *Lomatiphora aquila, ?Austrochoristites levisulcatus* gen. et sp. nov., and *Kitakamithyris occidua* sp. nov., together with the longer-ranging *Rugosochonetes obtectus* sp. nov., and *Cleiothyridina gloveri* Thomas. In section 114, in the southeastern part of Spirit Hill (Appendix 2, Fig. 11), *Lomatiphora aquila, Cleiothyridina gloveri,* and a rhynchonelloid are present in quartz sandstone at locality 114/9. A short distance northwards in section 115, *Lomatiphora aquila* and *Cleiothyridina gloveri* are present in quartz sandstone at locality 115/7. Both of these localities lie above a well-marked disconformity (Veevers & Roberts, 1968, fig. 58). *Lomatiphora aquila* is present on either side of the disconformity in section 116, at the northern end of the amphitheatre in southeastern Spirit Hill, suggesting that the hiatus was brief and confined within the *aquila* Zone, or at the most, within the range of *Lomatiphora aquila* (Fig. 5); abundant specimens of *aquila* are recorded from locality 116/1, and poorly preserved specimens from 115 feet above the base of the section.

Ferruginized and silicified outcrops of the Burt Range Formation near the Carlton to Legune track crossing at Sandy Creek (Appendix 2, Fig. 12) contain a fauna characteristic of the upper part of the *aquila* Zone: *Schuchertella dorsiplana* Thomas, *Rugosochonetes obtectus* sp. nov., *Lomatiphora aquila* gen. et sp. nov., *Rhytiophora* sp. cf. *R. raricostata* (Herrick), *Cleiothyridina gloveri* Thomas, *Ectochoristites? arenatus* Thomas, and *Prospira incerta* Thomas. From the same general locality Thomas (1970) describes *Prospira incerta* Thomas, *Ectochoristites? arenatus* Thomas, and *Cleiothyridina gloveri* Thomas, and Veevers (1959b) refers a rhynchonelloid to *Camarotoechia septima* Veevers; I consider the rhynchonelloid to be closer to *Septemirostellum simplex* sp. nov., from the Burt Range Formation, than to *S. septimum,* which is restricted to the Septimus Limestone and Zimmermann Sandstone. Thomas's specimens of *Ectochoristites? arenatus* are poorly preserved, and cannot be compared adequately with *Austrochoristites levisulcatus* gen. et sp. nov. from the Burt Range Formation. *Prospira incerta* Thomas appears earlier in the northeastern part of the basin than in the Burt Range area, being recorded in the *aquila* Zone at Sandy Creek, and in quartz sandstone identified as Burt Range Formation at locality 123/14, an isolated hill 11.5 miles northeast towards Legune Homestead from the Sandy Creek locality (Appendix 2, Fig. 12). Evidence in support of the earlier appearance of *P. incerta* in the northeast is provided by its occurrence with *Unispirifer laurelensis* (Thomas) and *Septemirostellum simplex* sp. nov. at Sandy Creek; both of these species are older than the specimens of *Prospira incerta* recorded from the Burt Range area (Fig. 5). Similarly, *Rhytiophora* sp. cf. R. *rariscostata* (Herrick) occurs slightly later in the Burt Range area (Fig. 5).

The *Unispirifer laurelensis* Zone is confined to the Burt Range area of the southeastern platform. As well as being in the uppermost beds of section 101, part of the composite section of the Burt Range Formation (Veevers & Roberts, 1968, fig. 41), it is present in the lower parts of sections 109 and 128 near the middle of Enga Ridge (Appendix 2, Fig. 10). In section 109 the zone extends from the base to locality 109/3 at 280 feet (Appendix 1, Fig. 8), and more than 100 feet of the base of section 128 belongs to the *laurelensis* Zone (Appendix 1, Fig. 9). The boundary between the *laurelensis* and overlying *eganensis* Zones in section 128 is not accurately defined because of lack of fossils, but lies between 100 feet and 250 feet above the base.

The *Grammorhynchus eganensis* Zone is widespread in the Burt Range area, mainly along the length of Enga Ridge. In section 109 it is identified between 280 feet and 725 feet in the uppermost beds of the Burt Range Formation and lowermost beds of the Enga Sandstone; the brachiopod fauna and its distribution are given in Appendix 1, Figure 8. The zone occupies a similar stratigraphic interval in sections 128 (Appendix 1, Fig. 9) and 129 (Appendix 1, Fig 10). At the southern end of the Burt Range Amphitheatre the zone is tentatively identified in calcareous sandstone between 425 feet and 500 feet above the base of section 113, at the top of the Burt Range Formation (Appendix 1, Fig. 11).

Faunas of the *Septemirostellum amnicum* Zone are identified from widely separate areas in the basin. The zone is recognized in the Enga Sandstone along the length of Enga Ridge (*see* especially Appendix 1, Figs 7, 13), and from an isolated outcrop at locality 78/3, 3 miles northeast of Mount Septimus (Appendix 2, Fig. 8). An unnamed dolomite breccia of Tournaisian age deposited on the shore of the Waggon Creek valley (Veevers & Roberts, 1966) contains *Schizophoria* sp. cf. *S. resupinata* (Martin), *Rhipidomella michelini?* (Léveillé), *Leptagonia analoga* (Phillips), *Protoniella? waggonensis* sp. nov., *Magnumbonella prolata* sp. nov., *Spinauris cristata?* sp. nov., *Rugosochonetes obtectus* sp. nov., *Septemirostellum amnicum* Veevers, *Prospira incerta* Thomas, *Eomartiniopsis costata* sp. nov., *Cyrtina* sp., *Composita* sp., and *Cleiothyridina* sp., and is identified as *amnicum* Zone. Section 116 at Spirit Hill (Appendix 2, Fig. 11) contains a grey sandy limestone overlying, probably unconformably, brown dolomitic sandstone containing the *aquila* Zone. The grey limestone, numbered locality 136/1, contains *Rhipidomella michelini?* (Léveillé), *Rugosochonetes obtectus* sp. nov., *Cleiothyridina* sp., *Austrochoristites levisulcatus* gen. et sp. nov., *Punctospirifer plicatosulcatus* Glenister, *Eomartiniopsis costata* sp. nov., and *Septemirostellum* sp., and may belong to either the *eganensis* or *amnicum* Zones; Veevers & Roberts (1968) correlated the grey limestone with the basal part of the Enga Sandstone.

Apart from the type section of the Septimus Limestone (Appendix 1, Fig. 14), the *Schellwienella australis* Zone is recognized in the lowermost 190 feet in section 110 at Mount Zimmermann (Appendix 1, Fig. 16). It cannot be positively identified in either section 150 at the southwestern end of Mount Septimus, or in section 112 at Langfield Point (Appendix 1, Figs 15, 17).

The succeeding *Syringothyris langfieldensis* Zone provides a correlation within the Septimus Limestone throughout the central Burt Range area. A time line given by Veevers & Roberts (1968, fig. 47), correlating sections of the Septimus Limestone, utilized an abundant fauna (Thomas, 1962a, assemblage c) just above the base of the *langfieldensis* Zone; this fauna could of course be ecologically controlled. The zone is identified in Septimus Limestone from 190 feet to possibly 375 feet above the base in section 110 at Mount Zimmermann; from 125 feet to 325 feet in section 112 at Langfield Point, and in section 119, half a mile east of Langfield Point (Appendix 2, Fig. 10). At the southwestern margin of the Burt Range Amphitheatre, in section 111, Enga Sandstone is overlain unconformably by Septimus Limestone of the *langfieldensis* Zone; section 111 contains fossil localities 138/1 and 138/2. Dolomitic and calcareous sandstone at the top of section 416, 1 mile east of Mount Septimus, is tentatively referred to the *langfieldensis* Zone.

The *Spirifer spiritus* Zone is not known south of Mount Septimus, but has been identified by Thomas (1962a) in his localities 6 and 13, scattered outcrops of Septimus Limestone, near Spirit Hill. Veevers & Roberts (1968, figs 47, 48) interpreted the irregular upper surface and progressive thinning southwards of the Septimus Limestone in the central Burt Range as an erosional break before the deposition of the overlying Zimmermann Sandstone. Brachiopod species in the Zimmermann Sandstone are not sufficiently diagnostic to indicate either an age younger than or a lateral equivalence to the *spiritus* Zone. Because of the erosional surface, section 110 of the Zimmermann Sandstone is placed on top of section 104 of the Septimus Limestone to give the most complete section in the chart showing ranges of brachiopod species in the upper part of the Tournaisian (Fig. 6).

The *Punctospirifer pauciplicatus* Zone is recognized throughout the Utting Calcarenite at Utting Gap, and in an isolated locality near Tanmurra Creek

(Veevers & Roberts, 1968, pl. 1). No other occurrences of the zone have been identified.

Late Visean shoreline gravels and part of the succeeding paralic quartz sandstone are correlated, on the basis of the *Anthracospirifer milliganensis* Zone, with interbedded calcarenite and calcareous sandstone deposited farther offshore. The *milliganensis* Zone is recognized throughout the Burvill Beds and from the base of the Point Spring Sandstone in outcrops extending from Milligans Hills to the northwestern end of the Weaber Range; in the Tanmurra Formation penetrated by Bonaparte No. 1 and Kulshill No. 1 Wells; in the upper part of the Bonaparte Beds in Bonaparte No. 1 Well; and possibly in the Waggon Creek Breccia on the margin of Waggon Creek valley. The fauna from outcropping parts of the zone is listed in Table 1 of Appendix 1. The zone is identified at locality 301, Milligans Hills, in sections 25, 435, and 436 in the southern scarp of the Weaber Range (Appendix 2, Fig. 14), section 453 at Burvill Point, locality 306, three-quarters of a mile north-northwest of Burvill Point, locality 5/2, three-quarters of a mile south of Ningbing Homestead (Appendix 2, Fig. 6), and locality 108/7 at Utting Gap (Appendix 2, Fig. 13). In the subsurface, the *milliganensis* Zone is identified tentatively in Bonaparte No. 1 Well in core 4 (1257 feet to 1262 feet), core 5 (1561 feet to 1566 feet), and core 6 (1840 feet to 1851 feet), and in Kulshill No. 1 Well in core 27 (7752 feet to 7763 feet). In Bonaparte No. 1 Well, core 4 (Tanmurra Formation) contains *Schizophoria* sp. cf. *S. resupinata* (Martin), *Punctospirifer* sp., *Composita bonapartensis* Thomas, and *Schellwienella* sp.; core 5 (Tanmurra Formation) contains *Anthracospirifer* sp. cf. *A. milliganensis* Thomas; and core 6 (Bonaparte Beds) contains *Lingula* sp., *Rhipidomella michelini?* (Léveillé), *Schellwienella weaberensis* Thomas, *Dictyoclostus? funiferus* sp. nov., *Cleiothyridina?* sp., *Anthracospirifer milliganensis* Thomas, and an Elythidid gen. et sp. A species tentatively referred to *Marginatia* sp. from cores 6 and 7 in Bonaparte No. 1 Well may also be present in the *milliganensis* Zone in outcrop; insufficient material is available to compare this form with *Dictyoclostus? funiferus,* another productacean in the *milliganensis* Zone. In Kulshill No. 1 Well, core 27 contains *Dictyoclostus funiferus* sp. nov., *Echinoconchus gradatus* Campbell, *Rugosochonetes* sp., *Tangshanella?* sp., Elythidid gen. et sp., *Cleiothyridina* sp., *Spirifer* sp., and *Podtsheremia? thomasi* sp. nov.

Veevers & Roberts (1966) correlated the Waggon Creek Breccia with the Burvill Beds. The brachiopod fauna from the basal pebbly sandstone of the breccia (Appendix 1, Table 1) does not contain sufficient numbers of diagnostic species to definitely substantiate this correlation, but it does suggest a Visean age. Bivalves from the same bed, *Pterinopecten* sp., *Phestia* sp., *Anthraconeilo* sp., *Parallelodon* sp., and *Sanguinolites* sp., are recorded in the Burvill Beds also, and hence support the original correlation.

The probable occurrences of the *Echinoconchus gradatus* fauna outside of the Weaber Range are given on page 22.

CORRELATION WITH OTHER SEDIMENTARY BASINS IN WESTERN AUSTRALIA

Devonian and Carboniferous marine sediments are known in the Canning Basin, 300 miles to the southwest, and in the Carnarvon Basin, more than 1000 miles to the southwest, of the Bonaparte Gulf Basin (Fig. 8). Although most of the brachiopod faunas from these sediments have recently been described (Coleman, 1951; Glenister, 1955; Veevers, 1959a; 1959b; Thomas, 1970), there is difficulty in correlating precisely between the basins because the stratigraphic ranges of the species have either not been determined, or are obscured by intricate stratigraphy

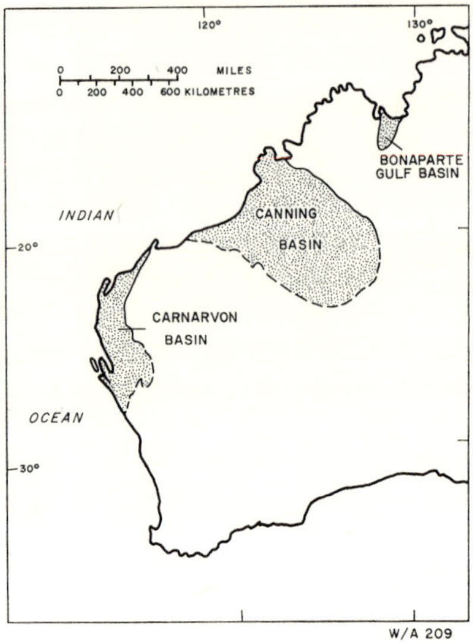

Figure 8. The major Palaeozoic sedimentary basins in the western part of Australia.

within reef-complexes. Correlation charts of the Bonaparte Gulf, Canning, and Carnarvon Basins, based on all available faunal evidence, are given by Johnstone et al. (1967) for the Devonian, and Banks et al. (1969) for the Carboniferous (the charts are combined in Fig. 9).

Canning Basin

The *Crurithyris apena* Zone provides the only correlation based on brachiopods between the late Devonian rocks of the Canning and Bonaparte Gulf Basins. On the basis of the *apena* Zone, the Kununurra, Hargreaves, Westwood, and possibly the Abney Members of the Cockatoo Formation correlate with the Frasnian limestone in the Bugle Gap area (probably reef), near Longs Well, and in the Oscar Range area (Veevers, 1959a, p. 15).

None of the succeeding Frasnian and Famennian brachiopod zones recognized by Veevers (1959a) in the Canning Basin are identifiable in the Bonaparte Gulf Basin. Veever's *saltica, torrida,* and *scopimus* Zones are in a reef-complex of similar morphology, and in the case of the *scopimus* Zone age, to the Ningbing Limestone; and the late Famennian *proteus* Zone is recognized both in platform limestone (Fairfield Formation) and in back-reef Pillara Limestone (Playford & Lowry, 1966). Famennian brachiopods are, with the exception of *Ptychomaletoechia lucida* Veevers, endemic to either the Bonaparte Gulf or Canning Basins, and hence suggests that at this time there was no exchange of brachiopod species between the basins. Species of conodonts and ostracods, and the bivalve *Buchiola* sp., which inhabits basinal sediments, are common to the Famennian parts of both basins (Roberts et al., 1967), and indicate a marine connexion between them. The barrier to migration of the reef-dwelling brachiopods is interpreted as the absence in the Famennian of a reef-complex around the Precambrian Kimberley Block between the Canning and Bonaparte Gulf Basins. Frasnian and Tournaisian faunas, on the other hand, are in many ways comparable, and suggest an extensive interchange of species, and possibly the presence of interconnecting reefs in the Frasnian and early Tournaisian.

The Tournaisian Laurel Formation, included by Playford & Lowry (1966) in the Fairfield Formation, is used here in the sense of Thomas (1959). Detailed sections through the Laurel Formation and ranges of species described by Thomas (1970) and presently being described by G. A. Thomas and M. Garrett from the Fairfield Formation are needed to establish the faunal succession, and the relationship with the underlying Fairfield Formation. Thomas (1959), Jones (*in* Veevers & Wells, 1961), Banks et al. (1969), and Roberts & Veevers (in press) suggest an unconformity between the two formations. A continuous section between them is not known in outcrop, and hence it is difficult to prove a hiatus. The youngest known outcropping beds of the Fairfield Formation are toV-toVI in age (Glenister & Klapper, 1966), and the oldest outcropping beds of the overlying Laurel Formation are early but not earliest Tournaisian; the Fairfield Formation contains brachiopods of Veevers's *proteus* Zone, and the oldest outcropping beds of the Laurel Formation contain brachiopods of the *teicherti* Zone. Hence the hiatus, if present, is brief and equivalent to the latest Famennian (upper part of toVI) and earliest Tournaisian (used here in the sense of the German ammonoid zones — CuI).

In the subsurface in BMR Laurel Downs No. 2 Bore, Jones (*in* Veevers & Wells, 1961) suggests a hiatus equivalent to the lowermost part of the Tournaisian between the two formations. Roberts & Veevers (in press), however, point out a barren interval, 300 feet thick, between the fossiliferous late Devonian and early Carboniferous which could represent this time.

Thomas (1970) described most of the brachiopods in the Laurel Formation. His collection came from the upper 750 feet of the formation, particularly from 260 feet to 625 feet above the base of his section II (Thomas, 1959). Section I of Thomas (1959), the lower part of the Laurel Formation, contains the *Acanthocosta teicherti* Zone in its lower part, and is correlated with the lower to middle part of the Burt Range Formation in the Bonaparte Gulf Basin. Species in the lower part of the section include *Acanthocosta teicherti* gen. et sp. nov., *Rugosochonetes obtectus* sp. nov., *Schuchertella* sp., and *Cleiothyridina gloveri* Thomas. The upper part of this section contains *Grammorhynchus eganensis* (Veevers) and *Magnumbonella prolata* sp. nov., and probably belongs to a younger zone. Section II, the upper part of the Laurel Formation, contains the *Unispirifer fluctuosus* zonal assemblage (Thomas, 1970); included in this assemblage are *Rhipidomella michelini?* (Léveillé), *Schizophoria* sp. cf. *S. resupinata* (Martin), *Schuchertella dorsiplana* Thomas, *Marginatia mimica* sp. nov., *Rhytiophora* sp. cf. *R. raricostata* (Herrick), *Grammorhynchus eganensis* (Veevers), *Cleiothyridina minilya* Thomas, *Unispirifer fluctuosus* (Glenister), and *Unispirifer laurelensis* (Thomas). Within this fauna are recognized elements of the *Unispirifer laurelensis, Grammorhynchus eganensis,* and *Septemirostellum amnicum* Zones, and hence the upper part of the Laurel Formation is correlated with the upper part of the

Burt Range Formation, and the Enga Sandstone. Conodonts identified by E. C. Druce (Roberts & Veevers, in press) suggest an equivalence with the lower part of the Enga Sandstone. Detailed sections and determination of the ranges of species within the Laurel Formation are required before closer correlations can be made between the Laurel and Burt Range Formations.

Most species from the *fluctuosus* zonal assemblage are common to both the Bonaparte Gulf and Canning Basins, but it is noteworthy that the nominate species *Unispirifer fluctuosus* (Glenister), *Schellwienella* sp. cf. *S. minilyensis* Thomas, *Cleiothyridina? fitzroyensis* Thomas, and *Composita hendersoni* Thomas are restricted to the Canning Basin. Those species present in both basins include *Rhipidomella michelini?* (Léveillé), *Schuchertella dorsiplana* Thomas, *Unispirifer laurelensis* (Thomas), *Punctospirifer plicatosulcatus* Glenister, *Cleiothyridina minilya* Thomas, *Grammorhynchus eganensis* (Veevers), *Rugosochonetes obtectus* sp. nov., *Magnumbonella prolata* sp. nov., *Protoniella? waggonensis* sp. nov., and *Spinauris cristata* gen. et sp. nov.

Carnarvon Basin

Fewer correlations based on brachiopods can be made with the late Devonian and early Carboniferous sequence in the Carnarvon Basin. The Gneudna Formation of latest Givetian and Frasnian age (Veevers, 1959b; Condon, 1965) contains three brachiopods — *Spinatrypa prideri prideri* (Coleman), and probably *Ladjia saltica* Veevers and *Productella occidua* Veevers (? equals cf. *Productella* sp. of Veevers, 1959a) — found also in the Canning Basin, but none from the Bonaparte Gulf Basin. The indirect evidence of age equates the Gneudna Formation with part of the Cockatoo Formation in the Bonaparte Gulf Basin (Fig. 9; Johnstone et al., 1967, fig. 2). The overlying Munabia Sandstone and Willaraddie Formation are nearly unfossiliferous (Condon, 1965), and cannot be assigned precise ages.

Thomas (1970) recognizes his *Unispirifer fluctuosus* zonal assemblage from 460 feet to 730 feet above the base of the type section of the Moogooree Limestone. The fauna from this interval is: *Rhipidomella michelini?* (Léveillé), *Schellwienella minilyensis* Thomas, *Unispirifer fluctuosus* (Glenister), *U. laurelensis* (Thomas), *Kitakamithyris moogooriensis* Thomas, *Syringothyris spissa* Glenister, *Punctospirifer plicatosulcatus* Glenister, *P. mucronatus* Thomas, *Cleiothyridina minilya* Thomas, *Composita carnarvonensis* Thomas, and *Septemirostellum amnicum* (Veevers). Of these, *R. michelini?, U. laurelensis, P. plicatosulcatus, P. mucronatus, C. minilya,* and *S. amnicum* are known from the Bonaparte Gulf Basin also. *S. amnicum* is first recorded from 140 feet above the base of the Moogooree Limestone, and is probably longer ranging than in the Bonaparte Gulf Basin. Another possible discrepancy is the occurrence of *Punctospirifer mucronatus* at 460 feet above the base of the Moogooree Limestone and in the *spiritus* Zone in the Bonaparte Gulf Basin. Disregarding these possible discrepancies, the fauna from 460 feet to 730 feet in the Moogooree Limestone, although not diagnostic of a particular zone in the Bonaparte Gulf Basin, is probably equivalent to the *laurelensis, eganensis,* and *amnicum* Zones. Roberts & Veevers (in press) suggest that the fauna from 710 feet in the Moogooree Limestone could be similar to part of the fauna in the Septimus Limestone. Detailed correlations can be made only when the stratigraphic ranges are determined for the species in the Moogooree Limestone, and there is a fuller understanding of the ranges of

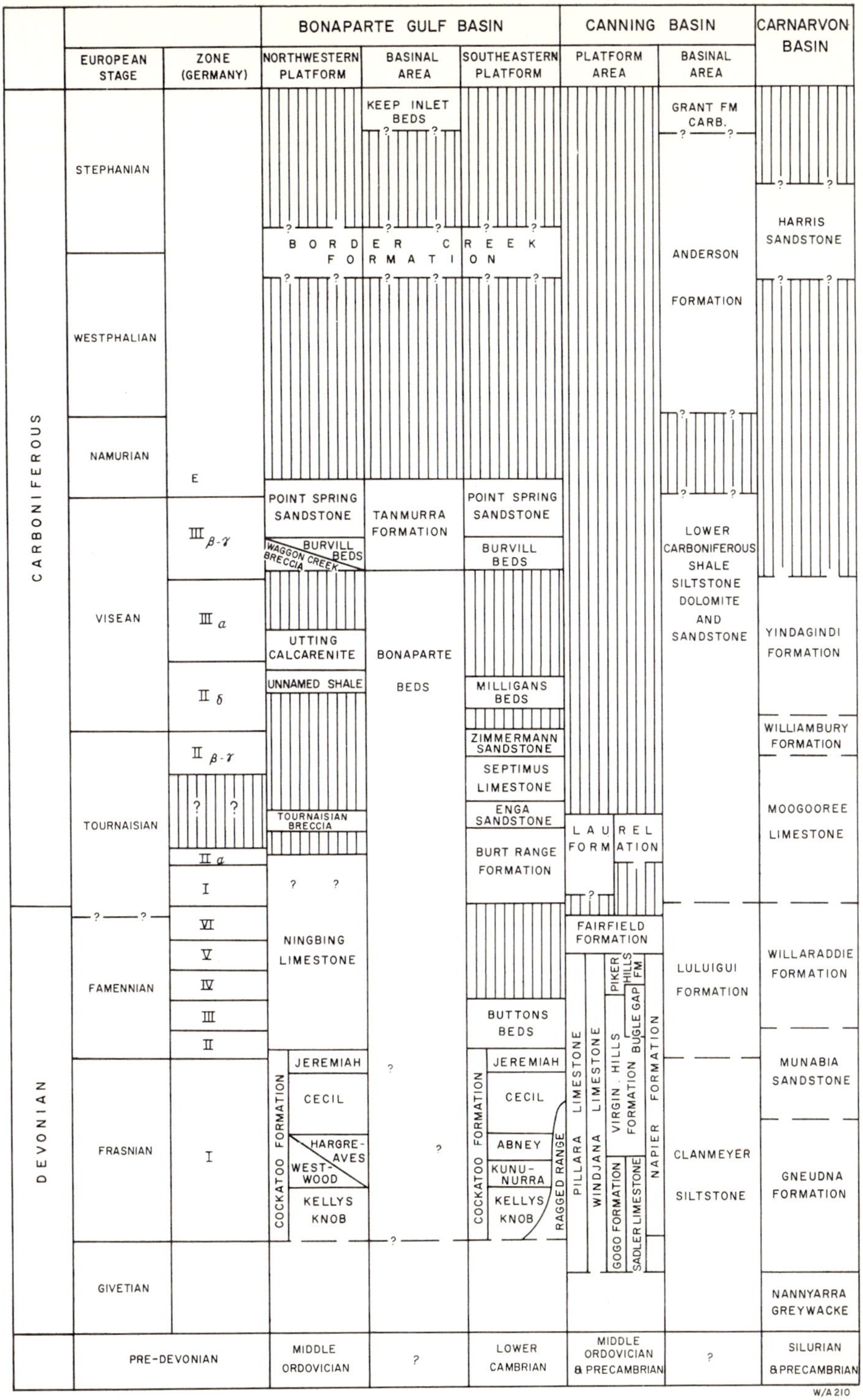

Figure 9. Correlation chart of the late Devonian and Carboniferous formations in the Bonaparte Gulf, Canning, and Carnarvon Basins. Data partly from Playford (1967), Johnstone et al. (1967), and Banks et al. (1969).

Grammorhynchus eganensis and *Septemirostellum amnicum* in the Canning and Carnarvon Basins. Evidence from conodonts could be used to determine the local ranges of both species, and then to compare these with the ranges in the Bonaparte Gulf Basin; this may give an insight into the time of possible migrations of faunas into the Bonaparte Gulf Basin from the Canning and Carnarvon Basins.

Younger Carboniferous formations in the Carnarvon Basin cannot be correlated precisely with the sequence in the Bonaparte Gulf Basin. The Williambury Formation is unfossiliferous (Thomas, 1962b; Condon, 1965), and the overlying Yindagindy Formation contains a sparse brachiopod fauna which Thomas (1962b; 1970) believes to suggest a Visean age.

CORRELATION WITH EASTERN AUSTRALIA

The late Devonian brachiopod faunas of eastern Australia are not as yet generally known in detail, and hence it is difficult to determine their relationships with the widespread and abundant late Devonian faunas of Western Australia. Late Devonian species from Queensland described by Maxwell (1950; 1951; 1954) do not appear closely related to any Western Australian species. Work on productoids from Queensland, presently being undertaken by R. G. McKellar, and on the faunas of the Cania region by J. F. Dear, may bring to light some faunal similarities.

In New South Wales, the late Devonian to possibly early Carboniferous Lambie Group, a post-orogenic sequence of mainly nonmarine, but in part marine, lithic sandstone, siltstone, shale, and conglomerate, contains a *Cyrtospirifer* fauna (Benson, 1922; Maxwell, 1950; Mackay, 1964). A small number of species have so far been described from New South Wales, and one of these was referred by Mackay (1964) to *Cyrtospirifer gneudnaensis* Glenister; this identification is difficult to verify from Mackay's illustrations. Apart from this one possibly related species, the late Devonian fauna of New South Wales cannot be closely compared with that of Western Australia. Veevers (1959a, p. 32) suggested that the differences between the eastern and Western Australian species reflected an inherent difference in the faunas, an inference made also by Teichert (1943, p. 168). Two other species of *Cyrtospirifer* described by Mackay (1964) were referred to American species, reinforcing the theories of Benson (1922) and Gregory (1930) that the Lambian faunas are similar to those in eastern North America.

Carboniferous brachiopods are distributed in two zoogeographic provinces within Australia; a western province in the intracratonic basins of Western Australia, and an eastern province in the New England Geosyncline (Campbell, fig. 6.2, *in* Brown, Campbell, & Crook, 1968).

The western province existed from the Tournaisian to the Namurian, at which time there was uplift, erosion, and the subsequent deposition of late Carboniferous nonmarine sediments, whereas the eastern province existed throughout nearly the entire Carboniferous period; an analysis of the fauna of the eastern province is given by Campbell & McKellar (1969). Faunas of each province contain many species of cosmopolitan brachiopod genera, as well as endemic forms, but there is only a small number of long-ranging cosmopolitan species common to both areas, and direct detailed correlations based on brachiopods cannot be made between them. Species common to both provinces include *Leptagonia analoga*

(Phillips), and species morphologically very close to *Rhipidomella michelini* (Léveillé) and *Schizophoria resupinata* (Martin) in the Tournaisian and Visean, and *Echinoconchus gradatus* Campbell, which is close to *E. elegans* (M'Coy), in the Visean and Namurian. One other species, *Prospira incerta* Thomas from the Bonaparte Gulf Basin, is morphologically close to *P. prima* Maxwell from the Tournaisian of Queensland.

Genera which on present knowledge are endemic to the western province include the new genera *Dorsoscyphus, Schistochonetes,* possibly *Acanthocosta, Lomatiphora, Spinauris, Grammorhynchus, Cardiothyris,* and *Austrochoristites.* Tournaisian to early Namurian species of the following genera are present in the western province, but have not been recorded in the eastern province: *Serratocrista* Brunton, 1968, *Globosochonetes* Brunton, 1968, *Spinocarinifera* gen. nov., *Magnumbonella* Carter, 1967, *Rhytiophora* Muir-Wood & Cooper, 1960, *?Protoniella* Bell, 1929, *Pleuropugnoides* Ferguson, 1966, *Septemirostellum* gen. nov., *Anthracospirifer* Lane, 1963, *Cyrtina* Davidson, 1858, *?Tangshanella* Chao, 1929, *Eomartiniopsis* Sokolskaya, 1941, and *Torynifer* Hall & Clarke, 1894. *Podtsheremia* Kalashnikov, 1966, may be represented in Queensland by *Spirifer* aff. *duplicostatus* Phillips, identified by McKellar (1967).

Genera endemic to the eastern province include *Tulcumbella* Campbell, 1963, *Asyrinxia* Campbell, 1957, and *Balanoconcha* Campbell, 1957. The Tournaisian to early Namurian part of the eastern province contains species of the following genera which have not been recognized in the west: *Streptorhynchus* King, 1850, *Lissochonetes* Dunbar & Condra, 1932, *?Tornquistia* Paeckelmann, 1930, *Quadratia* Muir-Wood & Cooper, 1960, *Eomarginifera* Muir-Wood, 1930, *Antiquatonia* Miloradovich, 1945, *Fluctuaria* Muir-Wood & Cooper, 1960, *Inflatia* Muir-Wood & Cooper, 1960, *Gigantoproductus* Prentice, 1950, *Marginirugus* Sutton, 1938, *Krotovia* Fredericks, 1928, *Eumetria* Hall, 1864, *?Merista* Suess, 1851, *Athyris* M'Coy, 1844, *Acuminothyris* Roberts, 1963, *Alispirifer* Campbell, 1961, *Spiriferellina* Fredericks, 1924, *Crurithyris* George, 1931, *Perissothyris* Carter, 1967, *?Retzia* King, 1850, and *?Rotaia* Rzhonsnitskaya, 1959 (Maxwell, 1954; Campbell, 1956; 1957; Campbell & Engel, 1963; Roberts, 1965; McKellar, 1967; Campbell & McKellar, 1969). No species of these genera have as yet been recorded in the western province.

Brachiopod zones established by Roberts (1965) and Campbell & Roberts (1969) from New South Wales, and the faunas of McKellar (1967) from Queensland, are combined in a sequence of zones extending from the Tournaisian to the Stephanian in a chart showing correlations of the Carboniferous of Australia (Banks et al., 1969; Campbell & McKellar, 1969). The zones, recognizable in the eastern province only, are dated by brachiopods, and ammonoids (where they are available) in terms of the German ammonoid zones. The faunas from the western province are referred to the German zones also, on the basis of conodont, brachiopod, ostracod, and foraminifer evidence, and hence it is possible to make generalized correlations throughout the marine Carboniferous of the continent. A comparison of the zones from the Bonaparte Gulf Basin with those in the eastern province shows that in the Tournaisian there are a greater number of zones in the Bonaparte Gulf Basin; this may be a reflection of the abundantly fossiliferous platform sediments compared with the more sparsely fossiliferous geosynclinal sediments in the east.

The Tournaisian sequence of nine zones from the *adunata* to the *spiritus* Zones is equivalent in age to the *Tulcumbella tenuistriata, Spirifer sol,* and part of the *Schellwienella* cf. *burlingtonensis* Zones; the *pauciplicatus* Zone to the *Orthotetes australis* Zone; and the *milliganensis* Zone to the *Rhipidomella fortimuscula* and possibly part of the *Marginirugus barringtonensis* Zones. Equivalents of the Westphalian *Levipustula levis* Zone and the Stephanian *Syringothyris bifida* Zone are not recognized in the western province because of late Carboniferous nonmarine sedimentation in the platform areas.

Marginirugus was identified by Öpik (*in* Noakes et al., 1952) from the Septimus Limestone, and is listed in that fauna by Traves (1955) and Thomas (1962a). Despite an exhaustive search in field seasons in 1963 and 1965, no further specimens have been found, and I suggest that the original reference to *Marginirugus* was a misidentification of a large specimen of *Marginatia mimica* sp. nov.

The faunas of the eastern and western provinces were distinct probably because of their geographical isolation on two margins of the Australian continent rather than because of climatic differences; palaeolatitude maps for the Devonian and lower Carboniferous of Australia given by Irving (1964) suggest there was slight change only in latitude, and hence probably a stable climate throughout the late Devonian and early Carboniferous. Continuous reef growth in the Bonaparte Gulf Basin during this time is supporting evidence, and infers continuation of a tropical or subtropical climate. A similar climate is envisaged for the Tournaisian and Visean of eastern Australia by Campbell & McKellar (1969).

OVERSEAS FAUNAL AFFINITIES

Geographically, the faunas of the Bonaparte Gulf Basin are closest to those of southeast Asia, notably Malaya, Vietnam, Burma, Thailand, and India. The latter faunas, however, are mostly poorly known, and with one exception in the Visean it is not possible to establish a direct link between faunas from Australia and elsewhere in the world.

Relationships between species given in the following discussion are dealt with in detail in the systematic portion of this Bulletin, and, in addition, references are given in the discussion on zones.

Frasnian faunas in the Bonaparte Gulf Basin are small, but can be readily related to a world-wide Frasnian 'platform' fauna best known in Europe, USSR, and North America. Excluded from the 'platform' fauna are assemblages from post-orogenic regions, such as the Catskill Delta in northeastern USA, which are characterized by different assemblages of species (Hall, 1852; 1867; Greiner, 1957). The closest affinities in the Frasnian lie with faunas in North America and the USSR which contain species morphologically similar to *Tenticospirifer columnaris* sp. nov., *Spinatrypa prideri larga* subsp. nov., and *Desquamatia (Synatrypa) kimberleyensis* (Coleman) (p. 13). To the north of Australia, the rich Frasnian fauna of China described by Grabau (1931a) contains species of *Productella* Hall, 1867, *Spinatrypa* Stainbrook, 1951, *Desquamatia* Alekseeva, 1960, and *Cyrtospirifer* Nalivkin, 1924, but none of these are related to forms in the Bonaparte Gulf Basin.

The Famennian fauna, which from present knowledge appears distinct from other Australian assemblages, is similarly closest to the 'platform' faunas of North

America and the USSR. *Ptychomaletoechia lucida* (Veevers) is related morphologically to *'Camarotoechia' sobrina* Stainbrook from the Percha Shale of New Mexico; *Cyrtospirifer ningbingensis* sp. nov. to *C. pamiricus paralis* (Reed) from the Pamirs, USSR, and to specimens referred to *C. whitneyi* (Hall) from the Ouray Limestone, New Mexico; and *C. depressus* sp. nov. to two species from USSR (p. 14). Species of the genera *Nayunella* Sartenaer, 1961, *Rugaltarostrum* Sartenaer, 1961, *Sentosia* Muir-Wood & Cooper, 1960, and *Leioproductus* Stainbrook, 1947, are restricted to the Famennian, and are known in the Bonaparte Gulf Basin and in North America; *Nayunnella* is recorded also from the Famennian of the USSR and China.

Carboniferous faunas of the Bonaparte Gulf Basin contain more taxa than those from the Devonian. In the Tournaisian, and to a lesser extent in the Visean, there is a substantial number of endemic forms (p. 33), but most species can be closely compared with others from various parts of the world. Cosmopolitan species such as *Leptagonia analoga* (Phillips), *Schizophoria resupinata* (Martin), and *Rhipidomella michelini?* (Léveillé) range throughout most of the lower Carboniferous.

Despite endemic forms, the Tournaisian species indicate faunal relationships with Europe, USSR (mainly the Kazakhstan-Siberian province of Einor et al., 1965), and North America; details of these relationships are given throughout the discussion on zones and in the systematics. Perhaps the most puzzling feature is the similarity between the fauna from the Chappel Limestone of Texas, a thin bioclastic limestone, and that from the early Tournaisian reef in the Ningbing Limestone. It is noteworthy that most of the species from the reef are absent from back-reef limestone of equivalent age on the southeastern platform of the Bonaparte Gulf Basin, and hence it is difficult to determine the significance of the faunal relationship. Species from elsewhere in the Tournaisian which have affinities with North American forms include *Spinocarinifera adunata* gen. et sp. nov., which may be compared with *S? arcuata* (Hall) from the Kinderhook of Missouri; *Unispirifer laurelensis* (Thomas) with *U. platynotus* (Weller) from the Kinderhook of Iowa, and possibly *U. minnewankensis* (Shimer) from the upper Banffian of Alberta; *Cleiothyridina minilya* Thomas with *C. obmaxima* (McChesney) from the Fern Glen Formation of Illinois and the Keokuk of Iowa; *Septemirostellum simplex* sp. nov. with *S? chouteauensis* (Weller) from the Chouteau Limestone of Missouri; and *Schellwienella australis* Thomas with *S. burlingtonensis* Weller from the Burlington Limestone in the Mississippi Valley area. In addition, *Rhytiophora calhounensis* (Moore) from the Chouteau Limestone of Missouri, *R. raricostata* (Herrick) from the Logan Formation of Ohio, and *Brachythyris peculiaris* (Shumard) have closely comparable species in the Tournaisian of the Bonaparte Gulf Basin.

As well as the cosmopolitan species listed earlier (pp. 32-33), European Tournaisian faunas contain a number of species morphologically comparable with forms from this basin. *Spinocarinifera adunata* gen. et sp. nov. is close to *S. niger* (Gosselet) from the early Tournaisian of Belgium, France, and Germany; *Septemirostellum simplex* sp. nov. to *S. mitcheldeanensis* (Vaughan) from the Tournaisian of Great Britain; *Tylothyris transversa* sp. nov. to *T. laminosa* (M'Coy) from the Tournaisian of Belgium and the Tournaisian and early Visean of Great Britain; *Spirifer spiritus* Thomas to *S. attenuatus* Sowerby from the late Tournaisian and early Visean of Britain and Belgium; and possibly *Septemirostellum amnicum*

(Veevers) to *S? acutirugatum* (de Koninck) from the Tournaisian of Belgium. Soviet faunas also contain specimens of *S? acutirugatum.* Species close to those from the Kuznetsk Basin in the Kazakhstan-Siberian province of Einor et al. (1965) are: *Septemirostellum simplex* sp. nov. to *S? biplex* (Tolmachoff); *Spinocarinifera adunata* gen. et sp. nov. to *S. inflata* (Sololskaya); *Marginatia mimica* sp. nov. to *M. deruptoides* Sarycheva; and *Unispirifer laurelensis* (Thomas) to *U. ussiensis* (Tolmachoff). *Unispirifer septimus* Thomas is close to *Spirifer ventricosus* Sokolskaya from the Moscow Basin, and to *S. mediocris* Tolmachoff and *S. similis* Tolmachoff from the Kuznetsk Basin; and *Torynifer? dorsiseptatus* Thomas to *T. pseudolineatus* (Hall), also from the Moscow and Kuznetsk Basins; the Moscow Basin is in the Europe-Tian Shan faunal province of Einor et al. (1965).

In Asia, the Tournaisian fauna from the Kitakami Mountains in northeast Japan described by Minato (1951; 1952) has only the cosmopolitan species *Schizophoria resupinata* (Martin) and *Leptagonia analoga* (Phillips) in common with that from the Bonaparte Gulf Basin. Chinese faunas, which are not well known, also contain few comparable species: *'Camarophoria' moshihlingensis* Ozaki and *'Camarotoechia' kinlingensis* Grabau may be comparable to *Septemirostellum amnicum* (Veevers); *Cleiothyridina obmaxima* (McChesney), described by Yang (1964), to *C. minilya* Thomas; and Ozaki (1939) records *'Avonia'* cf. *niger* (Gosselet) which could resemble *Spinocarinifera adunata* gen. et sp. nov.

The Visean and early Namurian faunas of the Bonaparte Gulf Basin are closest to those of Europe and North Africa, and to a lesser extent have similar elements to the Visean faunas, mainly from the Europe-Tian Shan faunal province, of the USSR. So far as can be determined from the available literature, they are close also to faunas in parts of China, parts of Japan, and southeast Asia, particularly Malaya.

In addition to long-ranging cosmopolitan species, *Productina margaritacea* (Phillips) and *Megachonetes zimmermanni* (Paeckelmann) are common to Europe and the Bonaparte Gulf Basin. Specific relationships between forms from the two areas include: *Pleuropugnoides* sp. with *P. pleurodon* (Phillips); *Echinoconchus gradatus* Campbell with *E. elegans* (M'Coy) (a comparable species, *E. biseriatus* (Hall), is also known from the late Mississippian in USA); *Delepinea uttingi* Thomas with *D. destinezi* (Vaughan) and *D. carinata* (Garwood); *Podtsheremia? humilicostata* sp. nov. and *P? thomasi* sp. nov. with the *Spirifer duplicostatus* Phillips group of species; *Anthracospirifer milliganensis* Thomas with the *Spirifer bisulcatus* (Sowerby) group of species; *Schellwienella weaberensis* Thomas with *S. ornamenta* Demanet; *Rugosochonetes macgregori* sp. nov. with *R. silleesi* Brunton, and the holotype group of *R. celticus* Muir-Wood; and *R. ustulatus* sp. nov. with *R. transversalis* Brunton. The genera *Serratocrista* Brunton, 1968, and *Globosochonetes* Brunton, 1968, have their only known representatives outside Great Britain in the Bonaparte Gulf Basin.

In the USSR, *Productina margaritacea* (Phillips) is recorded from the Donetz Basin, *Megachonetes zimmermanni* (Paeckelmann) from the Russian Platform and the Tournaisian of the Kuznetsk Basin, and *Echinoconchus elegans* (M'Coy) from the northeastern part of European USSR (Kalashnikov, 1966). *Stegacanthia strigis* sp. nov. may be close to *S. sibirica* (Sarycheva) from the late Tournaisian and early Visean of the Kuznetsk Basin; *Schellwienella weaberensis* Thomas resembles *S. reprinki* Sokolskaya from the Moscow Basin; and *Anthracospirifer milliganensis* Thomas is morphologically related to *Spirifer parabisulcatus* Semichatova.

A Visean fauna from Malaya described by Muir-Wood (1948), geographically the closest fauna to that of the Bonaparte Gulf Basin, contains *Echinoconchus elegans* (M'Coy), which is close to *E. gradatus* Campbell, *Schizophoria* sp., probably comparable with *S.* sp. cf. *S. resupinata* (Martin), *Leptagonia* cf. *analoga* (Phillips), and *Productina* cf. *margaritacea* (Phillips). *Pugnax asiaticus* Muir-Wood, which in many aspects resembles *P. acuminatus* (Martin), may be the same as one specimen of an acuminate rhynchonellid from the Utting Calcarenite.

The Visean faunas from China are not fully described, but some resemblances are seen with the assemblage from the Bonaparte Gulf Basin. Chao (1927) illustrated *Echinoconchus elegans* (M'Coy), which is close to *E. gradatus* Campbell, and a *Striatifera*-like species referred to *Kansuella maxima* (M'Coy), which resembles the Gigantoproductinid in the Burvill Beds.

Mid-continental North American Visean and early Namurian faunas have little in common with those of this basin, Europe, or the USSR. The less well-known faunas of the Cordilleran belt in western North America, however, may have somewhat closer relations with the widespread European type of fauna. Three species from the Bonaparte Gulf Basin can be compared with forms from the Cordilleran region: *Podtsheremia? humilicostata* sp. nov. and *P? thomasi* sp. nov. with *'Anthracospirifer' shoshonensis* (Branson & Greger); and *Spirifer* sp. with *Spirifer haydenianus* Girty. Much of the Cordilleran fauna is still awaiting detailed description.

The overseas faunal affinities may be summarized:

1. Frasnian and Famennian faunas have much in common with 'platform' faunas in Europe, USSR, and North America.

2. The Tournaisian fauna contains a large number of endemic forms, which implies isolation from major migratory routes, but nevertheless can be related to faunas in Europe, USSR, and North America.

3. Visean and early Namurian faunas are extremely close to those of Europe and North Africa, and to a lesser extent those of USSR (mainly the Europe-Tian Shan province); with the exception of *Echinoconchus gradatus* they have no known similarity with the faunas of continental North America. This contrasts with eastern Australia, where Campbell & McKellar (1969) demonstrate affinities with faunas of eastern and central Asia, and suggest intermittent connexions with central USA.

SYSTEMATIC DESCRIPTIONS

Order ORTHIDA Schuchert & Cooper, 1932

Superfamily ENTELETACEA Waagen, 1884

Family ENTELETIDAE Waagen, 1884

Subfamily SCHIZOPHORIINAE Schuchert & Le Vene, 1929

Genus SCHIZOPHORIA King, 1850

Type species: Conchyliolithus (Anomites) resupinatus Martin, 1809, by original designation of King, 1850.

SCHIZOPHORIA sp. cf. S. RESUPINATA (Martin)

(Pl. 1, figs 6-32)

Description: External. Shell dorsally biconvex, and transversely elliptical to elliptical in outline; greatest width at midlength; hinge approximately half as wide as shell; commissure usually rectimarginate, but ranging from uniplicate to sulcate; costellae increasing by bifurcation and intercalation, having a density of thirteen to twenty per 5 mm at 20 mm from umbo; primary costellae higher than later-developing costellae, and together with occasional costellae of other orders bearing raised anteriorly facing apertures along their length; concentric ornament of well-marked growth halts.

Pedicle valve moderately convex, having greatest convexity at umbo, becoming flatter towards margins; valve highest at one-third its length; umbo small, gently incurved over delthyrium, and forming an apical angle of 100-110°; sinus shallow, broad, developed on anterior half of some valves, but absent in others; interarea strongly apsacline to catacline, convex immediately beneath umbo to flat near hinge, ornamented with horizontal growth lines, and separated from umbonal shoulders by well-defined cardinal ridges; delthyrium open, with a delthyrial angle of 60-70°.

Brachial valve higher and more strongly convex than pedicle valve; greatest height between one-third length and midlength, greatest convexity at umbo; umbo small, but well defined by concave umbonal shoulders, straight to occasionally incurved, extending slightly behind hinge; lateral slopes moderately convex, becoming flat on posterolateral margins; fold very poorly defined, represented by median dorsal deflection of anterior commissure only; some valves with well-defined sinus commencing at umbo, becoming wider and deeper towards front; dorsal interarea anacline to catacline, lower than ventral interarea, concave beneath umbo to flat at hinge, and ornamented by horizontal growth lines; notothyrium bordered by recessed brachiophore plates, having a notothyrial angle of 70-100°, and filled with cardinal process.

Internal. Pedicle valve. Diductor muscle scars elongate, bluntly pointed posteriorly, becoming wider anteriorly to about two-thirds of their length, and having blunt anterior terminations; lateral margins bounded by ridges extending anteriorly from bases of dental plates; muscle field extending one-half to one-third length of valve; median ridge arising from pedicle callist, highest anteriorly (up to 2 mm high in a specimen 48 mm wide), and terminating abruptly at front of muscle field; individual adductor muscle scars not observed; adjustor muscle scars on slightly raised platforms on inner margins of dental plates in pedicle cavity; dental plates divergent, extending down inner margins of delthyrium, distally giving off laterally directed wedge-shaped teeth, and having strong curved anterior extensions; mantle canals narrow and radially arranged in juveniles; in mature individuals two well-defined canals of vascula media arising from anterolateral margins of ridge bearing adductor muscle scars or possibly median anterior extremities of diductor muscle scars, slightly convex outwards, branching anteriorly, and extending to front of valve; other radially arranged lateral canals originating from extensions of dental plates bordering muscle field.

Brachial valve. Anterior adductor muscle scars pear-shaped, and posteriorly impressed; posterior adductor muscle scars in deep elongate bifid depressions on posterolateral margins of anterior scars, bounded laterally by ridges extending from front of brachiophore bases; median ridge low, commencing beneath cardinal process, and extending to front of muscle field; cardinal process a single lobe in juvenile specimens, multi-lobed in adults, with up to four low lateral lobes on either side of high bi- or trilobed median projection (Fig. 10); process bearing

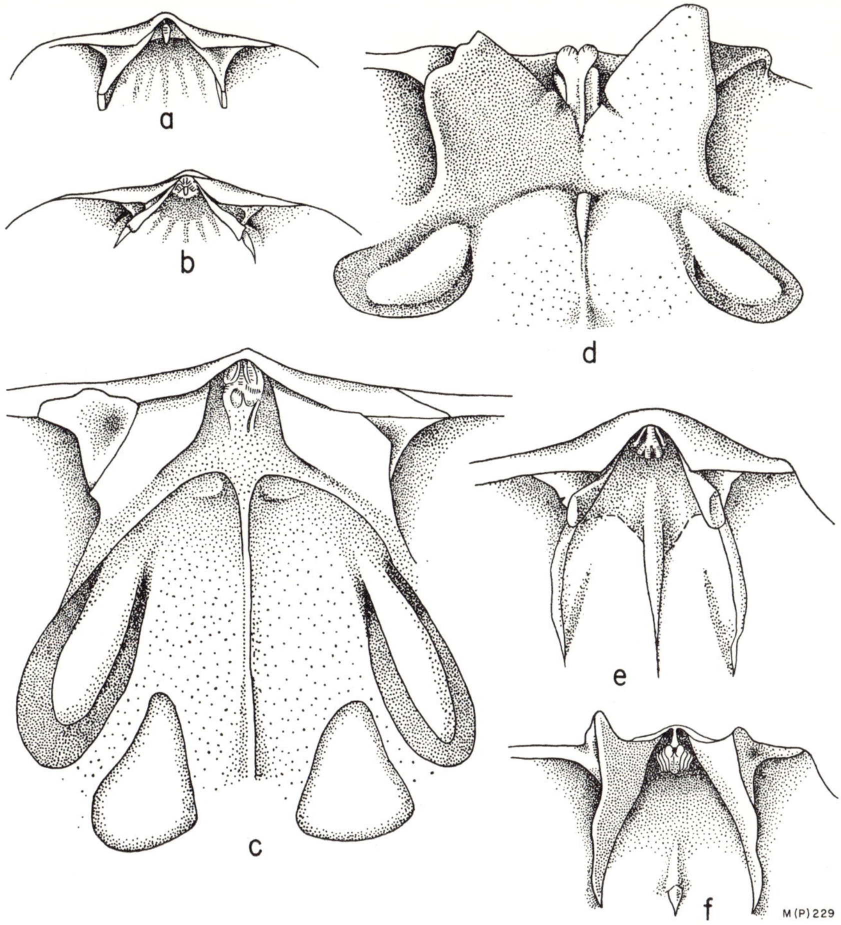

Figure 10. Ontogenetic development of the cardinal process in *Schizophoria* sp. cf. *S. resupinata* (Martin) from the single-lobed structure in juveniles to the multi-lobed structure in adults. a, b, c, and e are ventral views; d and f are anterior views of the apical parts of the brachial valve. a-CPC 11086, b-CPC 10801, c and d-CPC 10795, all from locality 108/0; e and f-CPC 10797, from locality 108/3; Utting Calcarenite, Utting Gap. × 3.

posteriorly facing lamellose myophores; in large individuals median lobes rising into stalk-like projections above apical callus; brachiophores erect, laterally divergent, and at an angle to supporting brachiophore plates; brachiophore bases diverging at 75-85°; sockets wide, supported by robust fulcral plates, posterior extremities infilled with callus; mantle canals narrow and radially arranged in juveniles; in mature individuals consisting of narrow linear vascula media (four

D

pairs) extending from inner margins of anterior adductor muscle scars, and vascula myaria (two pairs on each side of valve) extending from lateral margins of anterior adductor muscle scars or anterior extremities of posterior adductor muscle scars.

Measurements (in mm)

	Length (pedicle valve)	Length (brachial valve)	Width	Height
CPC 10796	—	21	27	9.5
CPC 10799	—	30.5	33.5	15.5
CPC 10800	—	15	19	5
CPC 10801	—	5	5.5	1.5
CPC 10802	5	—	6.5	2.5
CPC 10803	11	—	24.5	5
CPC 10804	9	—	11.5	3
CPC 10805	—	—	47.5	8.5
CPC 10807	18	—	21	7

Remarks: The pedicle valve of *Schizophoria* sp. cf. *S. resupinata* has a conservative morphology, and, apart from the variable development of a median sinus on the posterior of the valve, shows little intraspecific variation. The brachial valve on the other hand varies in convexity, thickness of the shell, and in the development of a fold and sinus. Most brachial valves are of moderate height and convexity. Some, which usually have a thick shell, are much higher and have a strongly convex venter; these are interpreted as gerontic individuals. Although the available material is fragmentary, it is clear that most brachial valves have an evenly convex venter, with the fold represented by a median dorsal deflection of the anterior commissure only. Two brachial valves from locality 108/0 are slightly more convex than average, and have a well-defined dorsal median sinus commencing at the umbo, and extending to the front of the valve. In all other respects they are identical with the typically moderately convex brachial valves, and hence are regarded as being within the range of variation of the species.

Most morphological features, and the variability in the shape of the shell, suggest a relationship with the European *S. resupinata* (Martin). *S. resupinata* has five varieties based mainly on differences in shape and outline (Demanet, 1934; Bond, 1941; Pocock, 1968). Most specimens from the Utting Calcarenite are morphologically close to *S. resupinata* s.s., differing only in the possession of a slightly shorter hinge, a consistently convex rather than resupinate pedicle valve, and brachiophore bases which are at an angle to the brachiophores. Some gibbous brachial valves from the Utting Calcarenite have an anterior profile closest to that of *S. resupinata pinguis* Demanet illustrated by Bond (1941, fig. 34), but have a rounded rather than a flat venter; their posterior profile is much lower than that of the specimens of *pinguis* illustrated by Demanet (1934, pl. 4, figs 9, 10) from the Visean (V3b) of Belgium. Internally, specimens from the Utting Calcarenite resemble *S. resupinata* in having weakly impressed ventral diductor muscle scars, and a narrow ventral median ridge.

S. resupinata dorsosinuata Demanet, described by Brunton (1968, pp. 11-17, pl. 2, figs 7-37) from the Visean of Fermanagh, Ireland, resembles this species from the Bonaparte Gulf Basin in having a consistently convex pedicle valve, and brachiophore bases which are at an angle to the brachiophores. Brunton's specimens of *S. resupinata dorsosinuata* are distinguished by constantly having a dorsal median sinus and subparallel anterior extensions of the dental plates along the margins of the diductor muscle scars; the brachial valve is less variable, and usually less convex than that of *S.* sp. cf. *S. resupinata*. Specimens of *dorsosinuata* described by Demanet (1934, pl. 3, figs 14-15) have resupinate pedicle valves and more strongly convex brachial valves than those described by Brunton; they are not as closely comparable with *S.* sp. cf. *S. resupinata*.

The eastern Australian species *S. verulamensis* Cvancara (1958, pp. 856-59, pl. 109, figs 14-16; pl. 110, figs 1-5) resembles this material in the length of hinge and the convexity of the pedicle valve. It is distinguished from *S.* sp. cf. *S. resupinata* by having more deeply impressed ventral diductor muscle scars, a broader ventral median ridge, a brachial valve which is less variable in convexity, and more deeply impressed mantle canals in the pedicle valve; the mantle canals in mature individuals of *S. verulamensis* have lateral distributory canals (Roberts, 1968) which have not been observed in specimens from the Bonaparte Gulf Basin.

Brunton (1968) described the ontogenetic development of *S. resupinata dorsosinuata* Demanet. The cardinal process in *dorsosinuata* began as a single longitudinal ridge, and ultimately in the adult formed a trilobed structure which consited of a bifid median lobe and two lower lateral lobes. The cardinal process in specimens from the Utting Calcarenite has a single longitudinal lobe in juveniles. This lobe tends to bifurcate, and also develops two low lateral lobes in slightly larger specimens; in the adult there are up to four lateral lobes on either side of a higher bifid or trifid median lobe (Fig. 10). Multiple lateral lobes are known also in *S. verulamensis* Cvancara (1958, p. 857).

A large convex brachial valve from locality 108/3 in the Utting Calcarenite (Pl. 1, figs 14, 15) which has been extensively bored by predators has pustule-like thickenings on the inner surface of the valve. The pustules are located over boring sites, and indicate secretion of callus deposits by the brachiopod to combat the activities of the boring organism.

Occurrence: Locality 7/1, Ningbing Limestone (Tournaisian part), Ningbing Range; locality 210/6, unnamed Tournaisian breccia, Waggon Creek Valley; locality 112/2, Septimus Limestone, Langfield Point; locality 138/1, Septimus Limestone, Burt Range Amphitheatre; localities 107/1, 108/0, 108/3, and 108/4, Utting Calcarenite, Utting Gap; locality 108/7, Burvill Beds, Utting Gap; locality 305, Burvill Beds, 1.5 miles northeast of Ningbing Homestead; and locality 435/0, Burvill Beds, 5 miles northeast of Point Spring in the Weaber Range.

Material: CPC 10795-10807, CPC 11086, CPC 10795-10797, 10800-10802, 10804-10806, 11086 from locality 108/0; CPC 10798-10799, 10806-10807 from locality 108/3. Many broken silicified specimens from the Utting Calcarenite, and several poorly preserved specimens from each of the other localities.

Age: Tournaisian and Visean.

Family RHIPIDOMELLIDAE Schuchert, 1913

Genus Rhipidomella Oehlert, 1890

Type species: Terebratula michelini Léveillé, 1835, by original designation of Oehlert, 1890.

Rhipidomella michelini? (Léveillé)

(Pl. 1, figs 1-5; Pl. 2, figs 1-5)

1970 *Rhipidomella michelini?* (Léveillé), Thomas, pl. 21, figs 1-16; pl. 25, figs 2, 5, 6.

Remarks: Thomas (1970) described specimens of *Rhipidomella michelini?* (Léveillé) from the Moogooree Limestone of the Carnarvon Basin, the Laurel Formation of the Canning Basin, and the Septimus Limestone of the Bonaparte Gulf Basin. *R. michelini?* has since been collected from the upper part of the Burt Range Formation, the Enga Sandstone, the Utting Calcarenite, the Burvill Beds, and the Bonaparte Beds.

Many specimens from locality 107/7 in the Utting Calcarenite are larger than those from other localities. They are pentagonal in outline, and have the greatest width at approximately three-quarters the length of the valve. The population from locality 107/7 also contains small and medium-sized individuals, and hence the large specimens are interpreted as gerontic individuals which lived in a particularly favourable environment. Specimens from elsewhere in the Utting Calcarenite have a similar size to *R. michelini?* in the other Tournaisian and Visean formations in Western Australia.

An addition to Thomas's description of *R. michelini?* is the morphology of the mantle canals in the pedicle valve. The canals of vascula media are irregular, arise from the front of the muscle field, and extend towards the front of the valve; the vascula genitalia comprise pitted areas on the margins of the vascula media.

Thomas (1970) suggested that his specimens were closest in relative width and the position of maximum width with *R. michelini divaricata* (M'Coy) described by Demanet (1934, pp. 41-3, pl. 2, figs 18-22). In other features, however, they resembled both *R. michelini divaricata* and *R. michelini* s.s. (Demanet, 1934, pp. 37-41, pl. 2, figs 1-9). The large specimens from the Utting Calcarenite, which are widest towards the front of the shell, are closer in shape to some specimens of *R. michelini* (*see* Demanet, 1934, pl. 2, figs 2, 6), but are considerably larger than either *R. michelini* or *R. michelini divaricata.*

Measurements

	Length	Width	Height
CPC 10790	29.5	32	12.5
CPC 10791	33.5	36.5	—

Occurrence: Locality 128/6, Burt Range Formation, middle part of Enga Ridge; locality 210/6, unnamed Tournaisian breccia, Waggon Creek valley; locality 102/4, Enga Sandstone; locality 136/1, a possible equivalent of the Enga Sandstone at Spirit Hill; localities 104/3, 104/7, 104/11A, 104/15, 104/17, and 150/5, Septimus Limestone, Mount Septimus; localities 110/3 and 110/4, Septimus Limestone, and locality 110/11, Zimmermann Sandstone, Mount Zimmermann; localities 112/1 and 112/2, Septimus Limestone, Langfield Point; localities 107/1, 107/6A, 107/7, 108/0, 108/3, 108/5A, and 143/1, Utting Calcarenite, Utting Gap; locality 25, Burvill Beds, 2.5 miles east-northeast of Point Spring in the Weaber Range; locality 306, Burvill Beds, Burvill Point; Bonaparte No. 2 Well, core 11, at 4929 feet in the Bonaparte Beds.

Material: CPC 10790-10794 from locality 107/7. Several hundred unaltered and silicified specimens.

Age: Tournaisian and Visean.

Order STROPHOMENIDA Öpik, 1934

Suborder STROPHOMENIDINA Öpik, 1934

Superfamily DAVIDSONIACEA King, 1850

Family SCHUCHERTELLIDAE Williams, 1953

Subfamily SCHUCHERTELLINAE Williams, 1953

Genus Schuchertella Girty, 1904

Type species: Streptorhynchus lens White, 1862.

Schuchertella peltata sp. nov.

(Pl. 3, figs 10-19)

Diagnosis: Shell rounded-quadrate to rounded-rectangular, and convexoconcave; pedicle valve convex posteriorly, concave anteriorly and laterally, having a prominent pointed umbo; brachial valve slightly flattened posteriorly, convex over remainder of valve, and with an obsolete umbo; apical angle 150-160°; costellae in four orders, with five to six primary and five to seven secondary costellae per 5 mm at 10 mm from umbo; primary costellae and some secondary costellae higher than those of other orders; concentric ornament closely spaced and nearly lamellose; socket plates diverging at approximately 100°.

Description: External. Shell rounded-quadrate to rounded-rectangular convexoconcave, brachial valve convex throughout, and pedicle valve convex at umbo but concave anteriorly and around lateral margins; greatest width at midlength; commissure slightly sulcate anteriorly in some individuals; costellae in up to four orders, with primary and in some cases secondary costellae higher than those of other orders; second to fourth order costellae established by intercalation, and starting as extremely fine lirae between primary costellae; costellae rounded apically, and together with intervening furrows ornamented by short closely spaced concentric lamellae; density of costellae 10 mm from umbo five to six primary, five to seven secondary, and four to five thread-like beginnings of later-order costellae per 5 mm; density of costellae of all orders fifteen per 5 mm at 20 mm from umbo on a shell 22 mm long and 28 mm wide.

Pedicle valve. Umbo pointed, convex, generally straight, jutting well behind hinge; umbonal shoulders sloping on to flat or slightly concave posterolateral margins; ventrally curved lateral and anterior margins forming concave region in front of umbo — greatest concavity slightly in front of midpoint; apical angle 150-160°, cardinal margins straight or slightly curved; interarea apsacline, inclined at 135° to plane of commissure, generally flat, bearing horizontal growth lines, 3.5 mm high on a valve 20 mm long and 21 mm wide; perideltidium broad, extending along nearly half the width of hinge on either side of umbo; pseudodeltidium convex, sharply pointed, at times twisted, and ornamented with horizontal growth lines; delthyrial angle 50°.

Brachial valve more transverse than pedicle valve, slightly flattened posteriorly, but moderately convex elsewhere; greatest height and convexity at midpoint; umbo small, and barely raised above posterolateral margins; interarea linear; shallow indistinct median sinus in some specimens commencing at umbo, and extending to anterior margin; chilidium short, divided by median furrow, and covering dorsal extremities of lobes of cardinal process.

Internal: Pedicle valve. Adductor muscle scars elongate, trigonal; diductor muscle scars narrow, pointed posteriorly, wider and rounded anteriorly, and in some valves divided by low median ridge; teeth long, sharply pointed, and extending from rounded dental ridges along inner margins of delthyrium; internal surface of valve bearing impressions of external ornament, but pustulose towards anterior and anterolateral margins.

Brachial valve. Socket plates joined with anterolateral margins of cardinal process, diverging at about 100°, with straight ventral surfaces; floor of sockets wide, flat to slightly concave, and recurving nearly to hinge; cardinal process projecting ventrally below socket ridges, and having a pointed median anterior boss; two divergent trough-like lobes separated by a median groove on external face of process, lobes triangular and bearing marginal lamellose myophores; muscle field not observed; median ridge absent; internal surface bearing impressions of external ornament.

Measurements

	Length	Width	Width (of hinge)
CPC 8465 Holotype	18	20.5	17
CPC 8466 Paratype	22	26.5	18 est.
CPC 8467 Paratype	22	28	21
CPC 8468 Paratype	25.5	27	22
CPC 8469 Paratype	17	21.5	17

Remarks: Schuchertella peltata sp. nov. is readily distinguished from the Devonian species *S. gratillica* Veevers (1959a, pp. 72-73, pl. 5, figs 1-10) and *S. dromeda* Veevers (1959a, pp. 69-72, pl. 5, figs 11-21) from the Sadler Formation and the Fairfield Formation, respectively, of the Canning Basin. *S. gratillica,* which resembles *peltata* in having inclined cardinal margins on the pedicle valve and divergent socket plates, is distinguished by its larger size, convex pedicle valve, higher ventral interarea, and stronger adductor muscle scars in the brachial valve; primary costellae are as high as those of later orders, and there are fewer intercalating costellae. *S. dromeda* is resupinate, and has inclined ventral cardinal margins. It is separated from *S. peltata* by its greater width, less concave pedicle valve, and an ornament in which the primary costellae are not well differentiated from the remainder of the radial ornament.

A comparison with *S. dorsiplana* Thomas, 1970, from the Laurel Formation in the Canning Basin and the upper part of the Burt Range Formation and the Enga Sandstone in the Bonaparte Gulf Basin is given on page 46.

S. subcancellata Hyde (1953, pp. 230-31, pl. 8, figs 1-14) from the Cuyahoga Formation and the Vinton Member of the Logan Formation, Ohio, has an ornament in which the primary costellae are more prominent than those of other orders, inclined cardinal margins on the pedicle valve, and a brachial valve with a broad shallow median sinus. *S. subcancellata* is distinguished from *S. peltata* by its more transverse shell, coarser concentric ornament, flat pedicle valve, and larger size.

S. portlockiana (Semenov) from the C_1tl (Visean) of the Moscow Basin, and figured by Sokolskaya (1954, pl. 7, figs 10-14), has primary costellae which are more prominent than those of other orders. *S. peltata* is distinguished from Sokolskaya's specimens by its denser costellae, more quadrate outline, concave pedicle valve, more convex brachial valve, and inclined cardinal margins on the pedicle valve. Specimens of *S. portlockiana* figured by Paeckelmann (1930, pl. 10, figs 4-9), from the Lower Carboniferous of Germany, have more low costellae between the primary costellae, and are much larger than *S. peltata.*

S. gueizhouensis Yang (1964, pl. 1, figs 1-6) from the Tournaisian of northeast Gueizhou, China, resembles *S. peltata* in having a comparable external ornament and a similar lateral profile in which the pedicle valve is convex at the umbo but concave towards the margins. *S. peltata* is distinguished from *S. gueizhouensis* by being less transverse and having a more strongly convex anterior on the brachial valve.

The specific name is derived from the Latin *pelta,* a small shield, and refers to the shield-shaped ventral aspect of the shell.

Occurrence: Localities 100/12, 100/18, 100/20, 100/24-100/30, in an area 6 to 7 miles northwest of Mount Septimus; localities 101/2 (the type locality), 101/7B, 101/7C, 101/9-101/11, 101/13-101/16, in an area between 4.5 and 6 miles west of Mount Septimus; all Burt Range Formation. Locality 7/1, Ningbing Limestone (Tournaisian part), Ningbing Range.

Material: CPC 8465-8469. Holotype CPC 8465, paratypes CPC 8466-8469. All from locality 101/2. Approximately 50 silicified specimens, and more than 20 specimens preserved in calcarenite.

Age: Lower Tournaisian.

SCHUCHERTELLA DORSIPLANA Thomas

(Pl. 3, figs 20-34)

1970 Schuchertella dorsiplana Thomas, pp. 34-36, pl. 16, figs 7-9; pl. 25, figs 1, 4; pl. 29, fig. 6.

Description: External. Shell low, gently biconvex; pedicle valve slightly convex posteriorly and flat anteriorly; brachial valve flat posteriorly and gently convex anteriorly; outline transversely subovate, lateral margins well rounded, greatest width near hinge or at midlength; costellae narrow, with rounded crests, separated by deep troughs of varying width, increasing by intercalation, with a density of thirteen to eighteen per 5 mm at 10 mm from umbo; costellae in three orders, with primary costellae slightly higher at margins; micro-ornament of poorly preserved concentric lirae.

Pedicle valve highest at or slightly behind umbo; umbo small, pointed, straight to slightly incurved, extending slightly behind hinge, and separated from flat posterolateral extremities by straight or slightly concave umbonal shoulders; apical angle 160-165°; interarea apsacline, flat, and 3 mm high on a valve 25 mm long and 40 mm wide; perideltidium wide, and ornamented with vertical striations and horizontal growth lines; delthyrium covered by pointed convex pseudodeltidium; delthyrial angle 50-60°.

Brachial valve nearly flat posteriorly, becoming gently convex anteriorly; juvenile specimens flat; umbo small, and slightly elevated; chilidium short, and covering dorsal extremities of lobes of cardinal process; dorsal interarea linear.

Internal. Pedicle valve. Diductor muscle scars subovate, narrow and impressed posteriorly, and wider anteriorly; adductor muscle scars narrow, elongate, and separated from one another by a median ridge; median ridge arising well in front of apex of valve, and strongest towards front of muscle field.

Brachial valve. Socket plates buttressing cardinal process, short, diverging from hinge at 105-120°, and recurved to hinge on floor of sockets; sockets deep and rounded; inner face of cardinal process supported by socket plates, and bearing prominent pointed median boss; two ventrally divergent trough-like lobes on external face of cardinal process, separated by a deep depression; internal surface of valve pitted; muscle scars not observed.

Measurements

	Length	Width
CPC 8324	8.5	14 est.
CPC 8461	12	15
CPC 8462	7.5	11.5
CPC 8464	27	37
CPC 8321	18.5 est.	25
CPC 8323	22.5 est.	28 est.

Remarks: Thomas (1970) recognized that *Schuchertella dorsiplana* was characterized by the possession of a flat brachial valve. His specimens came from the upper part of the Laurel Formation in the Canning Basin (Thomas, 1970, p. 36). Specimens of *S. dorsiplana* from the Bonaparte Gulf Basin differ in several small ways from those from the Canning Basin: the brachial valve is flat in small individuals but becomes gently convex at the front of adult specimens (brachial valves illustrated by Thomas are broken at the front); one pedicle valve has a slightly incurved rather than a straight umbo; and the ventral interarea is flat rather than convex. Thomas compared *S. dorsiplana* with *Orthis caduca* M'Coy (1844), from the lower Carboniferous of Ireland. M'Coy's species, which is unknown internally, has a similar outline, but the brachial valve is flat at the anterior, in contrast to the slightly convex anterior on specimens of *S. dorsiplana* from the Bonaparte Gulf Basin.

S. peltata sp. nov., from the underlying middle and lower parts of the Burt Range Formation, is smaller, and more quadrate in outline. It has a shorter hinge, a more convex brachial valve, and a costellate ornament in which the primary costellae are much higher than the costellae of other orders.

S. concentrica Roberts (1964, pp. 177-78, pl. 1, figs 1-7) from the Visean of New South Wales, is transverse, and has a nearly flat pedicle valve and a concentric micro-ornament comparable with that in *S. dorsiplana*. *S. dorsiplana* is distinguished from *S. concentrica* by its slightly finer costellate ornament, more acute apical angle, slightly flatter brachial valve, and socket plates which are more strongly divergent from the hinge; the concentric micro-ornament is weaker in *S. dorsiplana*.

Occurrence: Localities 103/8, 103/9, 103/12, 109/5, 128/1, middle part of Enga Ridge; locality 129/4, southern part of Enga Ridge; locality 113/5, Burt Range Amphitheatre; locality 133/2, Alpha Hill; locality 122/2, Sandy Creek; locality 123/14, near Legune Homestead; locality 124/1, Flapper Hill; all Burt Range Formation. Locality 102/4, northern end of Enga Ridge; localities 103/16, 103/17B, 109/7, 109/8, middle part of Enga Ridge; locality 129/7, southern end of Enga Ridge; all Enga Sandstone.

Material: CPC 8321-8324; CPC 8461-8464. Approximately 30 specimens. CPC 8321 and 8323, from locality 102/4; CPC 8322, 8324, 8461-8463 from locality 129/4; CPC 8464 from locality 103/17B.

Age: Tournaisian.

SCHUCHERTELLA sp. cf. S. GRATILLICA Veevers

(Pl. 2, figs 6-15)

Description: External. Shell unequally biconvex, and subelliptical in outline; greatest width near midlength; lateral margins meeting hinge at an obtuse angle; commissure rectimarginate to very slightly sulcate; ten to fifteen (usually thirteen to fourteen) costellae per 5 mm at 10 mm from umbo; costellae in up to three orders, increasing by intercalation; primary costellae, especially those on brachial valve, frequently stronger than those of other orders; micro-ornament of fine concentric lirae.

Pedicle valve convex, swollen at umbo, but becoming flatter towards lateral margins and anterior; umbo frequently asymmetrical, and usually poorly differentiated from lateral slopes; interarea variable in height (highest interarea 7 mm on a valve 25 mm wide and 22 mm long), apsacline, flat, and ornamented with horizontal growth lines; perideltidium not preserved; delthyrium covered by a convex pseudodeltidium; delthyrial angle approximately 40°.

Brachial valve moderately convex on midpart of valve, but flat at umbo and towards margins; umbo small, and separated from flat posterolateral margins by concave umbonal shoulders; chilidium short and convex.

Internal. Pedicle valve. Dental plates slender, extending along inner margins of delthyrium, and subtending sharp peg-like teeth; muscle field not observed.

Brachial valve. Socket plates initially extend anterolaterally from base of cardinal process, but then curve laterally and run parallel with hinge, enclosing deep pit-like sockets; cardinal process with two divergent posterior lobes, and supported anteriorly by low median ridge; muscle field not observed.

Measurements	Length (pedicle valve)	Length (brachial valve)	Width	Height (of ventral interarea)
CPC 8470	—	16.5	24	8.5
CPC 8471	19	14	20 est.	6.5
CPC 8472	—	17.5	24.5	—
CPC 8473	15	—	19	—
CPC 8474	—	14	19 est.	—
CPC 8475	—	9	14.5	—
CPC 8476	—	12.5	20	—
CPC 8477	18	—	23	—
CPC 8478	—	12 est.	19	3

Remarks: Specimens of *Schuchertella* from the Frasnian Kununurra Member of the Cockatoo Formation and the Famennian Ningbing Limestone (back-reef) are variable externally in the height of the ventral interarea, and in the asymmetry of the ventral umbo. The costellate ornament appears coarser in specimens from the Kununurra Member (Pl. 2, figs 11-15), but this is because they are better preserved than the partly decorticated specimens from the Ningbing Limestone. With this variation, and small collections, it is very difficult to differentiate between specimens from the two formations, and hence they are referred to the one species, *S.* sp. cf. *S. gratillica* Veevers.

In the pedicle valve of this species, the umbo is usually wrinkled and twisted to one side, probably as a result of attachment. Some specimens, however, have smooth straight umbos, and were probably unattached. Examples of this type of variation are seen in *S. gratillica* Veevers (1959a, pp. 72-73, pl. 5, figs 1-10) and in specimens of *S. devonica* (Keyserling) from Ferques which are housed in the collection of the Bureau of Mineral Resources. Davidson (1864, p. 80) commented on the wide range of variation in *S. devonica.*

In the Canning Basin, *S. gratillica* Veevers is restricted to the Frasnian *Ladjia saltica* Zone. It is replaced in the Famennian (probably not until the late Famennian) by *S. dromeda* Veevers (1959a), a characteristic species of the *'Avonia' proteus* Zone. *S. gratillica* compares closely with the specimens from the Bonaparte Gulf Basin in having a biconvex shell and the same density of costellae; it differs in the possession of a slightly more convex brachial valve and, in some specimens, a higher ventral interarea and longer ventral umbo.

S. prava (Hall), figured by Fenton & Fenton (1924, pl. 20, figs 21-28) from the Hackberry (Lime Creek) Formation of Iowa, has a comparable outline, but increases in convexity during the later stages of growth; it is distinguished also by its more robust socket plates, and wider delthyrium. Specimens of *S. prava* comparable with *S.* sp. cf. *S. gratillica* are figured by Stainbrook (1945, pl. 2, figs 19, 20) from the Independence Shale of Iowa, and by McLaren et al. (1962, pl. 15, fig. 30), as *S.* sp. cf. *S. prava* (Hall), from the Southesk Formation, Alberta Rocky Mountains, Canada.

Occurrence: Locality 28/10 in the Kununurra Member of the Cockatoo Formation, Matheson Ridge; locality 141/1, Ningbing Limestone (Famennian back-reef), Ningbing Range; and locality 443/13, Ningbing Limestone (Famennian back-reef), Jeremiah Hills.

Material: CPC 8470-8474 from locality 141/1; CPC 8475-8478 from locality 28/10. More than 40 specimens preserved in limestone, and as internal and external moulds in fine-grained sandstone.

Age: Frasnian and Famennian.

Genus SERRATOCRISTA Brunton, 1968

Type species: Serratocrista fistulosa Brunton, 1968, from the Visean (lower D Zone), Bunnahone, Ireland, by original designation of Brunton.

Diagnosis: See Brunton (1968, p. 39).

Remarks: Brunton, using silicified and partly silicified material, described the shell of *Serratocrista* as pseudopunctate. Specimens of *Serratocrista* from locality 108/0 in the Utting Calcarenite are coarsely silicified, but in three pedicle valves the inner part of the silicified shell is porous, having a network of small holes of comparable diameter to the punctae in *Punctospirifer* from the same locality. One brachial valve interior has the characteristic pustulose appearance of pseudopunctate shells, and hence it is likely that the 'punctae' in the inner part of the pedicle valves are impressions of unsilicified pseudopunctae which were removed when the specimens were extracted with acid.

Serratocrista sp. from the Utting Calcarenite has a well-defined dorsal muscle field, and in some specimens a dorsal median ridge. Both features are absent in *S. fistulosa,* the type species, but in other respects the two species are morphologically close.

SERRATOCRISTA sp.

(Pl. 3, figs 1-9)

Description: External. Shell of moderate size, transversely subovate and ventribiconvex; greatest width near midlength; costellae spinose (observed on one specimen only because of coarse silicification), separated by deep troughs, in two orders, and having a density of fourteen to fifteen per 5 mm at 5 mm from umbo; shell material probably pseudopunctate.

Pedicle valve gently convex, nearly evenly convex in lateral profile, highest at umbo, becoming flat on posterolateral margins; umbo blunt, projecting a short distance behind cardinal margins; interarea low, flat, apsacline, and having a delthyrium covered by a convex pseudodeltidium; delthyrial angle 60-70°.

Brachial valve gently convex to flat; umbo obsolete; chilidium short, and covering dorsal margin of bilobed cardinal process.

Internal. Pedicle valve. Dental plates forming anterolaterally directed flanges down inner margins of delthyrium, and distally bearing small teeth; muscle field triangular, and divided by low rounded median ridge; individual muscle scars not preserved.

Brachial valve. Cardinal process bilobed externally, having a rounded median boss internally, and supported anterolaterally by robust socket plates; socket plates diverging from hinge at 25-30°, slightly recurved distally, and enclosing deep concave sockets; muscle field impressed posteriorly, triangular to subovate, and in some valves divided apically by an obsolete median ridge; adductor muscle scars narrow, situated near front of ridge, and extending farther anteriorly than diductor muscle scars on to a slightly elevated area at front of muscle field; diductor scars triangular, and bounded posterolaterally by socket plates.

Measurements	Length (pedicle valve)	Length (brachial valve)	Width
CPC 10817	—	11	18.5
CPC 10818	—	—	17 est.
CPC 10819	12	—	16

Remarks: A comparison with the type species, *S. fistulosa* Brunton, is given under the remarks on the genus (p. 48).

Occurrence: Locality 108/0, Utting Calcarenite, Utting Gap.

Material: CPC 10817-10820, all from locality 108/0. Eleven silicified separate valves.

Age: Visean.

Family ORTHOTETIDAE Waagen, 1884

Subfamily DORSOSCYPHINAE nov.

Diagnosis: Orthotetidae with socket plates joined to the cardinal process, and prolonged into anteromedially directed plates forming an elevated cup-like muscle platform; muscle platform divided by a median septum, and with deeply impressed adductor muscle scars.

Remarks: The family Orthotetidae now contains three subfamilies distinguished essentially on the degree of anterior prolongation of the socket plates: the Orthotetinae in which the socket plates are recurved; the Derbyiinae in which the socket plates extend anterolaterally and enclose the posterior parts of the dorsal adductor muscle scars; and the Dorsoscyphinae in which the socket plates extend around the entire muscle field. In addition, the Dorsoscyphinae is characterized by a robust median septum dividing a deeply impressed muscle field. *Dorsoscyphus spinulosus* gen. et sp. nov. is so far the only known member of the subfamily Dorsoscyphinae.

Genus DORSOSCYPHUS nov.

Type species: Dorsoscyphus spinulosus gen. et sp. nov. from the Utting Calcarenite, Bonaparte Gulf Basin.

Description: Shell small, quadrate to subrectangular, ventribiconvex, and having a multicostellate ornament; costellae in two or three orders, increasing mainly by intercalation, but second order costellae sometimes bifurcating from primary costellae; concentric ornament lamellose, to sometimes strongly lamellose; ornament imbricated at growth halts; pedicle valve symmetrical to asymmetrical, with straight or incurved umbo, and high ventral interarea; pseudodeltidium convex, lamellose, and covering nearly all delthyrium; median septum robust, joining inner surface of pseudodeltidium in apex of valve; septum having a wide dorsal surface, probably bearing adductor muscle scars, and having an abrupt anterior termination at front of muscle field; diductor muscle scars triangular; brachial valve nearly flat posteriorly, and convex anteriorly; cardinal process with two laterally divergent lobes flanking a chevron-ribbed median area, and probably covered by a chilidium; small median boss on inner face of cardinal process; socket plates curving anteriorly from sockets, forming an elevated cup-like muscle platform; muscle platform divided by a wide anteriorly expanding median septum, and having spinose anterior margin.

Remarks: Dorsoscyphus is distinguished from other davidsoniacean genera by the manner in which the socket plates extend anteromedially to form a cup-like dorsal muscle platform, and by its smaller size. Specimens from the Utting Calcarenite are about 5 mm wide, and the largest are 7.5 mm wide and 5.5 mm long. They are interpreted as adults on the basis of their thick shell, which is ornamented with up to three orders of costellae, and has up to three periods of significant growth retardation. Callus deposits along the inner surface of the pseudodeltidium in the apex of the pedicle valve, a complex cup-like muscle platform in the brachial valve, thickened dorsal and ventral median septa, and numerous dorsal lamellose extensions of the ventral interarea are additional features characteristic of adult brachiopods. The pedicle valve of *Dorsoscyphus* has a median septum joined with the inner surface of the pseudodeltidium, and unsupported teeth, and hence is a typical orthotetidid. *Dorsoscyphus* is so far the only genus referable to the subfamily Dorsoscyphinae. Its shell structure is unknown because silicified specimens only are available for study.

The genus is named from the Latin *dorsum,* the back, and *scyphus,* a cup, for the cup-like muscle platform in the brachial valve.

DORSOSCYPHUS SPINULOSUS gen. et sp. nov.
(Pl. 4, figs 1-19)

Diagnosis: Costellae with a density of four to five per 1 mm at 3 mm from umbo; ventral interarea catacline or strongly apsacline in specimens with a straight umbo, shallowly apsacline in specimens with an incurved umbo; pseudodeltidium extending two-thirds to three-quarters length of delthyrium; delthyrial angle 45-55°; ventral median septum extending one-third to one-half length of valve; dorsal median sinus commencing just in front of umbo, and extending to front of valve.

Description: External. Shell small, quadrate to subrectangular, triangular in lateral profile, and ventribiconvex; pedicle valve much higher than brachial valve; greatest width at either hinge or midlength; shell multicostellate; costellae high, with rounded apices, separated by deep narrow grooves, and ornamented with closely spaced concentric lamellae; lamellae absent from furrows, usually slightly lamellose, but becoming thickened and flange-like at certain growth stages; costellae in two or three orders; second order costellae intercalate between or from close to lateral margins of primary costellae, or in some valves bifurcate from primary costellae; density of costellae four to five per 1 mm at 3 mm from umbo; ornament imbricate at growth halts.

Pedicle valve hemipyramidal, highest at or slightly in front of umbo, cardinal margins straight or slightly inclined, lateral slopes steep; umbo straight to moderately incurved, and level with or extending slightly behind hinge; some valves with straight umbos asymmetrical; ventral interarea varying in inclination from catacline to strongly apsacline in specimens with straight umbos, and shallowly apsacline in specimens with incurved umbos; interarea mainly flat, sometimes concave at umbo, and in some valves strongly lamellose dorsally; interarea ornamented with horizontal growth lines, and perideltidium with both horizontal lines and closely spaced vertical striae; pseudodeltidium convex, lamellose dorsally, and extending two-thirds to three-quarters length of delthyrium; delthyrial angle 45-55°.

Brachial valve flat to slightly convex posteriorly, and more strongly convex at front; umbo varying from nearly obsolete to slightly convex, extending a little way behind hinge, and flanked by flat posterolateral extremities; dorsal median sinus shallow, extending from just in front of umbo to anterior margin; chilidium probably covering a chevron-ribbed area between two laterally facing lobes on external face of cardinal process.

Internal. Pedicle valve. Median septum robust, joined with inner apical surface of pseudodeltidium, having a wide flat to slightly concave dorsal surface, probably bearing the adductor muscle scars, and terminating abruptly at between one-third and one-half length of valve; anterior extremity of septum blunt, but in one specimen passing into a low median ridge; diductor muscle scars triangular, slightly impressed, and extending to front of median septum; dental plates forming low

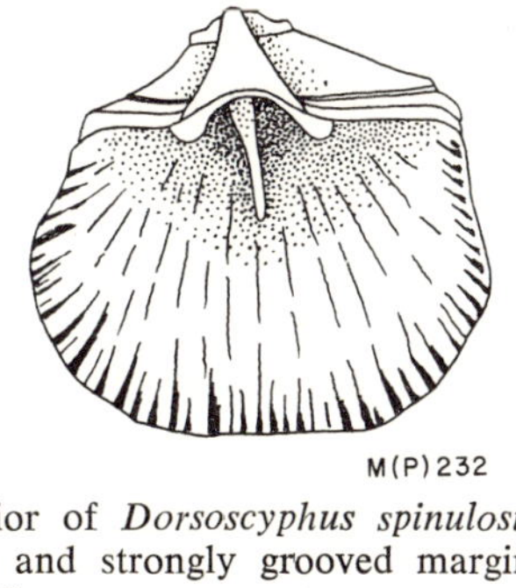

Figure 11. Pedicle valve interior of *Dorsoscyphus spinulosus* gen. et sp. nov. showing the robust median septum and strongly grooved margins. CPC 10829, locality 108/3, Utting Calcarenite. × 8.

flanges along inner dorsal margins of delthyrium, and supporting peg-like teeth; margins of inner surface of valve strongly grooved with follicle cavities (Fig. 11).

Brachial valve. External face of cardinal process has two laterally directed trough-like lobes flanking a median area bearing four chevron-shaped ribs (Fig. 12); external lobes of process supported internally by socket plates; small median

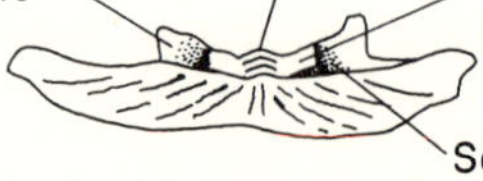

Figure 12. Posterior view of the brachial valve of *Dorsoscyphus spinulosus* gen. et sp. nov. showing the two lateral lobes of the cardinal process flanking a chevron-grooved ?myophore, the sockets, and the socket plates. CPC 10822, locality 108/0, Utting Calcarenite. × 8.

boss nearly continuous with the median septum on internal face of cardinal process; socket plates strongly divergent from hinge, curving anteromedially, extending to midlength of valve, and forming elevated cup-like muscle platform (Fig. 13); walls of platform highest medially, and spinose but still elevated above

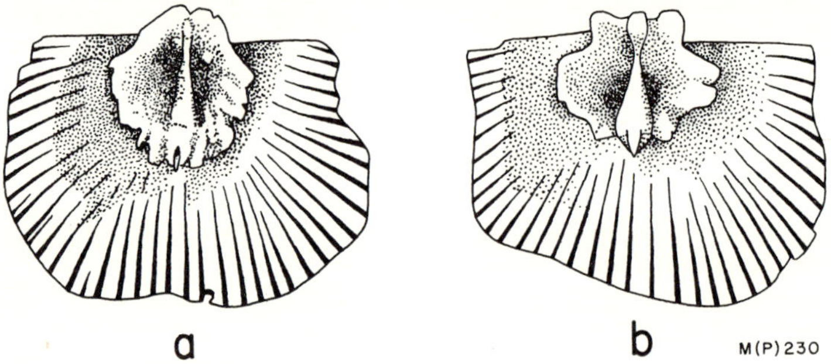

Figure 13. *Dorsoscyphus spinulosus* gen. et sp. nov. showing the anterior extension of the socket plates forming a cup-like muscle platform. The platform is divided by a thick anteriorly spinose median septum. a-CPC 10822, b-CPC 10821; from locality 108/0, Utting Calcarenite. × 8.

floor anteriorly; median septum commencing beneath cardinal process, dividing muscle field, pointed posteriorly, wider and rounded at front, and bearing high median anterior spine; adductor muscle scars triangular, and situated on floor of platform; ridge-like 'brachiophores' originating from anterolateral margins of cardinal process, and extending along ventral surface of socket plates; sockets small; inner margins of valve with strong follicle grooves.

Measurements	Length (pedicle valve)	Length (brachial valve)	Width	Height (of ventral interarea)
CPC 10821 Holotype	—	4.5	5.5	—
CPC 10822 Paratype	—	4.5	5.5	—
CPC 10824 Paratype	6	—	8	3.5
CPC 10825	5	—	5.5	2.5
CPC 10826 Paratype	—	4.5 est.	6	—
CPC 10827	5.5	—	6	—
CPC 10828 Paratype	6	—	6.5	2
CPC 10829	5.5	—	6	2

Remarks: Dorsoscyphus spinulosus cannot be compared with any previously described species.

The specific name refers to the spines on the front of the muscle platform in the brachial valve.

Occurrence: Localities 108/0 (the type locality) and 108/3, Utting Calcarenite, Utting Gap.

Material: CPC 10821-10829. Holotype CPC 10821, paratypes CPC 10822, 10824, 10826, 10828. All from locality 108/0, except CPC 10829, which is from locality 108/3. Seventeen silicified valves.

Age: Visean.

Family MEEKELLIDAE Stehli, 1954
Subfamily MEEKELLINAE Stehli, 1954
Genus SCHELLWIENELLA Thomas, 1910

Type species: Spirifera crenistria Phillips, 1836, by original designation of Thomas, 1910.

SCHELLWIENELLA (SCHELLWIENELLA) WEABERENSIS Thomas
(Pl. 2, figs 16-25)

1970 *Schellwienella (Schellwienella) weaberensis* Thomas, pp. 44-77, pl. 16, figs 1-5; pl. 17, figs 7-9.

Thomas (1970) described the internal morphology of the pedicle valve of *S. (Schellwienella) weaberensis* from serial sections. Internal and external moulds from localities 306 in the Burvill Beds and 436 in the Point Spring Sandstone provide additional information on both valves.

Pedicle valve. Ventral interarea flat, apsacline, with wide delthyrium covered by gently arched pseudodeltidium; diductor muscle scars flabellate to subquadrate, multi-lobate and dendritically marked posteriorly, and strongly ribbed anteriorly; adductor muscle scars narrow and separated by grooved median ridge posteriorly, wider medially, tapering again anteriorly, and bearing fine longitudinal striations; muscle field impressed posteriorly; anterior extensions of dental plates long, flanking smooth apical region and lateral margins of diductor muscle scars, and diverging along floor of valve at 80-85°.

Brachial valve. Muscle field in adults nearly round, deeply impressed, and divided by a median ridge; adductor muscle scars elongate, subovate, and bearing strong dendritic markings; median ridge obsolete in small specimens, but in adults supporting base of cardinal process, low and rounded, and in one large valve longer than 20 mm (Pl. 2, fig. 20); median boss immediately above median ridge on internal face of cardinal process; external face of process bilobed, with dorsal part covered by short convex chilidium; floor of valve adjacent to muscle field with pustulose vascula genitalia.

Measurements	Length (pedicle valve)	Length (brachial valve)	Width
CPC 10809	35 est.	—	57
CPC 10811	—	8	13
CPC 10814	42	—	53
CPC 10816	36	—	52

Remarks: The strong median septum in the brachial valve in one specimen of *Schellwienella weaberensis* Thomas (Pl. 2, fig. 20) resembles that in ?Orthotetinid gen. et sp. indet. of Brunton (1968, pl. 7, figs 4, 6). Additional features in Brunton's specimens comparable with *S. weaberensis* are a distinct concentric micro-ornament, socket plates which are parallel with the hinge, and the morphology of the cardinal process. ?Orthotetinid gen. et sp. indet. lacks a pedicle valve, and hence is difficult to refer to a genus; it comes from the Visean (low D Zone) of Bunnahone, Ireland.

The affinities of *S. weaberensis* have been discussed by Thomas (1970).

Occurrence: Locality 301, Milligans Hills; locality 305, 1.5 miles northeast of Ningbing Homestead; localities 306, 453/1, 453/7, Burvill Point; locality 25, 2.5 miles east-northeast of Point Spring in the Weaber Range; all from the Burvill Beds. At 265 feet above the base of section 436, Point Spring Sandstone, 2.5 miles north-northeast of Point Spring in the Weaber Range.

Material: CPC 10808-10816. CPC 10808-10811 from locality 306; CPC 10812-10816 from locality 436. Approximately 50 specimens.

Age: Visean.

Suborder CHONETIDINA Muir-Wood, 1955
Superfamily CHONETACEA Bronn, 1862
Family CHONETIDAE Bronn, 1862
Subfamily RUGOSOCHONETINAE Muir-Wood, 1962
Genus SCHISTOCHONETES nov.

Type species: Schistochonetes abruptus gen. et sp. nov. from the Utting Calcarenite of the Bonaparte Gulf Basin, Australia.

Description: Shell moderately large, concavoconvex, and auriculate; costellae high, rounded, and very coarse on posterior of shell, much finer or obsolete on front; change of ornament usually result of abrupt bifurcation, and lesser trifurcation and intercalation, at or nearly at a single growth stage between one-third and two-thirds length of adult shell; costellae obsolete on posterolateral margins; spinule apertures scattered; micro-ornament of fine concentric ribs on crests of costellae; pedicle valve moderately convex; interarea well defined, apex of delthyrium closed by small pseudodeltidium; cardinal spines at 25° to hinge; median septum high posteriorly, low anteriorly, extending one-third length of valve; adductor muscle scars flabellate and impressed; mantle canals well defined, straplike, and extending anteriorly from adductor muscle scars; papillose border on visceral disc; brachial valve moderately concave; interarea linear; cardinal process projecting ventrally, bilobed internally, and quadrilobate externally; median septum arising in front of a shallow alveolus, and extending one-half to two-thirds length of valve; anderidia diverging at 30-35°; accessory septa absent, inner adductor muscle scars subrectangular; outer adductor muscle scars club-shaped, and impressed posteriorly; inner socket ridges short and divergent; brachial ridges papillose, and extending from lateral margins of outer adductor muscle scars to front of inner adductor scars, enclosing elevated rounded-quadrate or triangular brachial discs; margins of valve papillose.

Remarks: Schistochonetes is distinguished from rugosochonetinid genera by its unusual external ornament, which is coarsely costellate posteriorly, and finely or indistinctly costellate anteriorly, the transition usually taking place at or nearly at a single growth stage. The most abrupt change from coarse to fine or obsolete costellae is seen in specimens from locality 107/1 (Pl. 5, figs 11-18). On some, the posterior costellae are simple, and on others they bifurcate before changing to finer or obsolete anterior costellae. The change towards the anterior is not as marked in specimens from locality 108/0, but most of these are of smaller size or have a broken anterior margin; many small individuals are broken near the change in the costellae, and the bifurcation of the posterior costellae is visible on the broken edge only. On other larger individuals (Pl. 5, fig. 9) the change is sometimes less abrupt than in specimens from locality 107/1.

A similar style of ornament is found in *Plebejochonetes* Boucot & Harper (1968, pp. 159-62, pl. 28, figs 11-16) from the Siegenian, Emsian, and Eifelian of Europe. The abrupt change from coarse to fine costellae in *Plebejochonetes* is well illustrated by Boucot & Harper (1968, pl. 28, figs 11, 12) on brachial valves of *P. semiradiatus* (Sowerby) and *P. plebejus* (Schnur); a pedicle valve of the latter species (Boucot & Harper, 1968, pl. 28, fig. 13) does not show such an abrupt change in the costellae. *Plebejochonetes* has a better defined dorsal interarea than *Schistochonetes,* an elongate posteriorly directed cardinal process, and, frequently, dorsal accessory septa.

Rugosochonetes Sokolskaya, 1950, is smaller than *Schistochonetes,* and has much finer capillae of nearly uniform size, hinge spines which diverge at about 60°, and, usually, capillate posterolateral margins. *Neochonetes* Muir-Wood, 1962, has a much finer external ornament in which the capillae are nearly of uniform size but sometimes become obsolete anteriorly. The type species of *Neochonetes, N. dominus* (King), is less concavoconvex, has stronger ridge-like mantle canals extending from the front of the ventral adductor muscle scars, a more papillose ventral visceral disc, and more deeply impressed ventral diductor muscle scars. In the brachial valve, *dominus* has a higher median septum and more prominent brachial markings. Except for the external ornament, these characters are unlikely to be of generic significance. *Plicochonetes* Paeckelmann, 1930, is much smaller than *Schistochonetes,* and has simple costae crossing the entire shell.

Coarse papillae in the pedicle valve of *Schistochonetes* border the visceral disc, especially on the posterolateral margins. The papillae are coarsest at the edge of the ears and the visceral disc, and extend on to the inner surface of the ears. They are interpreted as a protective device to prevent particles being swept into the shell in the feeding currents, as suggested by Grant (1968) for productoids. Protective papillae are recognized also in species of *Rugosochonetes* from the Bonaparte Gulf Basin: *R. obtectus* sp. nov., *R. macgregori* sp. nov., and *R.* sp. cf. *R. kennedyensis* Maxwell. No other species of *Schistochonetes* are known.

The name of the genus is derived from the Greek *schistos,* divided, and refers to the division of the costellae towards the anterior margin.

SCHISTOCHONETES ABRUPTUS gen. et sp. nov.

(Pl. 5, figs 1-18)

Diagnosis: Schistochonetes with seven to eight posterior costellae per 5 mm at 5 mm, and nine to eleven anterior costellae per 5 mm at 12 mm, both distances taken from the ventral umbo; approximately six spines at 25° to the hinge on either side of the umbo; other characters as in the diagnosis of the genus.

Description: External. Shell moderately large, transverse, rounded-rectangular, moderately concavoconvex, auriculate, and widest at hinge; costellae on posterior coarse, high, rounded, separated by deep furrows, seven to eight per 5 mm at 5 mm from ventral umbo; at or near a particular growth stage (at between one- and two-thirds length of valve) costellae suddenly increase in number by bifurcation and more rarely trifurcation and intercalation, producing a much finer ornament on front of shell, or suddenly becoming much weaker; costellae at front lower and usually narrower than posterior costellae, and at 12 mm from ventral umbo having a density of nine to eleven (averaging ten) per 5 mm; costellae obsolete on posterolateral extremities; spinule apertures scattered; concentric ribs on crests of costellae evenly spaced, nine to eleven per 1 mm.

Pedicle valve moderately convex; most convex at umbo, becoming flatter towards front, and highest at about one-third length of valve; umbo low, blunt, and barely incurved over ventral interarea; umbonal shoulders moderately to gently concave, and passing laterally on to wide flattened posterolateral extremities; cardinal margin obtuse; ventral interarea orthocline, gently concave beneath beak, becoming shorter and flatter towards lateral margins, and having a wide U-shaped delthyrium; interarea 2 mm long on a valve 16 mm long and 22.5 mm wide; narrow arched pseudodeltidium at apex of delthyrium; approximately six spines at 25° to hinge on either side of umbo.

Brachial valve having moderately concave visceral disc, and wide flat posterolateral extremities; most strongly concave at posterior, deepest at one-third length, and becoming flatter anteriorly; dorsal interarea linear.

Internal. Pedicle valve. Median septum arising from beneath rounded callus infilling apex of valve; septum high and blade-like posteriorly (for 2 mm on a valve 16 mm long and 22.5 mm wide), then changing abruptly into low rounded ridge extending to about one-third length of valve; adductor muscle scars narrow, elongate-oval, and situated on high platforms at posterior of valve; diductor muscle scars flabellate, pointed posteriorly, expanding anteriorly, and impressed into valve, especially at posterior; a narrow channel (perhaps associated with mantle canal system) arising immediately in front of median ridge, bearing spines or low median ridge, and extending to front of visceral disc; two strap-like canals of vascula media arising from front of adductor muscle scars, and extending along sides of median channel to front of visceral disc; another system of narrow linear mantle canals (?vascula genitalia) radiating from muscle field and covering remainder of visceral disc; these canals slightly irregular, and often linked transversely with one another, forming small pits on floor of valve; visceral disc bounded by coarsely papillose border; papillae largest near hinge, and finest at front of valve; encroaching on to visceral disc of small specimens, but confined to margins of large individuals; posterolateral margins flat to gently concave, bearing small papillae; teeth large, rounded, and transverse.

Brachial valve. Cardinal process bilobed internally, with short blunt ventrally projecting lobes, and situated immediately behind a shallow alveolus; external face quadrilobate, triangular (Fig. 14); lobes separated by rounded trough-like furrows of uniform width, two inner lobes expanding ventrally, and outer lobes of uniform width but inclined to midline of process; anterolateral margins of cardinal process supported by socket ridges; median septum arising from in front of alveolus, and extending to one-half to two-thirds length of valve; septum low and rounded posteriorly, higher in front of muscle field, and lower and narrower at distal extremity; anderidia short to elongate, arising in front of cardinal process, diverging at 30-35°, and not joined with median septum; inner adductor muscle scars narrow, rounded-rectangular, shallowly impressed, and separated from outer scars by anderidia; outer adductor muscle scars club-shaped, pointed posteriorly, much wider anteriorly, and having rounded anterior and lateral terminations; outer scars deeply impressed in front of socket ridges; brachial ridges spinose,

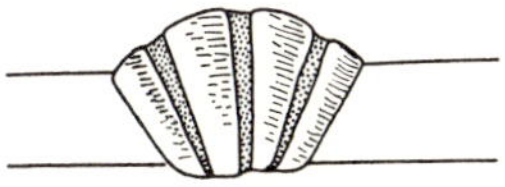

M(P)233

Figure 14. External face of the cardinal process of *Schistochonetes abruptus* gen. et sp. nov. CPC 10833, locality 108/0, Utting Calcarenite, Utting Gap. × 4.5.

originating on lateral margins of outer adductor muscle scars, extending anteriorly to about three-quarters length of valve, curving abruptly posteriorly, and terminating at front of inner adductor muscle scars; brachial ridges enclosing elevated rounded-quadrate or triangular brachial discs; sockets wide, rounded, anterolaterally directed, and bounded by short divergent socket ridges; anterior and lateral margins bearing longitudinal rows of moderately coarse papillae.

Measurements	Length (pedicle valve)	Length (brachial valve)	Width
CPC 10830 Paratype	14	—	—
CPC 10832 Paratype	—	13	19.5
CPC 10833 Paratype	—	11	17 est.
CPC 10834 Paratype	17	—	22.5
CPC 10835 Paratype	15.5	13	25 est.
CPC 10836 Paratype	17	14.5	26 est.
CPC 10837 Holotype	15.5	13.5	26.5

Remarks: Schistochonetes abruptus gen. et sp. nov. is not closely comparable with any other known species. Small specimens in which the costellae have not begun to divide are morphologically similar externally to a single specimen of *Plicochonetes crassistrius* (M'Coy) (1844, p. 119, pl. 20, fig. 10) from the Carboniferous of Ireland.

The specific name refers to the abrupt transition between coarse and fine costellae on most specimens.

Occurrence: Localities 107/1, 108/0 (the type localities), and 108/3, Utting Calcarenite at Utting Gap.

Material: CPC 10830-10837. Holotype CPC 10837, paratypes CPC 10830-10836 from a total of 46 free shells. CPC 10830-10836 from locality 108/0; CPC 10837 from locality 107/1.

Age: Visean.

Genus RUGOSOCHONETES Sokolskaya, 1950

1962 *Nix* Easton, p. 45, pl. 5, figs 14-20.

Type species: Orthis hardrensis Phillips, 1841 (in part) from the D of Hardraw, Yorkshire, England, by original designation of Sokolskaya, 1950.

Remarks: Rugosochonetes Sokolskaya was redefined by Muir-Wood (1962) from specimens identified as being from Phillips's original collection, and from comparable material (probably topotopic) from the Gayle Shale, Wensleydale, and the Hardraw Shale, Hardraw Force, Yorkshire. Muir-Wood selected a specimen, which was probably figured by Phillips, as lectotype. C. C. Branson's suggestion (1964, p. 96) 'that *Rugosochonetes* be left virtually dead in that it is tied to a type species which is little known and cannot be well defined' is rejected because the lectotype designated by Muir-Wood makes *R. hardrensis* (Phillips) a valid type species for *Rugosochonetes*. Phillips's specimens (Muir-Wood, 1962, pl. 8, figs 7, 9, 10, 13) are not particularly well preserved and do not appear to show all characters of the species; these are, however, visible on specimens identified by Muir-Wood as *R. hardrensis* (Phillips) from the Gayle Shale and the Hardraw Shale. In addition, Branson (1964) proposed that, assuming *Rugosochonetes* to be 'dead', *Neochonetes* Muir-Wood, 1962, should be enlarged to include Mississippian species having hinge spines set at a high angle (presumably *Rugosochonetes* species); he (1964, p. 96) considered that 'the sole clear distinction from *Neochonetes* is that the spines of *Rugosochonetes* rise from the hinge at a high angle'. This must be rejected also because *Rugosochontes* is the senior valid taxon (International Code of Zoological Nomenclature Article 67 (k)). I agree with Branson that there are few straightforward differences between *Rugosochonetes* and *Neochonetes,* and maintain that many characters used by Muir-Wood to differentiate the genera are so variable in chonetoid species as to be of very doubtful generic significance. Among these are the following criteria used by Muir-Wood to distinguish *Rugosochonetes* from *Neochonetes*: longer ventral median septum, less prominent ventral mantle canals, weaker ridges bounding the outer margins of the diductor muscle scars, less prominent brachial ridges and inner socket ridges, and spines at a high angle to the hinge. Characters which, in my opinion, may be important in distinguishing *Neochonetes* are its weaker convexity in the pedicle valve, flatter brachial valve, more transverse and frequently auriculate outline, more persistent and deeper ventral median sinus, capillae which become obsolete anteriorly, and the absence of dorsal accessory septa. The relationship between *Rugosochonetes* and *Neochonetes* (including species assigned by Sadlick (1963, p. 722) to *Quadrantes* Sadlick, a junior objective synonym of *Neochonetes*) will, however, be clarified only after an intensive systematic and biostratigraphic study of both Lower and Upper Carboniferous species.

The genus *Nix* Easton, 1962, is 'principally characterized by the absence of all brachial septa, and the interior of the brachial valve bears only a pit at the base of the cardinal process' (Easton, 1962, p. 45). Through the courtesy of Dr Mackenzie Gordon Jr of the U.S. Geological Survey I have been able to examine silicified topotype brachial valves of *Nix angulata* Easton, the type species, and all of these have well-defined median septa and short anderidia. One of Easton's figured specimens (pl. 5, fig. 20) has a short but well-defined dorsal median septum also. The absence of septa in the other brachial valves (pl. 8, figs 15, 16) is caused by the loss of the inner layer of shell by decortication (Mackenzie Gordon Jr, pers. comm.), and the type species is placed in *Rugosochonetes*.

RUGOSOCHONETES OBTECTUS sp. nov.

(Pl. 6, figs 1-26)

1970 *Rugosochonetes?* sp. A, Thomas, pp. 47-48, pl. 29, fig. 5.
1970 *Rugosochonetes?* sp. B, Thomas, pp. 48-50, pl. 29, figs 2-4.

Diagnosis: Shell large for genus, rounded-rectangular in outline, and moderately concavoconvex in small individuals to strongly concavoconvex in adults; greatest width at one-third to one-half the length of shell; eighteen to twenty-two capillae per 3 mm at 5 mm from umbo; spinules common; pedicle valve usually high in adult specimens, with a small bluntly rounded umbo; eight spines on each side of umbo; cardinal process quadrilobate externally; dorsal median septum extending one-half to two-thirds the length of valve; alveolus obsolete; anderidia diverging at 35-40°; inner socket ridges straight, diverging at 125°, and higher and wider distally; brachial ridges well defined for genus, and papillose anteriorly.

Description: External. Shell large for genus; rounded-rectangular, and less than twice as wide as long; greatest width at one-third to one-half length of shell; hinge about four-fifths width of shell; lateral margins well rounded and meeting hinge in an obtuse angle; fold and sinus present on a few large shells; commissure generally rectimarginate, but sometimes weakly uniplicate; body cavity narrow; capillae rounded, separated by narrow furrows, bearing many small spinules, and becoming obsolete on posterolateral margins; capillae increasing almost exclusively by bifurcation on pedicle valve, and by intercalation on brachial valve; eighteen to twenty-two capillae per 3 mm at 5 mm from umbo; micro-ornament of fine closely spaced concentric filae; growth stages lamellose, and increasing in number near anterior margins.

Pedicle valve moderately convex in small individuals to strongly convex in adults; flanks and front sloping gently to margins in small individuals, becoming steeper in large specimens, and resulting in a high valve; greatest height at one-third to one-half length of valve; umbo small, bluntly rounded, swollen in some larger specimens, moderately well differentiated from umbonal shoulders, and extending slightly behind hinge; umbonal shoulders concave, sloping on to flat posterolateral extremities; cardinal margins diverging at 160°, and bearing eight spines on each side of umbo; angle of spines to hinge not observed; interarea procline to apsacline, concave around delthyrium, flat laterally, and 1.5 mm long at umbo on a valve 20 mm wide and 15 mm long; delthyrium broadly U-shaped, sides forming angle of 65°, and bearing arched pseudodeltidium at apex; sinus shallow to completely obsolete; when present, starting a little in front of umbo, and becoming slightly broader anteriorly.

Brachial valve moderately concave in small specimens to more strongly concave in larger individuals, and deepest between umbo and midpoint of valve; posterolateral shoulders wide, and flat to slightly concave; interarea hypercline, flat, 0.75 mm high on a valve 20 mm wide and 15 mm long, convex chilidium covering base of cardinal process; fold low to completely obsolete; when present, starting at about one-third length of valve, and becoming slightly higher and broader towards front.

Internal. Pedicle valve. Median septum high and blade-like posteriorly, becoming abruptly lower a short distance from umbo (1.5 mm in a valve 20 mm wide and 13.5 mm long), angular to about one-quarter length of valve, and then wider and rounded to midlength; adductor muscle scars on narrow elongate ridges, or slightly impressed into floor, and extending along angular part of septum; diductor muscle scars trigonal, pointed posteriorly, with wide rounded anterior margins, and impressed along their posterolateral margins and at umbo; strong ridges separating posterior parts of adductor and diductor muscle scars; mantle canals arising from front of muscle field as a series of narrow linear canals, or in one or two broader canals of vascula media arising from front and sides of adductor muscle scars; teeth divergent from hinge, narrow and angular near umbo, becoming wider and more rounded laterally, and grooved along their inner margins; papillae coarsest on posterolateral and anterior margins, finest in front of muscle field, and commonly arranged in radial ribs.

Brachial valve. Cardinal process bilobed internally, with two high narrow lobes separated by a median furrow; external face of process triangular, and quadrilobate, having two long narrow median lobes separated by deep furrows from

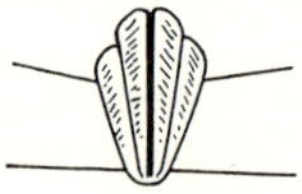

Figure 15. External face of the cardinal process on a paratype of *Rugosochonetes obtectus* sp. nov. CPC 10846, locality 128/6, Burt Range Formation, middle part of Enga Ridge. × 14.

slightly shorter lateral lobes (Fig. 15); alveolus absent; median septum commencing as rudimentary broadly rounded ridge at front of cardinal process, usually low posteriorly, higher and narrower anteriorly, and extending one-half to two-thirds length of valve; anderidia arising from near posterior of median septum, usually narrow, diverging at 35-40°, and extending two- to three-fifths length of septum; inner socket ridges straight, diverging at about 125°, and higher and wider distally; outer socket ridges not distinguishable from ridge along hinge; sockets becoming wider and deeper laterally, and with rounded floors; brachial ridges low, bordered internally by a depression, papillose anteriorly, arising from lateral extremities of inner socket ridges, and making a broad semicircular curve to front of median septum; papillae in radial rows on both sides of brachial ridges, at front of valve, and coarsest on ridge itself; margin of valve finely ribbed; muscle scars not observed.

Measurements	Length (pedicle valve)	Length (brachial valve)	Width
CPC 10839	—	7.5	11
CPC 10840	—	9.5	15
CPC 10841	—	8	14
CPC 10842	13.5	—	20
CPC 10843	14	—	21
CPC 10845 Holotype	15.5	14	21
CPC 10846 Paratype	—	9.5	13.5
CPC 10847 Paratype	13.5	—	20 est.
CPC 10849	—	13	20
CPC 10850	12.5	—	20 est.
CPC 10851	—	11	18.5
CPC 10854	14.5	12.5	19

Remarks: The rounded ridges separating the posterior parts of the ventral adductor and diductor muscle scars have been observed in silicified material only, and have not previously been reported in species of *Rugosochonetes*. Silicified specimens of

Permian chonetoids from Texas, kindly shown to me by Dr G. A. Cooper, have these and even more exaggerated ridges; the ridges are best interpreted as partitions between the musculature.

Rugosochonetes obtectus sp. nov. is distinguished from most early Carboniferous species by its larger size, transverse shell, high pedicle valve, and well-defined brachial ridges. Small specimens of *obtectus* from the lower and middle parts of the Burt Range Formation (localities 100/13, 28 and 29, and 101/21; *see* Fig.

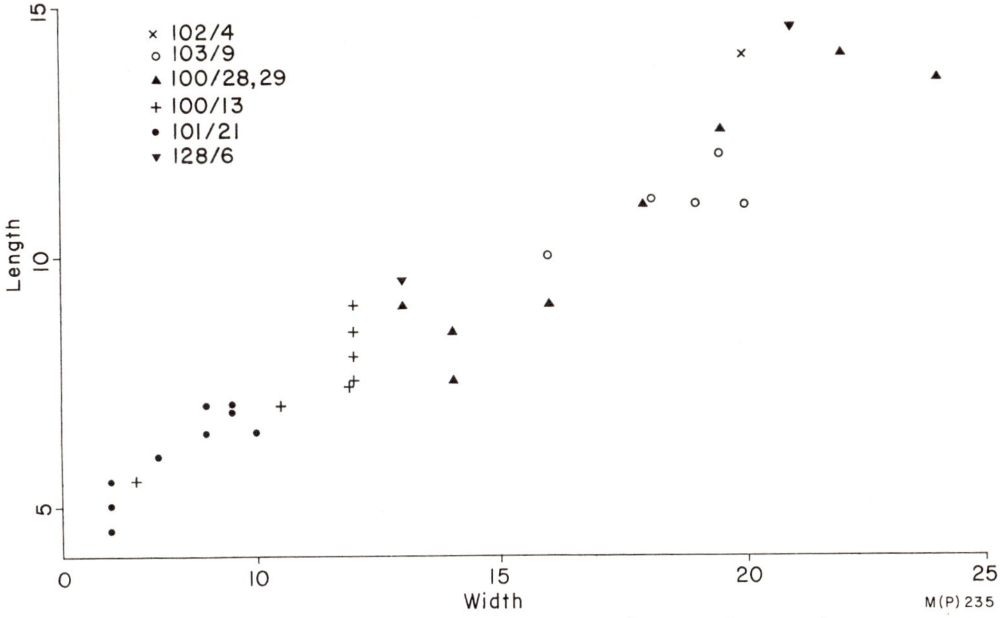

Figure 16. Plot of length/width ratios of brachial valves of *Rugosochonetes obtectus* sp. nov. Note the populations of different growth stages at localities 101/21, 100/13, and 103/9.

16) are comparable with *R. illinoisensis* (Worthen) figured by Weller (1914, pl. 8, figs 63-70) from the Burlington and Keokuk Limestones of Missouri. These specimens of *obtectus* have a similar finely capillate ornament, and a relatively low pedicle valve. Larger specimens, however, are much bigger than *illinoisensis,* and have a higher pedicle valve with a steep anterior and steep flanks, a more concave brachial valve, and more prominent brachial markings; some silicified specimens have better defined capillae. This comparison is based on specimens of *R. illinoisensis* from the U.S. National Museum collection, as well as those figured by Weller.

R. demini Fotieva (1961, pp. 101-2, pl. 9, figs 1-4) from the Tournaisian of the Timan-Pechora province, USSR, resembles *R. obtectus* in outline. *R. demini* is distinguished by its coarser costellate ornament, shorter dorsal median septum, less inflated pedicle valve, stronger ventral median sinus, and the possession of fewer cardinal spines.

R. laguessianus (de Koninck), figured by Muir-Wood (1962, pl. 8, fig. 12) from the lower Namurian of Britain, has the same density of capillae and a similar outline, but is smaller than *R. obtectus*. The specimens of *R. laguessianus* figured by Paeckelmann (1930, pl. 16, figs 1, 2) from the Carboniferous of Germany are smaller, have fewer spines along the cardinal margins, and a shorter ventral median septum. Russian specimens of *R. laguessianus* figured by Sokolskaya (1950, pl. 4, figs 1-33) from the C_1tl beds are, apart from being half the size of *R. obtectus,* quite close in shape, outline, morphology of the brachial ridges, and angle of divergence of the anderidia. Sokolskaya's specimens have a longer ventral median septum, and much weaker inner socket ridges.

R. loganensis (Hall & Whitfield), figured by Girty (1899, pl. 68, figs 5a-c) from the Madison limestone in the Yellowstone National Park area, USA, is similar in shape and external ornament to *R. obtectus.* The internal features of *R. loganensis* have not been described.

R. angulata (Easton) (1962, pl. 5, figs 14-20), from the Carboniferous Heath Formation, Montana, has a comparable size, outline, and external ornament. It is distinguished from *R. obtectus* by the possession of a slightly more prominent fold and sinus, more deeply impressed diductor muscle scars, a deeper alveolus, and shorter, more divergent anderidia and a shorter median septum in the brachial valve.

The specific name *obtectus* is Latin for covered over or concealed and refers to the obscure nature of the internal features of most specimens from the lower parts of the Burt Range Formation.

Occurrence: Localities 100/12, 100/13, 100/13C, 100/18, 100/25, 100/26, 100/28, 100/29 in an area 6 to 7 miles northwest of Mount Septimus; localities 101/1, 101/7B, 101/9, 101/21 in an area between 4.5 and 6 miles west of Mount Septimus; localities 103/7, 103/9, 109/2, 109/5, 128/2, 128/6 (the type locality) in the middle part of Enga Ridge; locality 117/1, Spirit Hill; locality 122/2, Sandy Creek; locality 123/14, 11 miles southwest of Legune homestead; locality 124/1, Flapper Hills; all in the Burt Range Formation. Locality 210/6, unnamed Tournaisian breccia, Waggon Creek valley. Locality 102/4, northern end of Enga Ridge in the Enga Sandstone; and locality 136/1, Spirit Hill, in a possible equivalent of the Enga Sandstone.

Material: CPC 10838-10854. Holotype CPC 10845; paratypes CPC 10846-10848. CPC 10838-10839 from locality 100/13; CPC 10840-10841 from locality 100/29; CPC 10842-10843, 10851-10852 from locality 103/9; CPC 10844, 10853 from locality 122/2; CPC 10845-10848 from locality 128/6; CPC 10849-10850 from locality 100/28; CPC 10854 from locality 102/4. More than 50 specimens.

Age: Tournaisian.

Rugosochonetes macgregori sp. nov.

(Pl. 7, figs 1-14)

Diagnosis: Shell of average size for genus, and rounded-rectangular in outline; pedicle valve moderately to strongly convex, with a well-defined broadly rounded umbo projecting behind hinge; three spines on either side of umbo projecting at 20-30° to hinge; capillae having a density of six to seven per 1 mm at 5 mm from umbo of the pedicle valve; cardinal process bilobed internally, and with six lobes externally.

Description: External. Shell rounded-rectangular, moderately concavoconvex, and widest at or slightly in front of hinge; capillae fine, rounded, bearing numerous spinule apertures, separated by narrow rounded furrows, and having a density of six to seven per 1 mm at 5 mm from ventral umbo; capillae increasing by bifurcation on pedicle valve, and mainly by intercalation, with some bifurcation towards margins, on brachial valve; capillae ornamented with regular concentric growth lines.

Pedicle valve moderately to strongly convex, with greatest convexity at umbo, greatest height at about one-third length; venter slightly flattened; umbo well developed, rounded, and projecting behind cardinal margins; umbonal shoulders gently concave, and passing on to wide convex non-auriculate posterolateral extremities; interarea orthocline, slightly concave beneath beak, becoming flatter towards lateral margins, and bearing wide U-shaped delthyrium; interarea 0.5 mm high at umbo on a valve 8.5 mm long and 10.5 mm wide; apex of delthyrium covered by a small arched pseudodeltidium; three spines on either side of umbo arising at 20-30° to hinge.

Brachial valve. Visceral disc moderately concave, deepest near midpoint of valve, rising on to wide gently concave posterolateral extremities; dorsal interarea hypercline, narrow, 0.3 mm long on a valve 7.5 mm long and 10.5 mm wide, and having a broad convex chilidium covering dorsal part of cardinal process; valve lamellose at anterior margin.

Internal. Pedicle valve. Median septum arising from beneath apical callus, high and blade-like initially, then passing abruptly anteriorly into a much lower ridge-like structure, and extending one-quarter to one-third length of valve; adductor muscle scars rectangular, slightly impressed into shell, and separated by median septum; diductor muscle scars flabellate, and impressed posteriorly; canals of vascula media arising in front of median septum and passing anteriorly along a shallow channel containing a low median ridge; anterior and lateral margins bearing rows of papillae; papillae largest and occasionally forming an incipient marginal ridge near posterolateral parts of valve, but finer at front; teeth wide, rounded, becoming broader and slightly ventrally inclined at lateral extremities.

Brachial valve. Cardinal process bilobed internally, with two narrow rounded lobes separated by a median furrow projecting ventrally above a wide shallow alveolus; external face broad, triangular, and having six narrow lobes arranged in three pairs; median pair longer than lateral pairs, with deep furrows between pairs; anterolateral margins of cardinal process supported by very broad inner socket ridges; median septum commencing well in front of alveolus, rounded posteriorly, more blade-like anteriorly, and extending to midpoint of valve; anderidia unconnected with median septum, arising from inner edge of inner socket ridges, diverging at 30°, and having a spine-like anterior termination near front of muscle field; inner adductor muscle scars subrectangular, diverging at same angle as anderidia, and slightly impressed into shell; outer adductor muscle scars subquadrate to subovate, and more deeply impressed into floor of valve; brachial ridges poorly defined, arising from anterolateral margins of outer adductor muscle scars, curving anteriorly, and then abruptly deflected back to front of inner adductor muscle scars; surface in front of visceral disc bearing rows of coarse papillae; sockets narrow, having rounded floors, and widening anterolaterally; outer socket ridges short and narrow, but inner socket ridges extremely broad — after curving around sockets they extend almost three-quarters of way along hinge.

Measurements	Length (pedicle valve)	Length (brachial valve)	Width
CPC 10855 Holotype	8.5	7	10.5
CPC 10856 Paratype	8.5	7	11.5
CPC 10857 Paratype	9	—	11.5 est.
CPC 10858 Paratype	9.5	—	13
CPC 10859 Paratype	10	—	13
CPC 10860 Paratype	—	9	12.5

Remarks: Rugosochonetes macgregori sp. nov. bears a close resemblance to *R. silleesi* Brunton (1968, pp. 55-62, pl. 8, figs 10-27) from the Visean (low D Zone) of County Fermanagh, Northern Ireland. Both species have a comparable external ornament and convex ventral umbo, and they are morphologically close in the interior of the brachial valve, especially in the median septum and the anderidia. *R. silleesi* has fewer lobes on the external face of the cardinal process, and seven pairs of hinge spines. The fine-ribbed holotype group of *R. celticus* Muir-Wood (1962, pl. 7, figs 4, 7) from the Visean of Flintshire, North Wales, is also close to *macgregori.* This group, one of three possible subspecies of *celticus* (Brunton, 1968, pp. 54-55), has a comparable size, shape, density of ornament, angle of inclination of the spines to the hinge, and an extremely close ventral interior. It is distinguished by having fewer lobes on the external face of the cardinal process. Another British Visean species, *R. delicatus* Brunton (1968, pp. 62-64, pl. 9, figs 13-15) from County Fermanagh, Northern Ireland, has a capillate ornament of comparable density to that of *macgregori. R. delicatus* is distinguished by its lower ventral umbo, weaker posterior capillae, greater number of hinge spines (four pairs, against three pairs in *macgregori*), and possibly by its weaker dorsal median septum. Some specimens of *delicatus* have the same outline as *macgregori* (*see* Brunton, 1968, pl. 9, fig. 4), but others have a wide hinge, and more extensive posterolateral extremities.

Specimens referred to *R. hardrensis* (Phillips) by Sokolskaya (1950, pp. 31-35, pl. 3, figs 1-16), from the C_1tsch (late Tournaisian) of the Russian Platform, have a comparable size and ornament. They differ from *macgregori* in having a weaker ventral umbo, and shorter and narrower inner socket ridges. Material from the Tournaisian of the Rudny Altai region of USSR referred to *R. hardrensis* by Grechishnicova (1966, pl. 3, figs 5, 6) also has a weaker ventral umbo.

R. kennedyensis Maxwell, described by Roberts (1964, pp. 181-82, pl. 2, figs 1-7) from the Visean of New South Wales, has a coarser capillate ornament, more weakly impressed dorsal adductor muscle scars, a weaker dorsal median septum, shorter anderidia, and narrower inner socket ridges.

This species is named after Mr Gordon McGregor in recognition of his assistance in the field in 1965.

Occurrence: Locality 107/1, Utting Calcarenite, Utting Gap: the type locality.

Material: CPC 10855-10860. Holotype CPC 10855; paratypes CPC 10856-10860. All from locality 107/1. Fifty-six free shells.

Age: Visean.

RUGOSOCHONETES USTULATUS sp. nov.

(Pl. 7, figs 15-25)

Diagnosis: Shell subrectangular to subovate, and auriculate; pedicle valve almost flat posteriorly, with a low obsolete umbo and wide posterolateral extremities, highest at one-half to two-thirds the length of valve, with a steep anterior margin; four spines on each side of the umbo diverging at 40-50° from hinge; capillae having a density of nineteen to twenty per 5 mm at 5 mm from ventral umbo; ventral median septum extending one-fifth the length of valve; brachial valve moderately to strongly concave; cardinal process bilobed internally and quadrilobate externally; median septum broad posteriorly, and extending two-thirds the length of valve; anderidia diverging at 40°; inner socket ridges wide, straight, and divergent from hinge; brachial impressions well defined.

Description: External. Shell transverse, subrectangular to subovate, auriculate, and moderately concavoconvex; greatest width at hinge; commissure rectimarginate; capillae rounded, separated by rounded furrows about half the width of each capilla, becoming rudimentary on posterolateral extremities, bearing occasional spinules, and having a density of nineteen to twenty per 5 mm at 5 mm from ventral umbo; capillae increasing mainly by intercalation on pedicle valve, and equally by bifurcation and intercalation on brachial valve; no micro-ornament observed.

Pedicle valve flattened posteriorly, rising gradually to highest and most convex area at one-half to two-thirds of its length, and then sloping steeply to anterior margin; venter flattened or medianly sulcate, and a very shallow sinus extending from near midpoint of valve to anterior margin; umbo low, and barely rising above posterior surface of valve; umbonal shoulders broad, gently concave, and extending on to pointed slightly convex ears; ears present in younger growth stages; four spines on each side of umbo diverging from hinge at between 40-50°; interarea orthocline to apsacline, flat along its entire width, and usually divided by longitudinal groove, coinciding with anterior margins of pseudodeltidium, at mid-length; interarea 1 mm long at umbo on a valve 18.5 mm wide and 10 mm long; delthyrium wide, U-shaped, and having a thick arched pseudodeltidium; valve lamellose anteriorly.

Brachial valve moderately to strongly concave, deepest at one-third to one-half length of valve, and having steep anterior margin crossed by extremely weak median fold; posterolateral margins abruptly separated from deep visceral disc, wide, moderately concave, and with sharply pointed ears; interarea hypercline, short, and having a small convex chilidium covering base of cardinal process.

Internal. Pedicle valve. Median septum arising from front of apical callus and extending one-fifth length of valve; septum rounded, broad and high posteriorly, becoming abruptly lower anteriorly, and widening at front; adductor muscle scars on narrow elongate ridges extending along and in front of median septum; diductor muscle scars flabellate, and impressed posteriorly; teeth wedge-shaped, inclined anterolaterally, and with subangular anterior extremities; floor of valve away from muscle field papillose; papillae coarsest and irregularly arranged on posterolateral regions and immediately in front of muscle field, finest and arranged on ridges or in longitudinal rows at front of valve.

Brachial valve. Cardinal process bilobed internally, with two narrow rounded lobes separated by a narrow median furrow projecting ventrally above a swollen anterior face; external face triangular, quadrilobate, with a narrow elongate median pair of lobes separated by wide expanding furrows from lower shorter lateral lobes; anterolateral margins of cardinal process supported by massive inner socket ridges; alveolus broad, shallow, and poorly defined; median septum arising in front of alveolus, extremely broad and rounded posteriorly, tapering and becoming lower between inner adductor muscle scars, and then thicker and blade-like for remainder of length to two-thirds length of valve; anderidia arising from same platform-like area as median septum, diverging at 40°, becoming narrower and higher anteriorly, and having elevated spinose distal terminations; inner adductor muscle scars trigonal, sharply pointed posteriorly, with wider oblique anterior margins, and bounded along most of their lateral margins by anderidia; outer adductor muscle scars small, trigonal, deeply impressed, and with a short anteriorly directed ridge arising from inner socket ridges in middle of each scar; sockets deep, subangular, widening anterolaterally, and bounded by broad, straight inner socket ridges; inner socket ridges high, well rounded, and extending anterolaterally past ends of sockets; outer socket ridges short and narrow; brachial ridges moderately well defined, and enclosing smooth elevated kidney-shaped brachial discs; ridges originating almost horizontally from front of outer adductor muscle scars, curving evenly anteriorly, and then abruptly making a hook-like curve to front of inner adductor muscle scars; two linear vascula media canals extending from near either side of anterior end of median septum to anterior margin; anterior and lateral margins with poorly defined rows of fine papillae.

Measurements	Length (pedicle valve)	Length (brachial valve)	Width
CPC 10861 Holotype	—	9	14
CPC 10862 Paratype	10	8.5	19
CPC 10863 Paratype	10	—	16 est.
CPC 10864 Paratype	10.5	9.5	18 est.

Remarks: Rugosochonetes ustulatus sp. nov. is extremely close morphologically to *R. transversalis* Brunton (1968, pp. 65-66, pl. 9, figs 16-26) from the Visean (low D Zone) of County Fermanagh, Northern Ireland. Both species are transverse, have the same density of capillae, a low ventral umbo, a short ventral median septum, a slightly sulcate body cavity, and an external face on the cardinal process with subparallel inner lobes separated by wide furrows from shorter divergent lateral lobes. *R. ustulatus* has more pronounced auricles in the adult, and a thickened brachial valve interior containing stronger, more divergent inner socket ridges, a more robust posterior shaft on the median septum, and well-defined brachial markings. The internal surfaces of the ears in *ustulatus* are tuberculate, whereas those in *transversalis* are pitted.

R. auriculus Roberts (1964, pp. 179-81, pl. 2, figs 8-13) from the Visean of New South Wales has an outline comparable with that of this species, long auricles, capillae of similar density, and the same number of spines along the hinge. *R. auriculus* is distinguished by its swollen ventral umbo, and a brachial valve interior which has shorter inner socket ridges, a deeper alveolus, a slightly shorter median septum, and poorly defined brachial markings.

R. planumbonus (Meek & Worthen) from the Keokuk Limestone of Missouri, illustrated by Weller (1914, pl. 8, figs 58-62), is close externally to the less auriculate individuals of *R. ustulatus,* having a comparable flattened ventral umbo and a steep anterior margin.

A single pedicle valve of *R. kinghiricus transversus* (Nalivkin) figured by Monakhova (1969, pl. 2, fig. 13) from the Visean of central Kazakhstan has a shape and weak median sinus similar to those of *ustulatus;* it is distinguished by the possession of coarser capillae.

The specific name *ustulatus* is Latin for scorched; long grass in the Utting Gap area was burned off to expose low outcrops of fossiliferous calcarenite, including the locality from which this species was collected.

Occurrence: Locality 107/1, Utting Calcarenite, Utting Gap: the type locality.

Material: CPC 10861-10864. Holotype CPC 10861, paratypes CPC 10862-10864. All from locality 107/1. Nine free shells.

Age: Visean.

RUGOSOCHONETES sp. cf. R. KENNEDYENSIS Maxwell

(Pl. 8, figs 11-22)

Description: External. Shell of average size for genus, rounded-rectangular to rounded-quadrate, moderately convex, and widest at or slightly in front of hinge; capillae high, flat-topped, bearing occasional spinules, separated by deep furrows, and having a density of sixteen to eighteen (usually sixteen) per 5 mm at 5 mm from ventral umbo; capillae increasing by bifurcation and occasional intercalation on pedicle valve, and almost always by intercalation on brachial valve; micro-ornament not observed.

Pedicle valve moderately convex, most convex at umbo, becoming gradually flatter towards front; valve highest at about one-third its length; umbo small, relatively narrow, and projecting slightly behind hinge; umbonal shoulders moderately concave and passing on to flat or slightly convex posterolateral margins; valve non-auriculate, lateral margins meeting hinge almost at right angles; venter evenly rounded, no median sinus; interarea apsacline, short, slightly concave at umbo, and having a small U-shaped delthyrium; apex of delthyrium bearing a small convex pseudodeltidium; two (possibly three) spines on each side of umbo arising at 70-80° to hinge, and then curving to almost right angles to hinge.

Brachial valve having a slightly concave visceral disc, and wide flat postero-lateral extremities; valve deepest at about midpoint; interarea very short, flat, and hypercline; no chilidium observed.

Internal. Pedicle valve. Median septum arising from apical callus, narrow, high posteriorly, lower anteriorly, and extending from one-fifth to one-half length of valve; in specimens with a short septum, adductor muscle scars form narrow elongate ridges along and in front of median septum; diductor muscle scars triangular, pointed posteriorly, wider anteriorly, and impressed into floor of valve, especially at umbo; narrow canals of vascula media extending down sides of adductor muscle scars and on to front of valve; teeth small, subrectangular, and slightly wider laterally; papillae coarse and scattered on posterolateral margins, finer, or completely absent, on the front of the valve and on longitudinal ribs; anterior margin of valve ribbed.

Brachial valve. Median septum originating in front of alveolus, low posteriorly, higher and wider anteriorly, and extending to one-half length of valve; alveolus wide and shallow; cardinal process poorly preserved, and probably with a quadrilobate external face; anderidia commencing in front of inner socket ridges, low and narrow, varying from short to more than half length of median septum, and diverging at 45-50°; inner adductor muscle scars poorly defined, very slightly impressed, and trigonal; outer adductor muscle scars separated from them by anderidia, deeply impressed, and with a rounded trigonal outline; inner socket ridges short, narrow, high, and diverging from hinge; sockets subrectangular, becoming wider laterally; brachial impressions not preserved; rows of coarse papillae on radiating ribs at front of valve; anterior and lateral margins strongly ribbed.

Measurements	Length (pedicle valve)	Length (brachial valve)	Width
CPC 10873	7	—	10
CPC 10874	—	7	12 est.
CPC 10875	8	—	12 est.
CPC 10876a	—	7	11.5
CPC 10877	—	5.5	9
CPC 10878	—	5.5	9 est.
CPC 10879	—	6	9.5
CPC 10880	—	5	7.5
CPC 10876b	7	—	8.5
CPC 10881	7.5	—	10
CPC 10882	7	—	10
CPC 10883	7	—	9.5 est.

Remarks: Rugosochonetes sp. cf. *R. kennedyensis* Maxwell is morphologically close to a number of species from Australia, Britain, and Europe. The closest of these, *R. kennedyensis* Maxwell (1954) from both Queensland and New South Wales, is similar in shape and external ornament, and has a comparable internal morphology. Maxwell's specimens from the late Tournaisian Pond Argillite in Queensland (Maxwell, 1954, pp. 20-21, pl. 2, figs 8-12) have a similar size and capillate external ornament, but may have a more swollen pedicle valve, a slightly more concave brachial valve, and a greater number of spines along the hinge of the pedicle valve. The specimens of *R. kennedyensis* Maxwell, described by Roberts (1964, pp. 181-82, pl. 2, figs 1-6) from the Visean of Greenhills and Trevallyn, New South Wales, are morphologically even closer to these specimens, differing only in being slightly larger, in lacking well-defined ribs on the inner lateral and anterior margins of both valves, and in having shorter anderidia than some individuals of *R.* sp. cf. *R. kennedyensis.*

R. kennedyensis magnus Maxwell (1954, pp. 44-45, pl. 5, figs 9, 10; 1961, pp. 87-88, pl. 19, fig. 1), which comes from a stratigraphically higher horizon than *R. kennedyensis* in the Tellebang Formation of Queensland, is much larger, and has a finer ornament than this species. Maxwell (1961) placed *R? werriensis* Campbell (1957) and *Chonetes gloucesterensis* Cvancara (1958) in synonymy with *R. kennedyensis magnus.* Although all three species are morphologically quite close, this action is premature until larger collections of well-preserved material, preferably from well-dated horizons, become available for study.

R. gloucesterensis (Cvancara) (1958, pp. 866-68, pl. 111, figs 1-7), from the Copeland Road Formation (*Rhipidomella fortimuscula* Zone) at Barrington, New

South Wales, has a finer capillate ornament, a more concave brachial valve, a longer ventral median septum, and a larger number of internal papillose ribs.

R. macgregori sp. nov., from the Utting Calcarenite, is distinguished from this species by its finer capillate ornament, more convex pedicle valve, stronger inner socket ridges, more deeply impressed dorsal adductor muscle scars, and better defined brachial impressions.

R. hardrensis (Phillips), the type species of *Rugosochonetes* (Muir-Wood, 1962, pp. 67-68, pl. 8, figs 5-9, 13) has possibly finer capillae, shorter anderidia, and a variable development of accessory septa (Muir-Wood, 1962, pl. 8, fig. 8).

R. mosensis (Demanet) (1938, pp. 64-65, pl. 5, figs 41-59; pl. 6, figs 1, 2), from the V3c inferieur of Belgium, has a shape and ornament comparable with that of *R.* sp. cf. *R. kennedyensis;* internally some individuals of *R. mosensis* possess elongate anderidia (Demanet, 1938, pl. 5, figs 55, 57; pl. 6, fig. 1). *R. mosensis* is distinguished by its larger size, and slightly weaker ventral umbo.

Occurrence: Localities 453/1 and 453/5, Burvill Point; locality 435/0, 5 miles northeast of Point Spring in the escarpment of the Weaber Range; locality 25, 2.5 miles east-northeast of Point Spring in the Weaber Range; all in the Burvill Beds.

Material: CPC 10873-10883, all from locality 453/5. About 50 specimens preserved as moulds in coarse-grained quartz sandstone.

Age: Visean.

Subfamily RETICHONETINAE Muir-Wood, 1962

Genus RETICHONETES Muir-Wood, 1962

Type species: Chonetes amatus Bouchard-Chantereaux *in* de Verneuil, 1854, by original designation of Muir-Wood, 1962.

Diagnosis: See Muir-Wood (1962, p. 63).

Remarks: When compared with the type species, the specimens described below are more rectangular, less concavoconvex, and have a wide fold and sinus. In the pedicle valve, the septum is longer, but usually commences in front of the apex as in *Retichonetes amatus,* and a furrow outlines the visceral disc. A pedicle valve from Ferques, France, in the U.S. National Museum collection, has a similar furrow. Muir-Wood (1962, fig. 13B) illustrates a crescentic line of pits at the posterior of the pedicle valve, opposite rounded crenulations in the brachial valve, which may be equivalent to the furrow. Differences in the brachial valve lie mainly in the well-defined brachial markings, the variable development of accessory septa, and the absence of rounded crenulations on the posterior of the valve lateral to the inner socket ridges. Muir-Wood (1962, p. 63) regards the brachial ridges as being absent from the type species, but faint furrows similar to those in this material are visible in specimens of *R. amatus* from Ferques in the collection of the U.S. National Museum. The accessory septa of *R. amatus* are delicate, and are almost worn away in some specimens from Ferques. Abrasion of the loose shells on the beach at Westwood Creek may explain their absence in most specimens of *R. arenarius* sp. nov.

RETICHONETES ARENARIUS sp. nov.

(Pl. 13, figs 1-18)

Diagnosis: Shell subrectangular to subquadrate, and moderately concavoconvex; density of capillae eighteen to twenty per 5 mm at 5 mm from umbo of pedicle valve; pedicle valve with a broad shallow anterior sinus bearing a low median fold; umbo low and projecting behind hinge; median septum extending to midpoint of valve; diductor muscle scars large, deeply impressed, and bounded posteriorly by an anteriorly curving ridge; visceral area outlined by a marginal furrow; brachial valve with a slight anterior geniculation; interarea hypercline; fold broad, and bearing a narrow median furrow; inner socket ridges smooth, short, and more or less parallel with hinge; median septum commencing near front of anderidia; anderidia diverging at 40°; ?accessory septa present in some valves as broad slightly divergent ridges extending anteriorly from inner adductor muscle scars; brachial markings well defined; brachial ridges arising from anterolateral margins of outer adductor muscle scars, extending to front of median septum, and enclosing elevated kidney-shaped brachial discs.

Description: External. Shell subrectangular to occasionally subquadrate, usually two-thirds wider than long, and moderately concavoconvex; greatest width at midlength; commissure rectimarginate or very slightly sulcate; density of capillae eighteen to twenty per 5 mm at 5 mm from umbo of pedicle valve; capillae rounded, separated by broad furrows, and crossed by fine concentric growth lines; capillae usually increasing by intercalation on pedicle valve, and by bifurcation on brachial valve.

Pedicle valve moderately convex, evenly convex in lateral profile, and highest at midpoint; umbo small, rounded, and projecting slightly behind hinge; posterolateral margins becoming flat near hinge; venter evenly rounded, but front of valve with broad shallow sinus bearing a low median fold corresponding with a depression on inside; cardinal spines not preserved; interarea orthocline to apsacline, short, gently concave beneath beak, and flat towards lateral extremities; delthyrium narrow, U-shaped, and possibly lacking a pseudodeltidium.

Brachial valve with a small anterior geniculation; valve deepest at midpoint; posterolateral margins flat to gently concave; interarea narrow, and hypercline; chilidium convex, and covering dorsal part of cardinal process; fold broad, and with a shallow furrow corresponding with a ridge in front of median septum on inner surface.

Internal. Pedicle valve. Visceral region deep, strongly concave, and outlined by a marginal furrow commencing from beneath teeth; teeth large, blunt, and closely spaced; median septum arising from a callus a short distance in front of apex of valve, high and blade-like posteriorly (1.5 mm on a valve 7 mm long and 9 mm wide), much lower medially, but becoming higher again anteriorly; septum extending to midlength of valve, and then merging laterally into a ridge around front of visceral disc, leaving a slightly concave area in median anterior part of valve; septum flanked by long narrow furrows; adductor muscle scars on small rectangular platforms at posterior of valve; diductor muscle scars large, lachrymose, and usually deeply impressed posteriorly; posterolateral margins of diductor scars bounded by prominent curving ridges; margins of inner surface covered with small pits.

Brachial valve. Cardinal process triangular, bilobed internally, and quadrilobate externally, the two median lobes resting against one another and separated from outer lobes by deep ventrally expanding channels; inner socket ridges supporting posterolateral parts of cardinal process, curving laterally around sockets, widening distally, and extending parallel with hinge; sockets closely spaced, deep, and having rounded floors; outer adductor muscle scars in deep, smooth, subovate to subrounded pits in front of inner socket ridges; inner adductor muscle scars on subrectangular platforms, separated from outer adductor scars by low divergent anderidia; anderidia short, diverging at about 40°, and supporting inner parts of inner socket ridges at front of cardinal process; some specimens with broad ?accessory septa extending anteriorly from inner adductor scars, diverging slightly from median septum; alveolus shallow, round, and situated immediately in front of cardinal process; median septum arising from near front of anderidia, extending to midlength of valve, narrow and blade-like for most of its length, but higher and swollen at front; area in front of septum usually swollen into a broadly rounded ridge extending to front of valve, and accommodated in depression in front of median septum in pedicle valve; brachial ridges arising from anterolateral margins of outer adductor muscle scars, enclosing smooth convex kidney-shaped discs, and meeting at front of median septum; brachial ridges frequently separated from discs by a narrow furrow; discs rising gradually anteriorly, and usually highest at front; fine radially arranged papillae on inner surface between brachial ridges and margins of valve.

Measurements	Length (pedicle valve)	Length (brachial valve)	Width
CPC 10911 Paratype	9	8	11.5
CPC 10912 Holotype	7	6	10.5
CPC 10913 Paratype	6.5	5.5	9
CPC 10915 Paratype	7.5	—	9.5
CPC 10916 Paratype	7	—	9
CPC 10919 Paratype	—	7.5	12
CPC 10920 Paratype	—	7	10 est.
CPC 10921 Paratype	7.5	—	10 est.
CPC 10922 Paratype	9	—	13 est.

Remarks: R. arenarius is compared with *R. amatus* (Bouchard), the type species, in the remarks on the genus (p. 69).

Plicochonetes macropatus Veevers (1959b, pp. 85-88, pl. 9, figs 4-10) from the Sadler Formation (Frasnian) of the Canning Basin, Western Australia, was questionably referred to *Retichonetes* by Muir-Wood (1962). *Plicochonetes macropatus* differs from *R. arenarius* in having a more strongly concavoconvex shell, a semi-oval rather than rectangular to subquadrate outline, a less well defined ventral umbo, more angular cardinal margins, and a broad median ventral fold. The brachial valve interior of *P. macropatus* is known from serial section only.

?Retichonetes douvillei (Rigaux) (1894, p. 104, pl. 1, fig. 1) from the Frasnian of Blacourt, France, has a shape and external ornament similar to that of *R. arenarius;* the pedicle valve is low and slightly convex, has a small pointed umbo extending a short distance behind the hinge, and moderately coarse capillae. The internal features of *?R. douvillei* have not been described.

A much shorter and lower ventral umbo and weaker dorsal muscle scars and brachial markings characterize the specimens of *R. vicinus* (Castelnau) from the Middle Devonian Hamilton Group of Ontario, Canada, figured by Muir-Wood (1962, pl. 7, figs 1, 2). Specimens of the same species described by Prosser (1913, pp. 155-57, pl. 12, figs 14-21) from the Middle Devonian of Maryland, USA, lack a fold and sinus, and have less well defined brachial markings and dorsal muscle scars. *R. marylandicus* (Prosser) (1913, pp. 157-58, pl. 13, figs 1-6), from the same area, is closer to *R. arenarius* in outline, but lacks a fold and sinus, and has a shorter ventral umbo; the interior is not figured.

The specific name *arenarius* (Latin, 'of sand') refers to the way in which these specimens were scooped up with handfuls of sand on the tidal flats near Westwood Creek.

Occurrence: Localities 37/2 (the type locality) and 12/4, Westwood Member of the Cockatoo Formation at Westwood Creek.

Material: CPC 10911-10922. Holotype CPC 10912, paratypes CPC 10911, 10913-10922. All from locality 37/2. Several hundred free specimens.

Age: Frasnian.

Family ANOPLIIDAE Muir-Wood, 1962, emend. Boucot & Harper, 1968

Genus GLOBOSOCHONETES Brunton, 1968

Type species: Globosochonetes parseptus Brunton, 1968, from the Visean (low D Zone) of County Fermanagh, Northern Ireland, by original designation of Brunton, 1968.

Diagnosis (from Brunton, 1968, p. 48): 'Small, strongly concavoconvex Anopliinae with strong ribbing and pair of ventrally serrated, anteriorly divergent septa in brachial valve.'

Remarks: Several recently described anopliid genera may be compared with *Globosochonetes*. *Eoplicanoplia* Boucot & Harper, 1968, from the Silurian (Ludlow) of North America, has finer capillae and a weak dorsal median septum instead of strong twin septa. *Plicanoplia* Boucot & Harper, 1968, from the Siegenian and Emsian of eastern North America and Columbia (Boucot & Harper, 1968) has a comparable arrangement of dorsal septa. When compared with *Globosochonetes* it has coarser rounded costae, a subtrigonal outline, wide posterolateral margins, a low ventral umbo, a denticulate ventral hinge, and smooth twin dorsal septa; the interior of the brachial valve lacks prominent radially arranged rows of papillae extending from near the posterior to the anterior of the valve, adjacent to the septa. *Caenanoplia* Carter, 1968, from the early Mississippian of USA, is more weakly capillate, and lacks well-defined septa in the brachial valve.

GLOBOSOCHONETES BURVILLENSIS sp. nov.

(Pl. 9, figs 1-8)

Diagnosis: Shell with fine ribbing, density of capillae seven to ten per 1 mm at anterior margin; ventral umbo small and blunt; delthyrium U-shaped, and closed apically by a small arched pseudodeltidium; three or four spines on each side of umbo, each approximately 70° to hinge; ventral median septum extending one- to two-fifths length of valve; dorsal septa low posteriorly, becoming higher anteriorly, and diverging at 20-30° (usually 25°); anderidia diverging at approximately 80°; inner adductor muscle scars subelliptical, and on low platforms; outer adductor muscle scars triangular, and deeply impressed; brachial markings arising from anterolateral margins of outer adductor scars, and extending to anterior third of septa.

Description: External. Shell small, strongly concavoconvex, and transversely subtrigonal; widest at hinge, which projects into small pointed auricles; lateral margins rounded except at acute angle of auricles; commissure rectimarginate;

capillae fine, separated by narrow grooves, and usually poorly defined on ventral umbo and posterolateral extremities; capillae occasionally increasing by bifurcation on pedicle valve (method of increase on brachial valve not observed), density seven to ten per 1 mm at anterior margin; micro-ornament not observed.

Pedicle valve moderately convex, highest near midpoint of valve, and most convex at umbo; umbo low, blunt, and projecting slightly behind hinge; umbonal shoulders gently concave, and sloping on to wide flat or slightly convex posterolateral extremities; venter evenly rounded, lateral and anterior slopes extending steeply to margins; interarea orthocline, short, flat, and with a small U-shaped delthyrium; apex of delthyrium bearing a small arched pseudodeltidium; three or four spines on cardinal margins on either side of umbo, each spine approximately 70° to hinge.

Brachial valve deeply concave, especially between umbo and midpoint of valve; posterolateral margins wide, flat to slightly concave, and separated by steep slopes from deep midpart of valve; dorsal interarea linear.

Internal. Pedicle valve. Median septum arising from apical callus, broad and high posteriorly, becoming lower and narrower anteriorly, and extending one- to two-fifths length of valve; adductor muscle scars obscure; diductor muscle scars triangular, and impressed at umbo; lateral and anterior parts of valve ornamented with fine radially arranged papillae; teeth not observed.

Brachial valve. Twin septa commencing at front of alveolus, diverging at 20-30° (usually 25°), and extending almost to anterior margin; septa low posteriorly, and becoming higher, thicker and in some cases serrated towards front; septa bordered on their inner and outer margins by much lower longitudinal papillose ridges commencing at various distances from the hinge; one or two low non-papillose 'median ridges' usually in midpart of valve; internal face of cardinal process small, pointed, rising ventrally above alveolus, and supported laterally by inner socket ridges; external face triangular, quadrilobate, with high dorsally broadened inner lobes separated by sharp furrows from much shorter, more divergent lateral lobes; alveolus deep and circular; anderidia extremely low, narrow, and short, commencing near front of alveolus and socket ridges, and diverging at approximately 80°; inner adductor muscle scars forming small sub-elliptical platforms; outer adductor muscle scars triangular, and deeply impressed between anderidia and inner socket ridges; sockets narrow, wide, and slightly divergent from hinge; inner socket ridges arising from lateral margins of cardinal process and extending more or less parallel with hinge a little way past sockets; brachial markings in form of depressions arising from anterolateral margins of outer adductor muscle scars, curving a short distance laterally, and then turning towards front (not ends) of twin septa; internal surface, particularly on anterior and lateral margins, bearing radial rows of fine papillae.

Measurements: Specimens of *Globosochonetes burvillensis* sp. nov. from the Burvill Beds are crowded together, and are figured in groups. Their dimensions can be determined from the illustrations on Plate 9.

Remarks: Globosochonetes burvillensis is distinguished from the type species, *G. parseptum* Brunton, by its finer capillae, greater number of spines on the hinge (three or four pairs instead of two pairs), and lower ventral umbo.

The specific name is taken from Burvill Point, a prominent bluff in the Weaber Range, 1 mile east of Ningbing Homestead.

Occurrence: Locality 305/1, Burvill Beds, 1.5 miles northeast of Ningbing Homestead.

Material: CPC 10884-10888. Holotype CPC 10886d; paratypes CPC 10884a, b, c; 10885a, b, c; 10886a, b; 10887a, b; 10888a, b, c, d. All from locality 305/1. More than 100 specimens preserved as casts and moulds in a fine-grained sandy limestone.

Age: Visean.

GLOBOSOCHONETES? MATHESONENSIS sp. nov.

(Pl. 8, figs 1-10)

Diagnosis: Shell subtrigonal and strongly concavoconvex; density of capillae ten per 1 mm at front of shell; at least two large spines projecting at 50-60° from lateral parts of hinge; ventral median septum extremely short; brachial valve deep; twin dorsal septa low, and probably accompanied by a median ridge; inner socket ridges straight and divergent from hinge.

Description: External. Shell subtrigonal, wider than long, and strongly concavo-convex; capillae fine, about ten per 1 mm at front of a shell 4 mm long.

Pedicle valve strongly convex medially, but with gently convex to almost flat posterolateral extremities; umbo small, blunt, poorly differentiated from lateral slopes and projecting slightly in front of hinge; median sinus absent; interarea orthocline, short, with small triangular delthyrium; at least two large spines projecting at 50-60° from lateral parts of hinge.

Brachial valve strongly concave medially, deepest near midpoint, and less concave and eventually flat towards posterolateral margins; fold absent.

Internal. Pedicle valve with a short blunt median septum; internal surface papillose; muscle scars not observed.

Brachial valve with two long divergent septa, and possibly a short median ridge; twin septa low, diverging at about 20°; inner socket ridges straight, divergent, and supporting base of cardinal process; alveolus deep; papillae in radial rows on front of valve.

Measurements	Length (pedicle valve)	Length (brachial valve)	Width
CPC 10865 Paratype	5.5	—	8.5
CPC 10866 Paratype	7	—	10.5
CPC 10867b	4.5	—	7
CPC 10868	—	5	6.5
CPC 10869	—	7	10
CPC 10870 Paratype	—	4	5.5
CPC 10871 Holotype	5 est.	4	9.5
CPC 10872a	—	3.5	5

Remarks: This species is described from indifferently preserved material and is tentatively referred to *Globosochonetes* Brunton, 1968. Features of *mathesonensis* prompting assignment to *Globosochonetes?* include the capillate ornament, short ventral median septum, and twin dorsal septa which possibly are separated by a median ridge. The divergent inner socket ridges are not typical of *Globosochonetes*.

G? mathesonensis differs from *G. burvillensis* sp. nov. from the Burvill Beds in having cardinal spines which diverge at a lower angle from the hinge, a more concave brachial valve, weaker dorsal septa, and inner socket ridges which diverge from the hinge.

Occurrence: Localities 28/10, Kununurra Member, Matheson Ridge (the type locality), 438/2, Kununurra Member at Abney Hill, and 406/1, Kununurra and Abney Members at Abney Hill near Kununurra township.

Material: CPC 10865-10872. Holotype CPC 10871, paratypes CPC 10865, 10866, 10870. CPC 10865-10867a, b, 10869-10871 from locality 28/10; CPC 10868, 10872 from locality 438/2. Approximately 20 specimens preserved in friable fine-grained quartz sandstone.

Age: Frasnian.

Family DAVIESIELLIDAE Sokolskaya, 1960

Subfamily DELEPINEINAE Muir-Wood, 1962

Genus DELEPINEA Muir-Wood, 1962

Type species: Productus comoides J. Sowerby, 1823, by original designation of Muir-Wood, 1962.

Remarks: The morphology of *Delepinea,* with particular reference to the cardinal spines and tubules, the muscle scars, the cardinal process and chilidium, and the brachial ridges, was discussed in detail by Campbell & Roberts (1964). Thomas (1965) outlined the characters of *Delepinea,* and described subdental pits in the pedicle valve of *D. uttingi* Thomas from the Bonaparte Gulf Basin in northwestern Australia; he noted that the pits had previously been recognized in species of *Daviesiella* Waagen rather than in *Delepinea.* The pits are again described and figured in *D. uttingi* and also recognized in *D. destinezi* (Vaughan) (Simpson, 1953, pl. 9, fig. 6). Brunton (1966b) emended Muir-Wood's diagnosis of *Delepinea,* and confirmed the interpretation of the brachial markings and cardinal spines given by Campbell & Roberts (1964).

The brachial ridges in *D. uttingi* Thomas are poorly preserved posteriorly, but probably arise from the posterior parts of the outer adductor muscle scars in the same manner as those in *D. aspinosa* (Dun), figured by Campbell & Roberts (1964, pl. 81, fig. 9). The ridges enclose broad elongate crescentic platforms, are discontinuous anteriorly, and then make an S-shaped curve to the front of the median septum as in *D. aspinosa,* and *D. destinezi* (Vaughan) figured by Brunton (1966b, p. 442). Both *D. aspinosa* and *D. destinezi* have shorter and more rounded brachial platforms than those of *D. uttingi,* and according to Brunton the brachial ridges arise from the front of the muscle scars in *D. destinezi.* The significance of this apparent difference is not known.

Furrows on the front and posterolateral parts of the brachial valve of *D. uttingi* are interpreted as mantle canals (Pl. 12, figs 1, 2). Those in the mid-anterior of the valve commence in front of the muscle field and extend as irregular linear canals to the front of the valve. On the brachial discs the canals are shallow, but curve parallel with the inner parts of the brachial ridges and pass through their discontinuous anterior wall, again to the anterior margin (Fig. 17). The canals on the posterolateral part of the valve (Fig. 18 and Pl. 12, fig. 9) extend from the margins of a small callus at the distal ends of the sockets and inner socket ridges; they are moderately wide linear canals, straighter than those on the front of the valve, and extend to the anterolateral margins. The pedicle valve has irregular linear canals extending from the muscle field to the margins, especially from the front of the anterior adductor muscle scars (Pl. 11, figs 4, 7).

Other species of *Delepinea* have similar but in many cases less well defined mantle canals. *D. aspinosa* (Dun) described by Campbell & Roberts (1964, pl. 81, figs 7, 8) has irregular linear canals on the front and lateral parts of the brachial valve but less well defined posterolateral canals; the pedicle valve (pl. 80, figs 3, 4, 7) has a pair of canals extending from the front of the anterior adductor muscle scars, and much weaker canals from the front of the diductor muscle scars. *D. gloucesterensis* Campbell & Roberts (1964, pl. 81, fig. 1; pl. 82, figs 1, 7) has a comparable but usually less well developed system. Brunton's illustrations of *D. destinezi* (Vaughan) (1966b, pl. 8, figs 3, 5) show a less well defined canal system than in *D. uttingi*. A pedicle valve of *D. carinata* (Garwood) figured by Cope (1935, pl. 3, fig. 2) from the Upper Michelinia Zone (C_2), Westmorland, England, has irregular linear mantle canals extending anteriorly from the front of the muscle field.

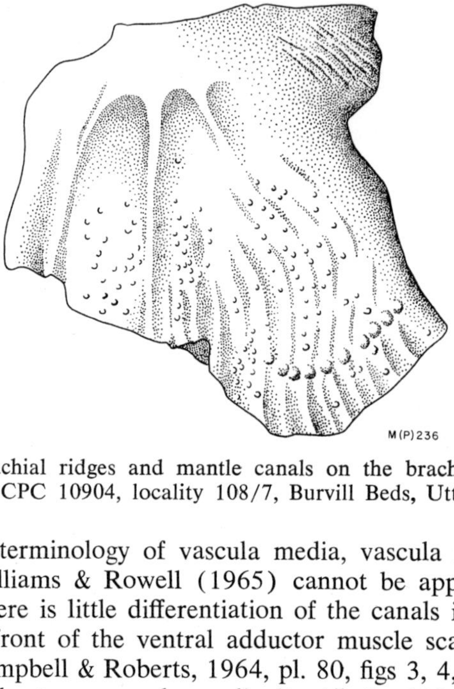

Figure 17. Papillose brachial ridges and mantle canals on the brachial valve of *Delepinea uttingi* Thomas. CPC 10904, locality 108/7, Burvill Beds, Utting Gap. × 1.15.

The mantle canal terminology of vascula media, vascula myaria, and vascula genitalia used by Williams & Rowell (1965) cannot be applied satisfactorily in *Delepinea* because there is little differentiation of the canals into specific systems. The large canals in front of the ventral adductor muscle scars (*see D. aspinosa* (Dun) figured by Campbell & Roberts, 1964, pl. 80, figs 3, 4, 7) could be termed vascula media, and the term vascula cardinalia (Öpik, 1934) could be used to describe the straight posterolateral canals in the brachial valve.

DELEPINEA UTTINGI Thomas

(Pl. 10, fig. 23; Pl. 11, figs 1-7; Pl. 12, figs 1-9)

1965 *Delepinea uttingi* Thomas, pl. 18A, figs 1-5.

Description: External. Shell large for genus, rounded-rectangular, approximately two-thirds wider than long, and moderately to strongly concavoconvex; costellae wide, separated by narrow furrows, wavy on front of shell, and increasing by bifurcation on both valves; density of costellae thirty-four to thirty-six per 10 mm at 20 mm from umbo.

Pedicle valve almost flat posteriorly, becoming moderately to strongly convex anteriorly; valve highest and most convex at or just in front of midpoint; umbo extremely low, barely differentiated from posterior of valve, and projecting a short distance behind hinge; posterolateral margins wide, and flat or gently convex; broad median fold on venter; low radial lateral plication on each flank, extending from umbonal shoulders to anterolateral margins of valve; anterior parts of strongly convex valves sloping steeply towards front, but less steeply inclined in other valves; cardinal margins obtuse, forming an apical angle of about 175°; interarea apsacline, flat, and 7 mm long at delthyrium on a valve estimated at 62 mm long and 135 mm wide; delthyrium widely triangular, sides making an angle of 90-110°, and apex filled with thick callus; one specimen (Pl. 12, fig. 5) with pseudodeltidium extending along whole of sides and possibly closing entire delthyrium; hinge spines not observed; cylindrical tubules through interarea inclined towards umbo, open to interior of valve, but not observed opening on to cardinal margins; tubules 0.5 to 0.7 mm in diameter, spaced at intervals of about 2 mm; more than twenty-seven tubules on either side of umbo on one large specimen 95 mm long.

Brachial valve moderately to strongly concave, flattened at umbo, deepest at midpoint, and gradually becoming shallower towards gently concave posterolateral extremities; shallow radial lateral furrows, opposite plications on pedicle valve, extending from near umbo to anterolateral margins; interarea hypercline, flat, 5 to 6 mm long on a valve estimated at 52 mm long and 135 mm wide; at least half of external face of cardinal process covered by large chilidium; chilidium extremely wide, flat or concave laterally, and highly arched medially over inner lobes of cardinal process.

Internal. Pedicle valve. Median septum arising from apical callus, and extending about one-third length of valve to slightly in front of adductor muscle scars; anterior tip of septum low and ridge-like (remainder not observed); adductor muscle scars dendritic, slightly impressed or on low platforms, and in two distinct pairs; a ?quadrate posterior pair, and a round anterior pair; diductor muscle scars flabellate to paddle-shaped, elongate, very coarsely striated, deeply impressed into floor, and extending well in front of adductor muscle scars; striations arising at posterior tips of diductor scars, extending both anteriorly and laterally, and usually coarsest at anterior margins; posterior margin of adductor muscle scars longitudinally ridged, and obscured by pointed callus; diductor muscle scars most deeply impressed on their posterolateral margins, and in some cases (Pl. 11, fig. 4) also deeply impressed into thick callus in front of adductor muscle scars; irregular rounded longitudinal ridges and depressions usually in front of adductor muscle scars, interpreted as mantle canals, extending towards front; teeth large, high, wedge-shaped, and situated immediately behind large subdental cavity; internal surface with irregular rounded ridges bearing small papillae; shell thickest around posterior of valve, particularly on posterolateral margins of diductor muscle scars.

Brachial valve. Median septum commencing well in front of cardinal process as a low rounded ridge, higher and more robust at posterior end of muscle field, and then tapering and becoming higher and blade-like for remaining length to midpoint of valve; anderidia low and narrow, commencing near base of cardinal process, diverging at 25-35°, and extending to about midlength of muscle field, dividing inner and outer adductor muscle scars; anderidia having a spinose projection at their junction with a low ridge around posterior margin of inner adductor muscle scars; adductor muscle scars well in front of umbo, and in two pairs — inner scars striated and trigonal, with broadly rounded posterior margins and wider straight anterior margins, and outer scars dendritic and subovate; inner face of cardinal process forming a massive ridge parallel with hinge, supported laterally by inner socket ridges, and bordered by a wide shallow alveolus; external face broadly triangular, and quadrilobate; median lobes high, ventrally expanding, and separated by a deep rounded furrow; lateral lobes lower, and separated by wider shallower furrows; sockets deep, rounded and bordered by an outer socket ridge and a massive rounded inner socket ridge; inner socket ridges straight, divergent, and apparently articulating with subdental pits in pedicle valve; brachial ridges poorly defined posteriorly, coarsely papillose and discontinuous anteriorly, forming a hook-like curve around a broad crescentic platform bearing indistinct canals and scattered papillae, and then recurving anteriorly to front of median septum (Fig. 17); canals on crescentic platform passing anteriorly through gaps between papillae on brachial ridge, and becoming deeper and wider towards anterior margin; surface of valve in front of muscle field on inner margins of brachial ridges also with canals between irregular papillose ridges; long linear ridges and canals arising from lateral margins of sockets, and extending towards lateral margins (Fig. 18).

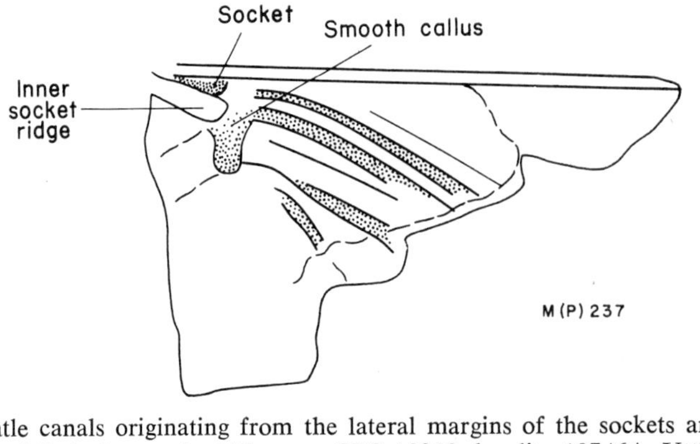

Figure 18. Mantle canals originating from the lateral margins of the sockets and inner socket ridges of *Delepinea uttingi* Thomas. CPC 10910, locality 107/6A, Utting Calcarenite, Utting Gap. × 1.2.

Remarks: The specimens described in this paper were collected from Thomas's localities L and M (Thomas, 1965, fig. 1), and have slightly finer costellae than Thomas's material. At 20 mm from the umbo there are thirty-four to thirty-six costellae per 10 mm, and in Thomas's specimens twenty-five to thirty-two costellae per 10 mm. Thomas (1965, p. 98) gave the density as sixteen to twenty-two per 10 mm towards the front, but did not specify the distance from the front of the umbo. Variation within the species may account for the different densities. Measurements on Thomas's original specimens were kindly made by Michael Garrett of the University of Melbourne.

In a detailed comparison with other species of *Delepinea,* Thomas (1965, pp. 100-101) considered *D. uttingi* to be closest to *D. destinezi* (Vaughan) and

D. carinata (Garwood). *D. destinezi* (Vaughan), described by Simpson (1953) and Brunton (1966b), has a similar size, outline, and external ornament to *D. uttingi,* but differs in lacking a ventral median fold, and in having wider diductor muscle scars and less well defined adductor muscle scars in the pedicle valve, a weaker and shorter dorsal median septum, and more elongate brachial discs. *D. destinezi* is known from the Visean (VIb) of Belgium, and the C_2S_1 of England and Ireland (Muir-Wood, 1962).

D. carinata (Garwood), described by Cope (1935), has a ventral median fold similar to that of *D. uttingi* but is distinguished by its higher and more convex pedicle valve and stronger ventral umbo (Thomas, 1965). *D. carinata* is known from the C_2 of England and the early Visean of Belgium.

Occurrence: Localities 107/1, 107/6A, 108/0, 108/3, 108/4A, and 143/1 in the Utting Calcarenite, and 108/7 in the Burvill Beds, at Utting Gap; locality 304/1 in the Utting Calcarenite, 14.3 miles bearing 348° from Ningbing Homestead (Thomas's locality K). Several very poorly preserved specimens of *Delepinea* with a coarsely costellate ornament and a convex pedicle valve having a strongly developed umbo have been collected from section 435 in the Burvill Beds on the southern scarp of the Weaber Range, 5 miles northeast of Point Spring.

Material: CPC 10898-10910, CPC 10898, 10900, 10902-10909 from locality 108/7; CPC 10899, 10901, 10910, from locality 107/6A. Thirty fragmentary specimens.

Age: Visean.

Genus MEGACHONETES Sokolskaya, 1950

Type species: Chonetes compressa Sibly equals *C. siblyi* I. Thomas, 1919, by original designation of Sokolskaya, 1950.

Diagnosis: See Muir-Wood (1962, p. 105).

MEGACHONETES ZIMMERMANNI (Paeckelmann)

(Pl. 10, figs 11-22)

1847 *Chonetes papilionacea* Phillips, de Koninck, pl. 19, fig. 2.
1930 *Chonetes (Chonetes) zimmermanni* Paeckelmann, pl. 17, fig. 23; pl. 18, fig. 1.
1950 *Chonetes (Megachonetes) zimmermanni* Paeckelmann; Sokolskaya, pl. 6, figs 7-13.

Description: External. Shell of moderate size for genus, rounded-rectangular, wider than long, and gently to moderately concavoconvex; shell widest at or near hinge; commissure rectimarginate; capillae rounded, separated by narrow furrows, and bearing numerous spinule apertures; capillae on both valves increasing mainly by bifurcation, more rarely by intercalation; density of capillae nineteen to twenty-one per 5 mm at 20 mm from umbo.

Pedicle valve moderately and evenly convex in lateral profile, highest at about midpoint of valve; umbo low, barely rising above umbonal shoulders, and extending a variable distance behind hinge (3 mm on a valve 21 mm long and 35 mm est. wide); umbonal shoulders slightly concave, passing on to wide flat or gently concave posterolateral extremities; venter evenly rounded to slightly flattened; cardinal margins forming an angle of approximately 165°, and bearing eight spines on either side of umbo; most spines extending at 45° to hinge, but some diverging at up to 80°; interarea apsacline, flat, and about 3 mm long on a valve 21 mm long and 35 mm est. wide; delthyrium triangular, very wide, and filled apically by an extremely thick convex pseudodeltidium.

Brachial valve with a gently concave visceral disc, and wide flat posterolateral extremities; interarea short, hypercline to catacline; chilidium convex, and covering dorsal half of external face of cardinal process.

Internal. Pedicle valve. Median septum low and ridge-like anteriorly, and extending one-third to one-half length of valve; adductor muscle scars faintly dendritic, narrow, pointed anteriorly, and impressed into floor; diductor muscle scars obscure; teeth large and subangular; papillae in radial rows on areas outside muscle field, particularly coarse and sub-imbricating on posterolateral parts of valve.

Brachial valve. Median septum slender, low, and extending two-thirds length of valve; inner socket ridges straight, and diverging from hinge; anterior and lateral surfaces bearing rows of small papillae; cardinal process widely triangular, having six lobes on external face; median lobes long, expanding ventrally, and separated by a deep ventrally widening furrow; the two pairs of lateral lobes progressively lower, shorter, and more divergent laterally from midline, separated from one another and median lobes by broad shallow ventrally expanding furrows; inner face of cardinal process, brachial ridges, and muscle scars not observed.

Measurements	Length (pedicle valve)	Length (brachial valve)	Width
CPC 10892	21	19	35 est.
CPC 10893	24.5	22	44
CPC 10895	20 est.	—	40 est.
CPC 10896	20.5	18	40 est.

Remarks: The six specimens available for study cannot be distinguished from Paeckelmann's material from the Visean of Poland (1930, pp. 272-74, pl. 17, fig. 23; pl. 18, fig. 1), being similar in size, shape, convexity of the pedicle valve, concavity of the brachial valve, number of spines, density of the costellae, and in the morphology of the external face of the cardinal process. The only difference I can detect in Paeckelmann's illustrations is the possible absence of the chilidium, which may have been broken. Russian material referred to *M. zimmermanni* by Sokolskaya (1950, pl. 6, figs 7-13) is characterized by a slightly less convex pedicle valve, but has a similar size, outline, and external ornament. *M. zimmermanni* is known from the Visean of Poland, Belgium, France, and the C_1mkh horizon of the Russian Platform, and is recorded from the Tournaisian of the Kuznetsk Basin, Siberia (Sarycheva et al., 1963).

These specimens are also morphologically close to one of the forms referred to *Chonetes papilionacea* Phillips by Davidson (1861, pl. 46, fig. 5), and questionably placed in *zimmermanni* by Paeckelmann (1930). Davidson's specimen comes from the Carboniferous of Settle, Yorkshire.

Megachonetes murchisoni (Julien) (1896, p. 76, pl. 7, fig. 14; pl. 8, fig. 1) from the Visean of Regny, France, has a similar external ornament, but is less transverse in outline.

Occurrence: Locality 108/7, Burvill Beds, Utting Gap.

Material: CPC 10892-10897. All from locality 108/7. Six specimens preserved in yellow silty limestone.

Age: Visean.

Chonetidid gen. et sp.

(Pl. 10, figs 1-10)

Description: External. Shell transverse, moderately concavoconvex, with rounded lateral margins and auriculate posterolateral margins; commissure rectimarginate to very slightly uniplicate; external surface very faintly capillate, and bearing fine concentric growth lines.

Pedicle valve with a low umbo, highest near midpoint, and a relatively steep anterior margin; greatest convexity at umbo, or at a point just in front of middle of valve; umbo barely extending behind cardinal margins; umbonal shoulders gently concave, passing laterally on to moderately wide convex to flat posterolateral extremities; venter evenly rounded in profile, but anterior margin in some specimens bears a faint median depression; three cardinal spines on each side of umbo diverge from hinge at about 40°; ventral interarea apsacline, flat, and 0.75 mm long on a valve 13 mm wide and 7 mm long; delthyrium U-shaped, ventral half covered by highly convex pseudodeltidium.

Brachial valve gently concave over visceral disc, but with abrupt anterior geniculation resulting in steep anterior and anterolateral margins; valve deepest at midpoint, slightly concave at umbo, posterolateral margins flat to slightly concave; some specimens with extremely low fold at front; interarea short, catacline to hypercline, with a convex chilidium covering dorsal part of cardinal process; cardinal process quadrilobate externally, lobes diverging and becoming wider ventrally; inner lobes separated by deep narrow slit, and inner from outer lobes by wider and shallower furrow.

Internal details not observed.

Measurements	Length (pedicle valve)	Length (brachial valve)	Width
CPC 10889	9	8	14
CPC 10890	7	6	13
CPC 10891	6.5	5.5	10.5

Remarks: The three specimens of Chonetidid gen. et sp. available for study are not closely comparable with any of the Carboniferous chonetoids so far described.

Occurrence: Locality 102/4 Enga Sandstone, northern end of Enga Ridge.

Material: CPC 10889-10891. All from locality 102/4. Three silicified shells showing external details only.

Age: Tournaisian.

Superfamily PRODUCTACEA Gray, 1840
Family PRODUCTELLIDAE Schuchert & Le Vene, 1929
Subfamily PRODUCTELLINAE Schuchert & Le Vene, 1929
Genus PRODUCTELLA Hall, 1867

Type species: Productus subaculeatus Murchison from the upper Devonian (late Frasnian) of Ferques, France, by subsequent designation by Oehlert, 1887.

Diagnosis: See Muir-Wood & Cooper (1960, pp. 146-47).

PRODUCTELLA WESTWOODENSIS sp. nov.

(Pl. 14, figs 1-16)

Diagnosis: Shell large for genus, and subrectangular to subquadrate; pedicle valve with a narrow concave interarea containing a small quadrangular delthyrium; rugae weak, but frequently extending across venter; brachial valve aspinose, and ornamented with low rugae; median septum extending one-half to two-thirds length of brachial valve; lateral ridge in front of sockets weak; adductor muscle scars faintly dendritic; brachial ridges oblique, arising from posterior tips of adductor muscle scars, and forming high hooked ridges around subquadrate brachial discs.

Description: External. Pedicle valve large, subrectangular to subquadrate, with poorly differentiated auricles, and rounded lateral margins; highest at midpoint of valve; valve moderately convex, most convex near umbo; umbo projecting behind hinge, but only slightly incurved; umbonal slopes steep and concave; flanks spreading; greatest width slightly in front of midlength of valve; hinge shorter than greatest width; interarea narrow, linear, and containing a small quadrangular delthyrium; body of valve with randomly scattered spine bases, 0.5 mm to 0.7 mm in diameter, bearing erect spines; more than six long erect spines in a poorly defined row near hinge; outer spines in this row curving laterally, and inner spines appearing to curve over umbo; rugae very weak but frequently extending across venter; micro-ornament of irregular growth lines.

Brachial valve moderately concave, deepest at midpart of valve, with a broad triangular depression at umbo; posterolateral margins flat to slightly concave and well differentiated from remainder of valve; interarea linear; valve aspinose, and ornamented with low rugae; rugae most prominent on posterolateral margins.

Internal. Pedicle valve interior not observed.

Brachial valve. Median septum low, narrow, arising from in front of an alveolus, and extending one-half to two-thirds length of visceral disc; septum highest between adductor muscle scars, becoming lower anteriorly; adductor muscle scars in a single pair, faintly dendritic, trigonal, with pointed posterior and rounded

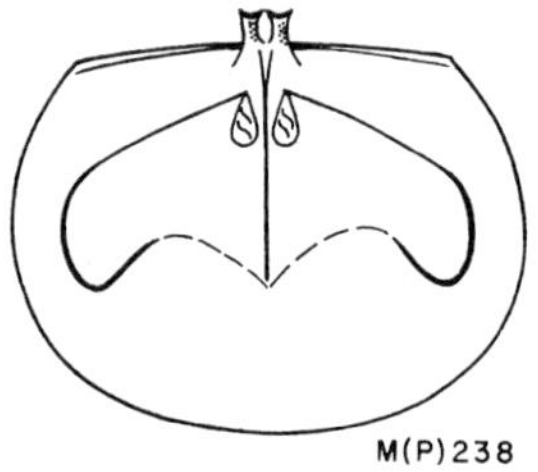

Figure 19. Brachial valve interior of *Productella westwoodensis* sp. nov. reconstructed from CPC 8615 and the exterior of CPC 8609. The greater part of the brachial ridges, including the extension (dashed) to the tip of the median septum, is from CPC 8609. Note the faintly dendritic adductor muscle scars.

anterior margins; posterior parts of muscle scars well separated from median septum, and anterior parts abutting against septum; cardinal process bilobate internally and externally; internal face with two long narrow lobes separated anteriorly by a deep alveolus, and supported anteriorly by two ridges joining base of median septum; inner lobes becoming wider posteriorly, expanding into two posterior-facing outer lobes; outer lobes deeply concave, joined medially, and tapering dorsally, forming a V-shaped posterior face; sockets laterally elongate and incised into lateral ridges; lateral ridges very low, extending along hinge, and in one specimen bordered anteriorly by a narrow furrow running obliquely across ears; brachial ridges arising from posterior tip of adductor muscle scars, extending obliquely towards lateral margins, higher towards front of valve, and enclosing broad subquadrate brachial discs; distal parts of brachial ridges possibly linked by very low ridges with anterior tip of median septum (Fig. 19); front of visceral disc finely spinose; posterolateral parts of valve pitted.

Measurements

Pedicle valve	Length	Length (of curvature)	Width (at hinge)	Width (of body)	Height
CPC 8609 Holotype	20.5	35	20.5	26	11.5
CPC 8610 Paratype	20.5	33.5	20 est.	22.5	11.5
CPC 8611 Paratype	15	21.5	14 est.	18.5	7

Brachial valve	Length	Width	Length (median septum)
CPC 8615 Paratype	22	25 est.	12.5

Remarks: Productella westwoodensis sp. nov. is larger than many species of *Productella*. The type species, *P. subaculeata* (Murchison), figured by Muir-Wood & Cooper (1960, pl. 32, figs 5-14) is smaller, less transverse, has a more prominent umbo, weaker rugae on the pedicle valve, and some spine bases on the body of the pedicle valve which obviously support sub-erect spines; the brachial valve exterior is marked by indentations which probably correspond to spine bases on the pedicle valve. The topotype brachial valve from Ferques (Muir-Wood & Cooper, 1960, pl. 32, figs 12, 13) has stronger ridges bounding the front of the sockets, and a shorter median septum.

P. occidua Veevers (1959b, pp. 23-26, pl. 1, figs 6-13) from the Gneudna Formation in the Carnarvon Basin, Western Australia, is much smaller and less transverse, with a more highly convex pedicle valve and a more concave brachial valve; the pedicle valve has a greater density of spines, and the spines themselves are finer than those on *westwoodensis*. A single specimen from the Sadler Formation in the Canning Basin, Western Australia, which was referred to Productidae gen. et sp. indet. (Veevers, 1959a, p. 85, pl. 9, figs 1-3) may belong to *occidua*. When compared with *westwoodensis* it is less transverse, has a more highly convex pedicle valve with a pronounced umbo, smaller ventral spine bases, and a number of poorly defined radial folds or 'pscudocostae'.

Externally closest to *westwoodensis* is a specimen referred to *P. subaculeatus* (Murchison) by Nalivkin (1930, pl. 6, fig. 5) from the Frasnian Voroneje Beds of the Russian Platform. Specimens referred to *P. herminae* Frech by Lyashenko (1959, pp. 211-12, pl. 81, figs 1, 2) from the Famennian (D^2_3el) of the Russian Platform resemble *westwoodensis* in shape, and in having a relatively small number of erect spines on the pedicle valve. The interior of *herminae* is not described, and a closer comparison cannot be made. Frech's original specimens (Frech, 1891, pl. 47, figs 3, 5, 6, 10, 11) from the Upper Devonian Iberger Limestone of Germany are smaller, globular, and have pronounced ears, and much smaller spine bases when compared with *westwoodensis*. *P. calva kosharica* Sokolskaya (1948, pl. 2, figs 8-10) from the Famennian of Koshari, USSR, has a comparable shape, but is distinguished by the possession of a large number of more elongate spine bases on the pedicle valve, and a shorter median septum and smaller muscle scars in the brachial valve.

Both *P. sheffieldensis* (Stainbrook) (1950, pp. 374-75, pl. 54, figs 10-13) and *P. applingtonensis* (Stainbrook) (1950, p. 375, pl. 54, figs 15-18) from the Famennian Applington Formation of Iowa are close in size to *westwoodensis*. Both American species are less transverse, and have a greater number of spines arising from elongate spine bases on the pedicle valve. In addition, *sheffieldensis* has larger ears, and a faint sinus on the pedicle valve.

The specimens of *Productus subaculeatus* Murchison = *fragaria* Sowerby figured by Davidson (1847, pl. 20, figs 1, 2) from Woolborough Quarry, England, have a similar transverse outline. They have more elongate spine bases and larger ears on the pedicle valve than *westwoodensis*.

This species is named after Westwood Creek.

Occurrence: Localities 12/4, 12/9, and 13/6 (the type locality), Westwood Member of the Cockatoo Formation, Westwood Creek. Locality 428/5, Hargreaves Member of the Cockatoo Formation, Hargreaves Hills.

Material: CPC 8609-8615. Holotype CPC 8609, paratypes CPC 8610, 8611, and 8615. All from locality 13/6. More than 150 free specimens.

Age: Frasnian.

Subfamily LOMATIPHORINAE nov.

Diagnosis: Productellidae with a costate ornament on both valves; pedicle valve having scattered spines arising from the costae, and a row of spines on the hinge; brachial valve aspinose, and with a weak reticulate ornament; pseudodeltidium, chilidium, teeth and sockets absent; cardinal process bilobed.

Remarks: Lomatiphora gen. nov. is so far the only genus in the subfamily, and occurs in the Lower Carboniferous (Tournaisian). The costate ornament on both valves, and the absence of interareas, teeth, and sockets distinguishes the Lomatiphorinae from the subfamilies Productellinae and Chonopectinae: the Productellinae is characterized by the possession of interareas, variably defined teeth and sockets, and a sparsely spinose ornament; and the Chonopectinae by the possession of interareas, well-developed teeth and sockets, and a variably spinose ornament. The morphopolgy of the cardinal process and median septum clearly places the Lomatiphorinae in the family Productellidae.

Genus LOMATIPHORA nov.

Type species: Lomatiphora aquila gen. et sp. nov. from the Burt Range Formation, Bonaparte Gulf Basin, Australia.

Diagnosis: Pedicle valve subelliptical, moderately to highly convex on the body of the valve, with a wide flattened flange normal to the body extending around anterior and lateral margins of adult specimens; ornament of rounded costae, a few weak rugae over visceral disc, and closely spaced concentric growth lines; costae increasing by bifurcation, and extending without a break on to flange; spines in a single row diverging from the hinge, and arising from costae and randomly scattered over the body of the valve; brachial valve geniculate, with a gently concave visceral disc and a flat recurved flange fitting against that on pedicle valve; visceral disc weakly reticulate; costae broad, increasing by bifurcation, and crossed by weak rugae; brachial valve aspinose; pedicle valve interior with triangular, radially striate diductor muscle scars, and a marginal ridge between the flange and body of the valve; teeth absent; brachial valve interior with a single pair of smooth adductor muscle scars; cardinal process erect, bilobed internally and externally, and buttressed anteriorly by two ridges or a platform; alveolus usually present; median septum extending two-thirds the length of visceral disc and low along its entire length; lateral ridges diverging from hinge, in some valves extending across ears; sockets absent; brachial markings oblique; front of valve with small radially arranged spines or pustules.

Remarks: Lomatiphora differs from other productellid genera by the possession of a costate ornament and marginal flanges on both valves. Marginal flanges are present in brachiopods from a number of diverse groups. Flat flanges are found in *Thomasella* Fredericks, 1928, *Institifera* Muir-Wood & Cooper, 1960, and this genus*; a ledge-like rim in *Marginicinctus* Sutton, 1938; and deep gutter-like structures in *Institella* Cooper, 1942, and *Setigerites* Girty, 1939. Many aulostegid genera also possess flange and gutter structures. Davidson (1880, p. 310) described a series of delicate recurved growth lamellae on a specimen of *Institifera* from the British Carboniferous, and judging from the plates in Muir-Wood & Cooper (1960, pl. 91, figs 24-26), *Thomasella* may also have had a number of successive flanges formed during the growth of the shell. In the other flange-bearing genera, the flanges are either in large specimens or in forms which appear to have prematurely adopted mature characters; G. A. Cooper (pers. comm.) interprets these flanges as adult or gerontic features. More recently Shiells (1968; 1969) has described flanges on species of *Kochiproductus* Dunbar, 1955, and *Pustula* Thomas, 1914, from the Carboniferous of the British Isles. Shiells (1968) postulates that the spinose flange on *Kochiproductus* traps sediment particles but allows food material to pass into the shell on the inflowing feeding current. The flange on *Lomatiphora* is smooth internally and hence probably did not have the same function as in *Kochiproductus*: I suggest the flange prevented larger, heavier specimens of *Lomatiphora* from sinking into a soft substrate.

In one horizon in the Burt Range Formation (locality 100/25A) there is a profusion of small specimens of *Lomatiphora* with flanges. These may be forms which have adopted adult characters because of an unfavourable environment. In other localities, the smaller specimens of *Lomatiphora* lack a wide marginal flange.

Mantle canals are rarely recorded in productoid brachiopods. In the brachial valve of *Lomatiphora* a pair of linear mantle canals extends from in front of the adductor muscle scars, along the median septum, and branches near the front of the valve (Fig. 20). Brunton's suggestion (1966a, p. 217) that depressions on either side of the median septum in *Overtonia fimbriata* (Sowerby) housed canals of vascula media is supported by the presence of these canals.

**Morganella* McKellar, 1970, which has *M. maxwelli* McKellar as type species, has a marginal flange similar to that on *Lomatiphora* gen. nov. *Morganella* resembles *Lomatiphora* in the morphology of the cardinal process, median septum, and brachial ridges. *Morganella* is distinguished by its mainly smooth shell, the only radial ornament being sparse peripheral radial lirae, and by the possession of small teeth, and short deep sockets. *M. maxwelli* is recorded from the Famennian Boulder Creek Grit in the Yarrol Basin, Queensland. (McKellar, R. G., 1970, The Devonian productoid brachiopod fauna of Queensland. *Geol. Surv. Qld Pub.* 342, *Pal. Pap.* 18, 1-40, pls 1-12.)

LOMATIPHORA AQUILA gen. et sp. nov.

(Pl. 15, figs 1-20)

Diagnosis: Lomatiphora with fifteen to eighteen costae per 10 mm immediately behind the ventral flange; ten rugae on the visceral disc of the pedicle valve, and ten to twelve on the visceral disc of the brachial valve; five to six spines on each side of the ventral umbo.

Description: External. Pedicle valve wider than long, subelliptical convex body surrounded (in adult form) by a wide marginal flange; in profile, body of valve having a moderately convex visceral disc and a less convex trail extending anteriorly into flange; greatest convexity near midlength; umbo small, incurved, extending slightly behind hinge, and well differentiated from gently concave umbonal shoulders; flange normal to body of valve, extending around anterior and lateral margins, and projecting a short distance behind hinge; flange up to 8 mm wide on a valve having a body 34 mm wide and a total width of 50 mm; flange flat on lateral margins, and usually gently folded at front, fold stopping at junction with body of valve; costae commencing at umbo, extending on to flange, crowded on its proximal parts, but on some valves becoming obsolete at its distal margins; costae rounded, variable in strength, and separated by intercostal furrows usually slightly wider than each costa; costae increasing by bifurcation, especially

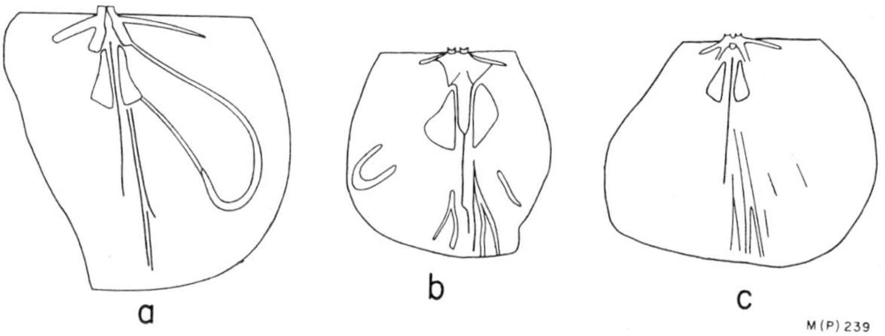

Figure 20. Brachial valve interiors of *Lomatiphora aquila* gen. et sp. nov. showing the buttressing ridges or platform in front of the cardinal process, the brachial ridges, and the mantle canals. a-CPC 8621a, b-CPC 8621b, c-CPC 8622b. Locality 126/5. Sorby Hills. × 1.5.

in front of spine bases, and having a density of fifteen to eighteen per 10 mm immediately behind flange; in some individuals two costae converge into a single rib; spines erect to sub-erect, in a single row diverging from hinge, five or six spines on either side of umbo, 0.5 mm in diameter at base of largest individual; other ovoid spine bases arising from costae and randomly scattered over body of valve (five spines observed on trail of one specimen); up to ten rugae, weak on visceral disc, slightly stronger on posterolateral margins; concentric growth lines prominent, particularly on flange, with a density in one instance of seven per 1 mm.

Brachial valve subquadrate, gently concave over visceral disc, genticulate anteriorly, and in adult having a marginal flange fitting against that of pedicle valve; valve deepest at two-thirds length, with a small rounded triangular depression at umbo; posterolateral margins broad, flat, smooth except for rugae, sloping gradually into concave visceral disc, but sharply differentiated from marginal flange; posterolateral margins ornamented with a row of round depressions opposite spines on pedicle valve; costae rounded, usually more than twice as wide as intercostal furrows, and increasing by bifurcation, mostly near geniculation of valve; density of costae twelve to fourteen per 10 mm at 20 mm from umbo; ten to twelve rugae lower than those on pedicle valve, extending across visceral disc, but not making a reticulate ornament with costae.

Internal. Pedicle valve. Diductor muscle scars slightly impressed, triangular, pointed posteriorly, rounded laterally and anteriorly, and radially striate; adductor muscle scars not observed; marginal ridge originating from posterior margin, and extending around valve between base of trail and flange.

Brachial valve. Cardinal process small, erect and bilobate; inner lobes high above alveolus and separated by a median furrow; outer face consisting of two triangular lobes separated by narrow median furrow, each lobe pointed dorsally, expanding into a rounded lobe ventrally, and having a median ventrally expanding trough; cardinal process buttressed by lateral ridges, and with two shorter anterior ridges or an anterior platform; lateral ridges rounded, variable in strength, curving anteriorly from hinge and at times cutting across ears of valve; two buttressing ridges at front of cardinal process slightly divergent, 2.5 mm long on a valve 22 mm long, and flanking an ovate or circular alveolus at base of cardinal process and narrow posterior extremity of median septum; in some specimens a semicircular platform in front of cardinal process replaces anterior ridges, obliterates alveolus, and extends anteriorly into broad posterior shaft of median septum; median septum commences between buttressing ridges or, more rarely, from front of platform; septum usually narrow and low along its entire length, best defined in trough between adductor muscle scars, and 12.5 mm long on a valve 22 mm long; septum thickened posteriorly in specimens with a platform buttressing cardinal process (the posterior 4 mm of septum 12 mm long up to 1 mm wide near front of platform); adductor muscle scars slightly elevated, smooth, and in a single pair; adductor scars irregularly triangular to crescent-shaped, elongate posteriorly with concave lateral margins, and a gently curved anterior margin extending obliquely towards median septum; posterior tips of muscle field impressed into anterior buttress ridges or platform at front of cardinal process; brachial ridges low, originating from buttress ridges, extending anterolaterally as low U-shaped ridges, and then returning to lateral margins of adductor muscle scars; faint linear mantle canals extending alongside median septum from near front of adductor muscle scars, and possibly branching towards front of valve (Fig. 20); internal ornament of small pustules or spines arranged radially on front of visceral disc and on trail; posterolateral shoulders pitted.

Measurements

Pedicle valve	Length (without flange)	Length (with flange)	Width (without flange)	Width (with flange)	Height
CPC 8634 Paratype	21	26	23.5	39.5	9.5

Brachial valve	Length (without flange)	Width (without flange)	Length (median septum)
CPC 8621a Holotype	21.5	27	12.5
CPC 8622 Paratype	20	22.5	9.5

Remarks: Lomatiphora aquila gen. et sp. nov. cannot be compared with any known species (see footnote on p. 85.).

The specific name *aquila* (Latin for blackish) refers to the well-preserved black iron-stained specimens from the Sorby Hills.

Occurrence: Localities 100/25-100/30 in an area between 6.5 and 7 miles north-west of Mount Septimus; localities 101/9, 101/10, 101/11, 101/13-101/15, 101/19 in an area between 4.5 and 5 miles west of Mount Septimus; locality 421/17, southern end of Enga Ridge; localities 114/9, 115/7, 116/1, 116/135, and 117/1 Spirit Hill; localities 122/2 and 122/3 at Sandy Creek; 123/14, 11 miles southwest of Legune homestead; locality 126/5, Sorby Hills (the type locality); and possibly at locality 124/1, Flapper Hill; all in the Burt Range Formation.

Material: CPC 8621-8635. Holotype CPC 8621a; paratypes CPC 8621b, 8622a, 8623, 8628, and 8634. CPC 8621-8623, 8628, 8631, 8634 from locality 126/5; CPC 8624-8625, 8630, 8632 from locality 100/30; CPC 8626-8627 from locality 100/27; CPC 8629 from locality 115/7; CPC 8633, 8635 from locality 101/13.

Age: Tournaisian.

Subfamily CHONOPECTINAE Muir-Wood & Cooper, 1960

Genus STEINHAGELLA Goldring, 1957

Type species: Leptaena membranacea Phillips, 1841, from South Petherwin, Cornwall, England, by original designation of Goldring, 1957.

Remarks: In Europe, *Steinhagella* is confined to Stages V and VI of the upper Devonian (Goldring, 1957), whereas in Australia the genus is known from the Frasnian Sadler Formation in the Canning Basin (Veevers, 1959a) and the Frasnian Westwood Member of the Cockatoo Formation in the Bonaparte Gulf Basin. Mansuy (1912, pl. 8, figs 5a-d) illustrated a species of the genus from Yunnan, and considered it to be Middle Devonian in age; this requires confirmation before extending the range of the genus below the Frasnian.

STEINHAGELLA sp.

(Pl. 16, figs 22-26)

Description: External. Pedicle valve rounded-rectangular, gently to moderately convex, and evenly curved in profile; umbo low, poorly differentiated from ears, not incurved, and barely extending behind hinge; ears broad, flat to slightly convex, and separated from umbo by gently concave umbonal shoulders; flanks gently convex; trail short, and in one specimen bearing a shallow median sinus; valve widest at midlength or at hinge, and highest at between one-third and one-half its length; rugae narrow, wavy, crowded at posterior of valve, more widely spaced anteriorly, and bearing or crossed by short prostrate spines; seven rugae between 10 and 15 mm from umbo; spines short, crowded, and quincuncially arranged on umbo, and longer but fewer in number at front; more than eight erect spines of larger diameter in one or two rows diverging from hinge, and in a group of four or more on ears; rugae ornamented with fine concentric growth lines.

Brachial valve mainly flat, but with a shallow concavity at umbo, and gently concave ears; ornament of rugae and prostrate spines as on pedicle valve.

Internal. No internal details are known.

Remarks: Steinhagella numida Veevers (1959a, pp. 77-79, pl. 9, figs 11-19), from the Sadler Formation (Frasnian) of the Canning Basin, is distinguished from this species by the possession of a flatter pedicle valve, a larger number of prostrate spines (particularly on the anterior and lateral margins of the pedicle valve), more closely spaced rugae on both valves, and a more concave brachial valve.

Steinhagella sp. is, as far as can be judged from the available illustration, morphologically close to *S. membranacea* (Phillips) (1841, p. 60, pl. 25, fig. 101), having comparable wavy rugae, and radially arranged prostrate spines. A single pedicle valve from the Pamirs referred to *Productella membranacea* (Phillips) by Reed (1922, pp. 87-88, pl. 13, fig. 22) has an ornament comparable with that on *Steinhagella* sp.

The specimen from the Middle Devonian of Yunnan referred to *Strophalosia productoides* Davidson by Mansuy (1912, pp. 52-53, pl. 8, figs 5a-d) has a stronger umbo which extends well behind the hinge of the pedicle valve, a less well defined concentric ornament, and a more concave brachial valve.

Occurrence: Locality 12/3, Westwood Member, Westwood Creek; locality 406/1, Kununurra Member, Abney Hill, and possibly at locality 28/10, Kununurra Member, Matheson Ridge; all in the Cockatoo Formation.

Material: CPC 8298-8301, consisting of three pedicle valves and one brachial valve all preserved in limestone. All from locality 12/3.

Age: Frasnian.

Family LEIOPRODUCTIDAE Muir-Wood & Cooper, 1960
Subfamily LEIOPRODUCTINAE Muir-Wood & Cooper, 1960
Genus LEIOPRODUCTUS Stainbrook, 1947

1947 *Leioproductus* Stainbrook, pp. 307-8.
1947 *Bispinoproductus* Stainbrook, p. 311.
1960 *Leioproductus* Stainbrook, Muir-Wood & Cooper, pp. 168-70.

Type species: Productella coloradoensis var. *plicatus* Kindle, 1909, from the Percha Shale, New Mexico, by original designation of Stainbrook, 1947.

LEIOPRODUCTUS BUTTONENSIS sp. nov.
(Pl. 16, figs 1-21)

Diagnosis: Shell subrectangular, and usually widest at front; pedicle valve with a shallow sinus on visceral disc; ears small and flat; trail long, gently recurved, and lacking a median spine-bearing ridge; spine bases of small diameter on trail, but in large individuals giving rise to ridges extending towards the front of valve; brachial valve bearing quincuncially arranged pits, as well as rugae, on visceral disc; trail long, and ornamented with faint longitudinal ridges, concentric growth lines, and occasional pits; diductor muscle scars smooth posteriorly and striated anteriorly; brachial valve interior with lateral ridges extending along hinge.

Description: External. Shell subrectangular, widest at front, especially in large specimens; pedicle valve most strongly convex posteriorly, with a blunt tapering incurved umbo extending a short distance behind hinge; visceral disc flat or with a shallow median sinus extending on to posterior part of long gently curving trail; ears small, flat, and well differentiated from flanks and concave umbonal shoulders; flanks steep, but flaring slightly at their dorsal extremities; visceral disc crossed by low, rounded rugae bearing irregularly placed spines; rugae strongest on umbonal shoulders, lower anteriorly, and replaced by closely spaced concentric growth lines on trail; spines on trail usually arranged in concentric rows, arising from small round to ovoid bases subtending low ridges down the length of trail; spine ridges pronounced in many large specimens; spines also in a curved row up flanks, and in a row along hinge; concentric growth lines frequently disrupted by spine bases at front of valve.

Brachial valve geniculate, with a moderately convex visceral disc, and a steeply inclined trail; ears wide, flat to slightly concave, separated from visceral disc by strong umbonal ridges; visceral disc ornamented by strong rounded rugae and large quincuncially arranged pits, and trail by concentric growth lines, poorly defined longitudinal ridges, and occasional pits.

Internal. Pedicle valve. Diductor muscle scars elongate, tapering posteriorly, broad anteriorly, and situated in shallow pits; posterior region of diductor muscle scars smooth, triangular with serrated anterior and inner margins; serrated parts join anteriorly with subquadrate scars bearing coarse longitudinal striae; adductor muscle scars on low ridges between diductor muscle scars.

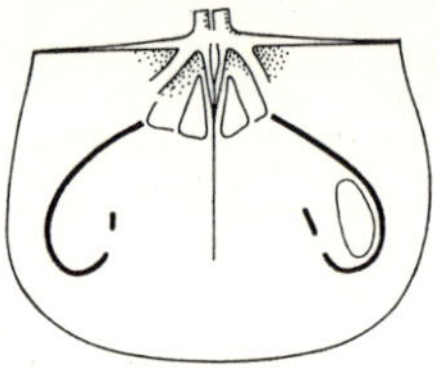

Figure 21. Reconstruction of the brachial valve interior of *Leioproductus buttonensis* sp. nov. showing the posteriorly divided median septum, the two pairs of adductor muscle scars, and the detached distal parts of the brachial ridges.

Brachial valve (Fig. 21). Cardinal process sessile, straight, bilobed internally, and supported by lateral ridges extending along hinge, by low rounded buttressing ridges along posterolateral margins of muscle field, and by median septum; alveolus not observed; posterior of median septum with two broad rounded ridges separated by a shallow elongate furrow; septum becoming narrower and higher at midpart of muscle field, extending as a narrow ridge to middle of visceral disc; adductor muscle scars in two pairs; inner scars triangular, pointed posteriorly, wider anteriorly; outer scars rectangular, and on posterolateral margins of inner scars; brachial ridges arising from posterolateral margins of adductor muscle field, extending obliquely at about 40° and enclosing elevated brachial discs; distal ends of brachial ridges detached; front of visceral disc and trail bearing many small sessile endospines.

Measurements

Pedicle valve	Length	Length (of curvature)	Width (of hinge)	Width (of body)	Height
CPC 8286 Holotype	27	53	20 est.	25.5	17.5
CPC 8288 Paratype	18	33.5	12	16.5	7

Brachial valve	Length	Width	Length (median septum)
CPC 8292 Paratype	—	22.5	12
CPC 8293 Paratype	18	11	—

Remarks: Leioproductus plicatus (Kindle), the type species, is smaller and more subquadrate than *L. buttonensis*. The pedicle valve of *L. plicatus* has larger ears, a spine-bearing median longitudinal ridge, a shorter trail bearing larger diameter spine bases, and lacks ridges at the front of the spine bases on the trail. The brachial valve has slightly divergent lateral ridges, and a longer trail.

L. buttonensis is externally close to *Magnumbonella prolata* sp. nov. (Pl. 25, figs 1-21), which in the Bonaparte Gulf Basin ranges from the Burt Range Formation (locality 100/14) into the Enga Sandstone (locality 103/19). *L. buttonensis* is distinguished from *M. prolata* by the possession of a posteriorly divided median septum, straight lateral ridges and a bilobed cardinal process in the brachial valve, and a row of spines up the flanks on the pedicle valve. Specimens of *M. prolata* used in this comparison are from the Enga Sandstone because the internal characters of those from the Burt Range Formation are unknown.

The specific name refers to Buttons Crossing on the Ord River, near the fossiliferous parts of the Buttons Beds.

Occurrence: Localities 105/525, 105/620 (the type locality), 105/735, 105/830, and 105/895 in the Buttons Beds in the bed of the Ord River near Kimberley Research Station, and probably at locality 141/1 in the Ningbing Limestone (back-reef) near Ningbing.

Material: CPC 8286-8297. Holotype CPC 8286, paratypes CPC 8288, 8291, 8292, 8293, 8294, and 8297. All from locality 105/620. More than 100 specimens.

Age: Famennian.

Genus MESOPLICA Reed, 1943

Type species: Leptaena praelonga J. de C. Sowerby, 1840, from the Upper Devonian of Devonshire, England, by original designation of Reed, 1943.

Diagnosis: See Muir-Wood & Cooper (1960, p. 175).

Remarks: The morphology of the type species of *Mesoplica* is not fully known because the type material from Devonshire is poorly preserved. Features lacking adequate description are the external ornament, especially that of the pedicle valve, and many of the internal structures of the brachial valve. The specimen figured by Sowerby (1840) and reproduced by Davidson (1865, pl. 19, fig. 22) is an internal mould showing the finely striate inner surface of the pedicle valve. Other specimens figured by Davidson (1865, pl. 19, figs 23-25), as well as those figured by Whidborne (1897, pl. 20, figs 12, 13), are also internal moulds but are more coarsely ribbed, suggesting that the exterior was partially costate. Whidborne described the interior as having 'very numerous, reticulating, longitudinal threads, which show a tendency to group themselves into more or less incipient ribs'. The absence of well-defined external ornament on specimens of the type species was noted by Reed (1943, pp. 97-98), who defined the genus as having 'irregular low rounded ribs without? spines' on the pedicle valve, and 'more or less pustulated radial ribs' on the brachial valve.

Muir-Wood & Cooper (1960, pp. 175-76) redefined the genus using Reed's description and additional specimens (but still none showing the exterior of the pedicle valve) from the Devonian of England. They described the pedicle valve as having low, irregular costae which did not extend to the umbo, and the brachial valve as having costae crossed by rugae. Until well-preserved topotype material of *Mesoplica praelonga* becomes available, many of the characters of *Mesoplica* Reed will remain doubtfully known.

Material from locality 88/5 in the Ningbing Limestone (back-reef) at Jeremiah Hills is tentatively referred to *Mesoplica.* The only well-preserved pedicle valve collected from the locality has an external ornament of random spine ridges rather than costae, and rugae on the posterior of the valve which grade anteriorly into closely spaced concentric growth lines. The brachial valve lacks the rows of spines along the hinge described from the type species by Muir-Wood & Cooper (1960, p. 175).

MESOPLICA? JEREMIAHENSIS sp. nov.

(Pl. 14, figs 17-27)

Diagnosis: Pedicle valve with large flat ears; widest at hinge; median ridge low and narrow; longitudinal ornament of random spine ridges on visceral disc and posterior part of trail; rugae low but well defined on umbonal shoulders, grading anteriorly into concentric growth lines; brachial valve with moderately concave visceral disc and long trail; ornament of rugae, concentric growth lines and deep pits on visceral disc, and longitudinal ridges on trail.

Description: External. Pedicle valve rounded-quadrate; ears wide, flattened, giving valve greatest width at hinge; umbo broad, incurved at hinge, and moderately convex; visceral disc less convex, traversed by very shallow median sinus extending on to trail, bearing a narrow, low, median ridge; trail of moderate length, extending almost at right angles to visceral disc, and gently incurved; flanks steep; umbonal shoulders with a group of indistinct spine bases; about ten rugae, low, rounded, strongest on concave umbonal shoulders, and grading anteriorly into concentric growth lines; visceral disc and posterior part of trail ornamented with randomly arranged elongate spine ridges up to 4 mm long, and closely spaced concentric growth lines; front of trail with concentric growth lines only; density of growth lines on front of trail of a valve 24 mm long and 30 mm wide, four per 1 mm.

Brachial valve geniculate, having a moderately concave visceral disc, and a long steeply inclined trail; median furrow narrow on visceral disc, becoming wider and deeper on trail; visceral disc deepest towards front, rising posteriorly to flat ears; ornament of low rugae, concentric growth lines, and deep elongate or subcircular pits, roughly arranged in quincuncial pattern, on visceral disc; growth lines, low longitudinal ridges, and occasional pits on trail.

Internal. Pedicle valve with a smooth apical thickening; diductor muscle scars triangular, smooth, and barely impressed into apical thickening posteriorly; subquadrate, coarsely striated and moderately impressed anteriorly; adductor muscle scars dendritic, ovoid, slightly impressed, and separated by a low median ridge; interior of valve ornamented with closely spaced irregular longitudinal striae.

Brachial valve. Cardinal process sessile, with a swollen inner face, and either bilobed or trilobed; lateral ridges curving posteriorly at base of cardinal process, extending along entire length of hinge; median septum extending to base of cardinal process, broad posteriorly, possibly possessing an antron, and slender anteriorly; termination of median septum, adductor muscle scars, and brachial markings not observed; rows of fine endospines on floor in front of visceral disc.

Measurements

Pedicle valve	Length	Length (of curvature)	Width (at hinge)	Width (at midlength)	Height
CPC 8617 Holotype	25.5	35 est.	31 est.	26	10

Brachial valve	Length	Width (at hinge)	Width (at midlength)
CPC 8618 Paratype	25.5	30	26
CPC 8619 Paratype	25	—	31 est.

Remarks: Mesoplica? jeremiahensis sp. nov. differs from the type species, *M. praelonga* (Sowerby), in having spine ridges rather than costae, and wider ears on the pedicle valve, and in lacking a row of spines along the hinge of the brachial valve.

M. simplicior (Whidborne) (1897, pp. 169-70, pl. 20, figs 14, 15), from the Devonian of southwestern England, is smaller, and from impressions on the internal moulds appears to have regularly developed costae on the pedicle valve. The pedicle valve of *M. praelonga simplicior* (Nalivkin) from the Famennian of Kazakhstan, Asia, figured by Muir-Wood & Cooper (1960, pl. 44, figs 6, 7), has more prominent rugae, a greater number of discontinuous ridges or costae over the pedicle valve, and a stronger median longitudinal fold; the median sinus is either obsolete or absent. Specimens referred to *M. praelonga* (Sowerby) by Dehee (1929, pp. 41-43, pl. 12, figs 13-16), from the Etroeungt of France, have continuous costae and a much longer trail on the pedicle valve.

A form from the Famennian of Queensland placed in *Mesoplica?* by R. McKellar (pers. comm.)* is more coarsely plicate on the pedicle valve, but has a closely comparable ornament on the brachial valve, and is apparently close internally to *M? jeremiahensis.*

The name of this species is taken from the Jeremiah Hills.

Occurrence: Locality 88/5 (the type locality), and locality 88/2, Jeremiah Hills; locality 21/26, Ningbing Range; all Ningbing Limestone (Famennian back-reef).

Material: CPC 8616-8620. Holotype CPC 8617, paratypes CPC 8616, 8618, 8619, and 8620. All from locality 88/5. Seven specimens.

Age: Famennian.

Subfamily PRODUCTININAE Muir-Wood & Cooper, 1960

Genus PRODUCTINA Sutton, 1938

1928 *Thomasia* Fredericks, p. 790.
1931 *Thomasina* Paeckelmann, p. 181.
1938 *Productina* Sutton, p. 551.
1942 *Thomasella* Paul, p. 191.
1951 *Argentiproductus* Cooper & Muir-Wood, p. 195.
1960 *Argentiproductus* Cooper & Muir-Wood, Muir-Wood & Cooper, p. 182.
1966 *Productina* Sutton, Brunton, p. 208.

Type species: Productus sampsoni Weller, 1909, by original designation of Sutton, 1938.

Remarks: Brunton (1966a, p. 209) showed that the internal features of immature brachial valves of *Productina margaritacea* (Phillips), which had been referred to *Argentiproductus* by Muir-Wood & Cooper (1960), are the same as those in

*The species is *Mesoplica hillae* McKellar. (McKellar, R. G., 1970, The Devonian productoid brachiopod faunas of Queensland. *Geol. Surv. Qld Pub.* 342, *Pal. Pap.* No. 18, 1-40, pls 1-12.

the type species of *Productina, P. sampsoni* (Weller). The internal morphology of brachial valves of *margaritacea* from the Utting Calcarenite also resembles that of *P. sampsoni,* which adds further support to Brunton's opinion that the brachial valve interior from Vise, Belgium, figured by Muir-Wood & Cooper (1960, pl. 123, figs 17, 17a) as *Argentiproductus margaritaceus* does not belong to that species. It also supports Brunton's contention that *Argentiproductus* is a subjective synonym of *Productina* Sutton.

Prior to this study, mantle canals in productoids had been reported only once (Brunton, 1966a). One pedicle valve of *P. margaritacea* exhibits moderately well-defined mantle canals (Fig. 22). Traces of mantle canals are also recorded in brachial valves of *Lomatiphora aquila* gen. et sp. nov. (Fig. 20). The overall patterns are not sufficiently clear to enable comparison with those in other brachiopod groups. In the pedicle valve of *P. margaritacea,* the main set of canals, interpreted as vascula media, extends as strap-like branches from two short canals immediately in front of the muscle field to the front of the valve. Less well defined ?vascula genitalia appear to originate from the vicinity of these canals also, and consist of short irregular anterolaterally directed depressions which give a pitted appearance to the posterior of the valve.

PRODUCTINA MARGARITACEA (Phillips)

(Pl. 17, figs 11-28)

1836 *Producta margaritacea* Phillips, p. 215, pl. 8, fig. 8.

1861 *Productus margaritaceus* Phillips, Davidson, p. 159, pl. 44, figs 5-8.

1960 *Argentiproductus margaritaceus* (Phillips), Muir-Wood & Cooper, p. 182, pl. 123, figs 11-16 (not 17, 17a).

1966 *Productina margaritacea* (Phillips), Brunton, p. 209, pl. 8, figs 1-19; pl. 15, figs 1-8.

Description: External. Pedicle valve subelliptical, wider than long, widest at or near hinge; valve moderately convex, bearing a strongly incurved umbo and a less convex trail; venter evenly rounded; umbo extending well behind hinge (4.5 mm on a valve about 30 mm wide and 22 mm long), broadly rounded, tapering posteriorly, having a narrow incurved tip facing anterior margin of shell; umbonal shoulders gently concave, sloping on to wide flat ears; flanks spreading and having same inclination as moderately convex trail; costae regular, rounded, separated by intercostal grooves of equal width, fine on umbo, much coarser towards front; density of costae twelve to fifteen per 10 mm at 10 mm from umbo, and ten to eleven per 10 mm at 20 mm from umbo; costae increasing by bifurcation, particularly on flanks and at front of valve; six laterally directed spines of large diameter on posterior of valve, two on umbo, two on umbonal shoulders, and two near posterolateral margins; on one specimen, two spines arising from median costa (one at 10 mm and one at 20.5 mm from umbo), causing it to form a keel-like ridge; valve irregularly lamellose.

Brachial valve non-geniculate; visceral disc strongly concave, deepest at mid-point, becoming flatter anteriorly; concavity at umbo continuous with that of visceral disc, containing a prominent rounded boss at base of cardinal process; lateral shoulders steep and passing abruptly posterolaterally into flat ears; costae weaker than those on pedicle valve; density of costae varying from the same as on pedicle valve (one valve with fifteen costae per 10 mm at 10 mm from umbo) to much finer (one valve with twenty-one costae per 10 mm at 10 mm from umbo); concentric lamellae crowded posteriorly and becoming wider and overlapping towards front; valve aspinose.

Internal. Pedicle valve. Adductor muscle scars elongate, pointed posteriorly, expanding anteriorly, having rounded anterior terminations; posterior half of adductor scars separated by furrow, anterior half by very low median ridge; diductor muscle scars not as clearly defined, but occupying smooth broad triangular areas on lateral margins of adductor muscle scars; mantle canals arising in front of muscle field (Fig. 22); ?vascula genitalia consisting of short irregular canals extending anterolaterally from two short canals in front of muscle field, giving a pitted appearance to posterior part of valve in front of muscle field; vascula media extending from two short canals to front of valve, sometimes branching, consisting of a number of broad strap-like canals; ginglymus long, flat to concave, bearing narrow teeth-like ridges; ridges originating beneath umbonal shoulders, extending along inner margin of ginglymus, curving anterolaterally, and becoming obsolete on ears; spines on posterolateral part of valve opening into interior.

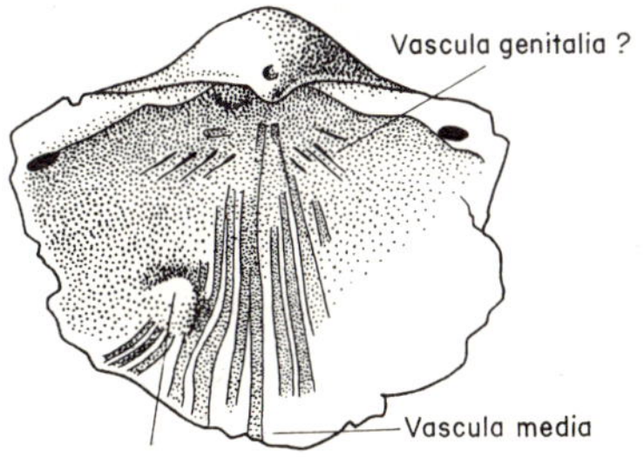

Figure 22. Pedicle valve interior of *Productina margaritacea* (Phillips) showing the mantle canals arising from two short canals in front of the muscle field, ?vascula genitalia posteriorly, and vascula media extending to the front of the valve. CPC 8657, × 1.5.

Brachial valve. Cardinal process sessile, resting on a thickened region at apex of valve; internal face having two widely diverging (approximately 50° from line of median septum) convex lateral lobes connected anteriorly by a high anteroventrally projecting bar having somewhat pointed anterolateral extremities; outer face ventrally inclined, with two deeply excavate lamellose lateral lobes bearing a quadrifid myophore; lateral lobes open or with margins almost enclosing central trough; platform in front of cardinal process supported by low anterolaterally directed ridge; median septum originating slightly in front of cardinal process, relatively wide and rounded posteriorly, becoming higher and blade-like at mid-length of muscle field; adductor muscle scars in a single ovoid pair, elevated above surface of valve; scars rounded posteriorly, becoming wider anteriorly, broadest at about three-quarters of their length, tapering again at their rounded anterior margins; a furrow corresponding to ridge on edge of ginglymus on pedicle valve diverging from hinge, extending from lateral margin of cardinal process to front of ears; furrow bounded by lateral ridges, one divergent from hinge, the other parallel with hinge; small anteriorly directed pustules on inner surface of valve, especially in front of muscle field.

Measurements

Pedicle valve	Length	Length (of curvature)	Width	Height
CPC 8656	22	39 est.	30 est.	11
CPC 8657	23	40	—	11

Remarks: Specimens of *Productina margaritacea* (Phillips) described by Brunton (1966a) from the Visean (D_1) limestone in County Fermanagh, Northern Ireland, have several minor differences from those in the Utting Calcarenite. In the Irish specimens, the pedicle valve has narrower and more deeply impressed diductor muscle scars, a slightly wider furrow between the adductor muscle scars, and sometimes more divergent articulatory ridges along the anterior margin of the ginglymus. In the brachial valve they have shorter adductor muscle scars, and a weaker median septum. These differences are minor, and are considered to be within the range of variation of the species. This is the first record of *P. margaritacea* in Australia.

Occurrence: Localities 107/1, 107/7, 108/0, Utting Calcarenite, Utting Gap.

Material: CPC 8656-8659. CPC 8656 from locality 107/1; CPC 8657-8658 from locality 108/0; CPC 8659 from locality 107/7. Four silicified specimens.

Age: Visean.

Family OVERTONIIDAE Muir-Wood & Cooper, 1960

Subfamily OVERTONIINAE Muir-Wood & Cooper, 1960

Genus SENTOSIA Muir-Wood & Cooper, 1960

Type species: Krotovia praecursor Stainbrook (1947, p. 313, pl. 46, figs 11-15) from the Percha Formation (Box Member), New Mexico, by original designation of Muir-Wood & Cooper, 1960.

Diagnosis: See Muir-Wood & Cooper (1960, p. 196).

SENTOSIA SUBQUADRATA sp. nov.

(Pl. 18, figs 19-30)

Diagnosis: Shell subquadrate to subovate; coarse spine bases arranged quincuncially at the posterior, and in rough rows on the anterior of the pedicle valve; small elongate pits and fewer spine bases on the brachial valve; concentric growth lines well defined; pedicle valve moderately convex, with an evenly rounded venter; brachial valve moderately concave, geniculate anteriorly, with flat to slightly concave posterolateral extremities.

Description: External. Shell concavoconvex; subquadrate to subovate, always wider than long; greatest width at one-third to one-half length of shell; ornament of rugae, spines, and concentric growth lines; rugae low, irregular, well defined on brachial valve and on lateral parts of pedicle valve, but obsolete on venter of pedicle valve; spine bases narrow, tapering posteriorly, about 1 mm long, bearing thin prostrate spines; spine bases quincuncially arranged posteriorly, but in rough rows towards front of pedicle valve; brachial valve with fewer spines, and bearing small elongate pits; both valves with slightly irregular, prominent, concentric growth lines; shell thin.

Pedicle valve moderately convex, with greatest convexity at umbo; highest at about one-third length of valve; umbo narrow, tapering posteriorly, projecting behind hinge, and strongly incurved; umbonal shoulders concave, extending on to small flat posterolateral extremities; venter evenly convex; sinus absent.

Brachial valve gently to moderately concave at umbo and over visceral disc; slightly geniculate anteriorly, deepest at midpoint of valve; ears broad, and flat to slightly concave.

Internal. Pedicle valve interior unknown.

Brachial valve. Cardinal process sessile, bilobed internally, two lobes separated by narrow furrow; lateral ridges short, narrow, and curving posteriorly at base of cardinal process; median septum extending to about midlength of valve; adductor muscle scars (from shell material of a partially decorticated valve) elongate and trigonal.

Measurements	Length (pedicle valve)	Length (brachial valve)	Width
CPC 8579	22	—	24 est.
CPC 8580 Holotype	19	—	—
CPC 8581 Paratype	—	21 est.	28.5
CPC 8582 Paratype	—	21	27 est.
CPC 8583 Paratype	—	23	24
CPC 8585 Paratype	—	23	30 est.

Remarks: No well-preserved interiors of *Sentosia subquadrata* sp. nov. are available for description.

Most of the other known species of *Sentosia* come from the Percha Formation of New Mexico. *S. praecursor* (Stainbrook), the type species, is distinguished from *subquadrata* by its more transverse outline, less convex pedicle valve (especially on the lateral margins), less concave brachial valve, and finer spine bases. *S. praecedens* (Stainbrook) (1947, p. 312, pl. 46, figs 24-26) appears to have a pedicle valve comparable with that of this species, but is usually smaller and has a more concave brachial valve.

A species of *Sentosia** from Queensland, currently being described by R. McKellar, has an outline similar to that of *subquadrata* but is much smaller, and is ornamented by a denser mat of spines. Photographs of the species from Queensland were kindly supplied by R. McKellar.

The specific name refers to the outline of the shell.

Occurrence: Localities 443/13 (the type locality), 443/22, and 443/23 in the Jeremiah Hills, and locality 11/2 in the Ningbing Range. All are from the Famennian part of the Ningbing Limestone; the Jeremiah Hills localities are in back-reef, and locality 11/2 is possibly reef.

Material: CPC 8579-8585. Holotype CPC 8580, paratypes CPC 8581-8585. CPC 8579 from locality 11/2; CPC 8580-8585 from locality 443/13. Forty specimens.

Age: Famennian.

**S. minuta deari* McKellar (McKellar, 1970). McKellar, R. G., 1970, *op. cit.*

Genus RHYTIOPHORA Muir-Wood & Cooper, 1960

Type species: Productus blairi Miller, 1891, from the Chouteau Limestone of Missouri, USA, by original designation of Muir-Wood & Cooper, 1960.

RHYTIOPHORA sp. cf. R. CALHOUNENSIS (Moore)

(Pl. 18, figs 1-13)

Description: External. Pedicle valve subrectangular; valve strongly convex, with greatest convexity at midlength; greatest width towards anterior margin; umbo small, narrow, not extending across hinge; umbonal shoulders broad and gently concave; posterolateral margins very broad, almost flat, produced into small convex ears; visceral disc strongly convex except for shallow median sinus; sinus broad, very shallow to almost obsolete, commencing at about one-quarter length of valve; flanks steep; anterior margin produced into short trail; rugae well defined on visceral disc, narrow near hinge, becoming broader and irregular across venter, having an uneven surface caused by round tubercular protuberances bearing very small spine bases; tubercles becoming larger and more elongate towards front of valve; two or three rows or a group of circular spine bases 0.3 to 0.4 mm in diameter along posterolateral margins; irregular growth lines on furrows between rugae; valve developing low longitudinal ridges on lateral and anterior margins.

Brachial valve subrectangular, geniculate, with a trail about 7 mm high on a valve 28 mm wide and 16 mm long; visceral disc moderately concave, deepest at front, crossed by a broad obsolete fold commencing just in front of umbo and extending on to trail; posterolateral margins flat, broad, clearly separated from visceral disc by rounded ridges; rugae more evenly arranged than on pedicle valve, having smooth rounded surfaces; intervening furrows marked with concentric growth lines; rugae commencing at hinge, strongest on posterolateral margins, extending across visceral disc; density of rugae twelve per 5 mm on midpart of a valve 28 mm wide and 16 mm long; scattered radially arranged pits on visceral disc passing anteriorly into longitudinal furrows on trail; valve aspinose.

Internal. Pedicle valve interior not observed.

Brachial valve. Cardinal process sessile, bilobate internally, supported by strong lateral ridges extending along hinge; inner lobes of cardinal process rounded, slightly divergent, separated by a wide furrow; alveolus small, situated immediately in front of base of process; remainder of internal structures not observed.

Measurements

Pedicle valve	Length	Length (of curvature)	Width (at hinge)	Width (body)	Height
CPC 8570	20	33	25	26	12
CPC 8572	15	—	21 est.	23 est.	8
CPC 8574	19	—	—	30 est.	—

Brachial valve	Length	Width
CPC 8571	16	25
CPC 8573	15	27 est.

Remarks: Rhytiophora sp. cf. *R. calhounensis* (Moore) differs from the type species, *R. blairi* (Miller), figured by Muir-Wood & Cooper (1960, pl. 51, figs 6-16), in having a more rectangular outline, more closely spaced and lower rugae on the visceral disc, weaker longitudinal ribs on the front of the pedicle valve, and in lacking a marginal frill.

Specimens of *R. calhounensis* (Moore) described by Carter (1967, pp. 294-7, pl. 21, figs 1-11) from the Chappel Limestone of Texas are extremely close to this species in shape and external ornament. Both Carter's specimens from Texas and the pedicle valves of *R. calhounensis* originally described by Moore (1928, pl. 10, figs 1-6) from the Chouteau Limestone of Illinois and Missouri are generally larger than the specimens from the Bonaparte Gulf Basin.

Occurrence: Locality 7/1, Tournaisian part of the Ningbing Limestone, Ningbing Range.

Material: CPC 8570-8574. All from locality 7/1.

Age: Tournaisian.

RHYTIOPHORA sp. cf. R. RARICOSTATA (Herrick)

(Pl. 13, figs 19-26)

Description: External. Pedicle valve subquadrate, moderately to strongly convex; greatest width at midlength; greatest convexity at umbo; umbo bluntly rounded, having an incurved weakly developed tapering tip; umbonal shoulders gently concave; posterolateral extremities short, and vertical to plane of commissure; hinge slightly shorter than greatest width; flanks gently convex, spreading, probably projecting dorsally a short distance behind hinge; venter broadly rounded, evenly convex towards umbo, bearing a shallow sinus on mid and anterior parts of valve; ornament of spine ridges at umbo (almost quincuncially arranged), passing anteriorly into obsolete to moderately well-defined costae bearing irregular concentric rows of spines; costae regularly developed on mid and anterior parts of valve, usually stronger behind each spine base; at front of some valves costae higher, broader, and forming narrow elongate folds in shell; costae rounded, separated by broad furrows twice as wide as one costa, and, on mid part of a valve 39 mm wide and 41 mm long, having a density of six per 10 mm; body spines prostrate to sub-erect, more numerous towards umbo, having bases 0.4 to 0.5 mm in diameter; spines also in a small group on posterior of flanks, and in a poorly defined row along hinge; more than twelve rugae, confined to visceral disc, prominent near hinge, usually weaker across venter (some completely obsolete); micro-ornament of fine closely spaced concentric growth lines.

Measurements	Length (pedicle valve)	Length (of curvature)	Width	Height (pedicle valve)
CPC 10923	27	45	28	16
CPC 10924	25	39	28.5	13
CPC 10925	41	64	39	21

Remarks: The pedicle valves of *Rhytiophora raricostata* (Herrick) figured by Hyde (1953, pl. 14, figs 1-15; pl. 15, figs 1-7) from the Logan Formation of Ohio are closely comparable in shape and external ornament. The costae on the pedicle valve are widely spaced, strongest at the front of the valve, and bear elongate spine bases; the rugae are restricted to the posterior of the valve. Hyde's specimens have slightly coarser costae than *R.* sp. cf. *R. raricostata*. Other features of *raricostata* cannot be compared because only pedicle valves have been collected from the Bonaparte Gulf Basin.

Occurrence: Localities 103/12, 128/2, 128/6, Burt Range Formation, middle part of Enga Ridge; locality 122/2, Burt Range Formation, Sandy Creek; and locality 102/4, Enga Sandstone, northern end of Enga Ridge.

Material: CPC 10923-10925. Pedicle valves. CPC 10923 from locality 103/12; CPC 10924-10925 from locality 102/4. Ten specimens.

Age: Tournaisian.

Genus SPINOCARINIFERA nov.

Type species: Spinocarinifera adunata gen. et sp. nov. from the Burt Range Formation of the Bonaparte Gulf Basin.

Description: Shell subquadrate; pedicle valve strongly and evenly convex, lacking a sinus, and with a moderately incurved umbo; greatest width at hinge or at four-fifths length of valve; posterolateral extremities flat; radial ornament of short spine ridges on posterior of umbo grading into well-defined spine-bearing costae on body of valve; roughly arranged concentric rows of sub-erect spines on body of valve, and a group or several rows of larger erect spines on flanks; concentric ornament of low rugae on posterolateral flanks (usually not extending across venter), and faint growth lines; brachial valve moderately concave, always geniculate, aspinose, reticulate posteriorly, costate anteriorly.

Pedicle valve interior with a low median ridge in posterior third of valve; adductor muscle scars smooth and narrow; diductor muscle scars flabellate and radially striated; cardinal process sessile, bilobed internally, trilobed externally, having a peg-like median lobe on external face; median septum narrow and blade-like, arising from in front of a shallow alveolus or platform, and extending two-thirds length of valve; adductor muscle scars in a single smooth pair; lateral ridges low, diverging from hinge near ears; brachial ridges poorly defined, diverging at 50-60° from anterolateral margins of adductor muscle scars; trail ornamented with rows of small endospines.

Remarks: Avonia I. Thomas, as redefined by Muir-Wood & Cooper (1960, p. 185) and discussed by Brunton (1966a, pp. 218-19), is distinguished from *Spinocarinifera* by its elongate-oval outline, shorter hinge, less geniculate brachial valve, and weaker and less continuous costae. In the brachial valve interior of *Avonia,* the median septum supports the cardinal process and extends one-third the length of the valve; the lateral ridges curve away from the hinge, and the cardinal process is bilobed. The pedicle valve lacks rows of spines on the flanks, and has a costate ornament formed by spine ridges. Concentric lamellae interrupt the ridges and separate the spines into bands. Brunton (1966a, p. 218) considered one specimen with well-defined costae on both valves figured by Muir-Wood & Cooper (1960, pl. 47, figs 22-24) as *Avonia youngiana* to be *Institifera tessellata* (de Koninck). Other specimens in the same plate have discontinuous spine ridges in the pedicle valve, and a lamellose brachial valve. Hence, the costae of *Avonia,* particularly on the brachial valve, are probably weaker than indicated by Muir-Wood & Cooper's description (p. 185), and weaker than costae on *Spinocarinifera.* Some of the brachial valves of *Avonia (Quasiavonia)* Brunton (1966a, pp. 219-20), with *Avonia aculeata* (J. Sowerby) as the type species, have a faintly costate ornament (Brunton, 1966a, pl. 10, fig. 11), but others are more lamellose (pl. 11, figs 14, 18).

Semicostella Muir-Wood & Cooper (1960, pp. 195-96, pl. 62, figs 1-7), a Mississippian genus from Oklahoma and Nevada, is distinguished from *Spinocarinifera* by the possession of a cincture between the trail and the visceral disc, and wider intercostal furrows on the pedicle valve. In the brachial valve *Semicostella* has a flatter visceral disc and a fine spinose ornament; the cardinal process is bilobate and supported by a broad median septum extending half the length of the valve, the lateral ridges are stronger, and the brachial ridges and impressions are better developed.

It is difficult to ascertain which of the previously described species, frequently referred to *Avonia,* should be placed in this genus because their internal characters are mostly unknown. Those which on external grounds may be referred to *Spinocarinifera* include:

Productus niger Gosselet, 1888, from the Etroeungt and Lower Tournaisian of Europe.

?Productus bassus Vaughan (1905, pp. 287-88, pl. 25, figs 1, 1a) from the K Zone of the Avon Gorge section in England.

Avonia inflata (Sokolskaya) (Sarycheva et al., 1963, pp. 138-39, pl. 14, figs 4-6) from the Lower Tournaisian of the Kuznetsk Basin, USSR.

?Avonia semicostata (Tolmachoff) (Sarycheva et al., 1963, pp. 142-43, pl. 14, figs 14-18) from the Tournaisian of the Kuznetsk Basin, USSR.

?Productus arcuatus Hall (Weller, 1914, pp. 107-8, pl. 13, figs 1-12) from the Kinderhook of Missouri and Iowa, USA.

Spinocarinifera, from the Latin *spino* — thorn, *carina* — keel or ridge, and *fero* — carry, refers to the spine-bearing costae on the body and trail of the pedicle valve.

SPINOCARINIFERA ADUNATA gen. et sp. nov.

(Pl. 19, figs 1-18; Pl. 20, figs 9-16)

Diagnosis: Spinocarinifera with a wide hinge and pronounced auricles; costae having a density of five to six per 5 mm at approximately 15 mm in front of the ventral umbo; five to six rugae on the ears and umbonal shoulders of the pedicle valve, and five to eight on the visceral disc of the brachial valve.

Description: External. Pedicle valve small, subquadrate, strongly and evenly convex, having an evenly curved venter; trail and flanks steep, ears small, flattened and well differentiated; greatest width at hinge or at about four-fifths length of valve; umbo bluntly rounded, incurved, projecting a short distance over hinge, with gently concave umbonal shoulders; spine ridges at posterior of umbo passing anteriorly into strongly defined costae on visceral disc and on trail; five to six costae per 5 mm at approximately 15 mm from umbo; costae occasionally increasing by bifurcation, and sometimes irregular on visceral disc; five to six rugae, confined to ears and umbonal shoulders; remainder of valve crossed by concentric growth lines; spines long, erect on flanks, recumbent to erect on body of valve, and arranged in two groups: (1) scattered recumbent to sub-erect spines, 0.2 to 0.3 mm in diameter at base, in irregular concentric rows over body of valve, spine bases always arising from costae; and (2) group or several rows of approximately ten larger erect spines, 0.5 mm in diameter at base, extending down flanks from hinge.

Brachial valve strongly geniculate, with moderately concave visceral disc and steep trail; visceral disc deepest at mid- or mid-anterior part of valve, joining a small depression at umbo; ears flat or slightly concave, abruptly separated from visceral region; rugae well defined on ears and usually crossing posterior of visceral disc, forming coarsely reticulate ornament; five to eight rugae; anterior half of visceral disc and trail costate, density of costae from five to seven per 10 mm in front of umbo; valve also bearing numerous dimples corresponding in position to spine bases on pedicle valve, especially on posterolateral margins; valve aspinose.

Internal. Pedicle valve. Low median ridge in posterior third of valve; adductor muscle scars smooth, narrow, elongate, extending the length of median ridge; diductor muscle scars slightly impressed, triangular to flabellate, radially striated, with irregular anterior and lateral margins; inner surface of trail faintly striated and with impressions of external ornament.

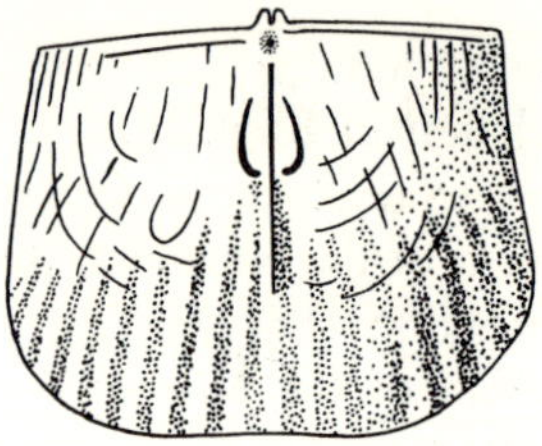
M(P)242

Figure 23. Brachial valve interior of *Spinocarinifera adunata* gen. et sp. nov. CPC 8567, locality 101/2, Burt Range Formation, 5.5 miles west of Mount Septimus. × 2.

Brachial valve (Fig. 23). Median septum arising from in front of shallow alveolus or platform, and extending two-thirds length of valve; septum low and sometimes rounded posteriorly, becoming high and blade-like in anterior third of its length; single smooth pair of adductor muscle scars, posterior margins pointed and anterior margins elevated; brachial ridges poorly defined, diverging at 50-60° from posterolateral margins of adductor muscle scars; cardinal process sessile, bilobed internally and trilobed externally; inner lobes convex and separated by deep median channel; outer face with lamellose peg-like median lobe surrounded by curving convex lateral lobes; lateral ridges variable in strength and usually diverging slightly from hinge near ears; visceral disc having small endospines roughly arranged in concentric rows, and impressions of external rugose ornament; trail and lateral margins bearing longitudinal rows of low-angle endospines less than 0.5 mm long, and having a density of six to eight per square mm on median part of trail.

Measurements

Pedicle valve exterior	Length	Length (of curvature)	Width (of hinge)	Width (of body)	Height
CPC 8564 Holotype	15	26.5	15.5	16.5	10.5
CPC 8562 Paratype	17.5	30.5	18.5	19.5	13
CPC 8563 Paratype	16	28	16 est.	16	10.5

Brachial valve interior	Length	Width (of hinge)	Width (of body)	Length (median septum)
CPC 8567 Paratype	14.5	15.5	16 est.	11.5

Remarks: Spinocarinifera niger (Gosselet) described by Dehee (1929, pp. 39-41, pl. 4, figs 1-6), from the Entroeungt and early Tournaisian of France, is morphologically very close to *S. adunata* in shape and external ornament. It is distinguished only by its slightly larger size, and more swollen umbo on the pedicle valve. Specimens of *S. niger* described by Paeckelmann (1931, pp. 98-100, pl. 4, figs 24, 25) from the Tournaisian of Aachen and Elberfeld, Germany, also have a similar shape and ornament; on the pedicle valve spines arise from the costae, and there is a group of spines near the ears; the brachial valve has a costate ornament. *S. adunata* is distinguished by its finer costate ornament, wider hinge, more pronounced auricles, and smaller size. The internal details of *S. niger* have not been figured. In Belgium, *S. niger* is recorded from the Tnla, Tnlb, and Tn2a (Demanet, 1958).

Spinocarinifera? bassa (Vaughan) (1905, pp. 287-88, pl. 5, figs 1, 1a) from the lower part of the K Zone of the Avon Gorge in Great Britain has stronger rugae on the posterior of the pedicle valve, more numerous costae, and flaring posterolateral margins; the brachial valve has not been described. Paeckelmann (1931) regards *bassa* as a subspecies of *S. niger* (Gosselet).

Spinocarinifera inflata (Sokolskaya) (Sarycheva et al., 1963, pp. 138-39, pl. 14, figs 4-6) from the early Tournaisian of the Kuznetsk Basin, USSR, has a similar style of ornament and convexity. Other close features include the median septum, which is obsolete anteriorly and commences some distance in front of the cardinal process, and the divergence and strength of the lateral ridges. *S. inflata* has a shorter hinge, smaller ears, less well defined costae on the pedicle valve, and is smaller than *adunata;* as illustrated in Sarycheva et al. (1963, fig. 47) *S. inflata* appears to have a bilobate external face on the cardinal process.

Spinocarinifera? arcuata (Hall) from the Kinderhook of Missouri and Iowa, described by Weller (1914, pp. 107-8, pl. 13, figs 1-12), is similar in size and shape to *S. adunata.* Specimens of the same species described by Hyde (1953, pl. 10, figs 1-5) have a slightly shorter hinge and a greater density of costae. Specimens of *Spinocarinifera? arcuata* (Hall) from the Kinderhook of Wassonville, Iowa, in the collection of the U.S. National Museum, are externally close to *S. adunata* except for the possession of slightly stronger rugae (up to nine) crossing the front of the visceral disc on the pedicle valve, slightly higher and sharper costae, deeper intercostal furrows, and a larger size (up to 25 mm long and 26 mm wide). In the brachial valve the median septum supports the base of the cardinal process, and there may be a second pair of poorly defined adductor muscle scars; the other features are close to those of *S. adunata.* Nalivkin (1937, pl. 6, figs 1-3) has also figured specimens of *S? arcuata* (Hall) from the Kassin Beds (Lower Tournaisian) of Kazakhstan, which appear to be externally similar to *S. adunata.*

'Avonia' proteus Veevers (1959a, pp. 80-84, pl. 8, figs 1-15), the index species of the uppermost Famennian Zone in the northern Canning Basin, Western Australia, resembles *S. adunata* in size, shape, morphology of the median septum, presence of an alveolus, and structure of the cardinal process and lateral ridges in the brachial valve. It is distinguished from *adunata* by the possession of spine ridges as opposed to true costae on the pedicle valve, a non-costate and more pitted visceral disc on the exterior of the brachial valve, and rudimentary teeth in the pedicle valve and sockets in the brachial valve. *'A.' proteus* is probably more closely related to *Spinulicosta* Nalivkin than to *Spinocarinifera.*

The specific name, from the Latin *adunatus* — united, refers to the median lobe on the cardinal process which is formed by the fusion of the inner margins of the lateral lobes.

Occurrence: Localities 100/13C, 100/14, 100/20, 7 miles northwest of Mount Septimus; localities 101/1A, 101/2 (the type locality), 101/5, 101/7B, 101/7C, 6 miles west of Mount Septimus; locality 421/5, southern end of Enga Ridge; all in the Burt Range Formation.

Material: CPC 8559-8569. Holotype CPC 8564, paratypes CPC 8562, 8563, and 8567. CPC 8559-8561 from locality 101/5; CPC 8562-8569 from locality 101/2. More than 100 specimens.

Age: Tournaisian.

Genus ACANTHOCOSTA nov.

Type species: Acanthocosta teicherti gen. et sp. nov. from the Burt Range Formation, Bonaparte Gulf Basin.

Description: Pedicle valve subcircular to subovate with strongly incurved umbo, rounded venter, and long trail; ornament of spine ridges on umbo passing anteriorly into narrow regularly spaced costellae; costellae low, occasionally increasing by bifurcation on posterior of trail, bearing elongate spine bases along their entire length; three to four rows of larger spines extending down flanks; rugae confined to visceral disc, frequently extending across venter; brachial valve subquadrate, moderately concave, and geniculate; ornament reticulate on visceral disc and costellate on trail; costellae bearing small spine bases; rugae extending across visceral disc; ventral diductor muscle scars flabellate, smooth posteriorly, striated anteriorly; median septum extending two-thirds length of brachial valve, having a posterior shaft divided by a shallow furrow, or arising from a platform in front of cardinal process; cardinal process bilobed internally and usually trilobed externally, having a wedge-shaped median lobe on external face; cardinal process in some specimens bilobate externally; lateral ridges parallel with and extending two-thirds along hinge; adductor muscle scars in two pairs: a smooth subovate to triangular inner pair, and a poorly defined triangular outer pair.

Remarks: Acanthocosta gen. nov. is characterized by spine ridges on the posterior and densely spinose costellae on the anterior of the pedicle valve. It is distinguished from *Tomiproductus* Sarycheva, 1963, by its larger size, profusion of small spines arising from the costellae on both valves, stronger rugae and a more pronounced reticulate ornament on the brachial valve, and the absence of spines on the hinge of the pedicle valve.

Acanthocosta gen. nov. is placed in the family Overtoniidae on the basis of its small sessile bilobate or trilobate cardinal process, and posterior spine ridges on the pedicle valve. The spine-bearing costellae on the front of the pedicle valve, and the posterior shaft of the median septum which frequently has a shallow antron, both suggest affinities with the Buxtoniidae.

Apart from the type species, no other species can be confidently assigned to *Acanthocosta.*

ACANTHOCOSTA TEICHERTI gen. et sp. nov.

(Pl. 23, figs 1-21; Pl. 24, figs 1-15)

Diagnosis: Shell moderately large, and subcircular to subovate; pedicle valve with spreading flanks, and evenly rounded venter; costellae low, separated by wide furrows, having a density of six to nine per 5 mm on the posterior of the trail, and giving rise to a large number of sub-erect to prostrate spines; fifteen to twenty larger erect spines in three or four rows on the flanks; brachial valve with well-defined rugae forming a reticulate ornament on the visceral disc.

Description: External. Pedicle valve subcircular to subovate having strongly incurved umbo, long evenly convex trail, spreading flanks, and evenly rounded venter; greatest convexity at umbo, and greatest width at dorsal projections of flanks immediately in front of hinge; umbo small, tapering, projecting only a short distance behind hinge, separated by concave umbonal shoulders from flat to slightly concave posterolateral margins; hinge slightly less than greatest width of valve; flanks steep, but flattening towards margins; costellae arising from spine ridges on umbo, extending over remainder of valve, but becoming slightly weaker anteriorly, and on large individuals obsolete at front of trail; costellae low and narrow, subparallel on flanks and trail, occasionally increasing by bifurcation on posterior of trail, and separated by wide rounded furrows; density of costellae on posterior of trail six to nine per 5 mm; spines arising along length of each costella, distributed over body of valve except at hinge and, in large individuals, at front of trail; body spines 0.2 mm in diameter at base, sub-erect to prostrate; larger erect spines 0.3 mm in diameter at base, in three or four rows containing a total of fifteen to twenty spines extending from posterolateral margins down flanks; usually ten to fifteen rugae (but as many as 20 on larger specimens), confined to visceral disc; rugae strongest at hinge, often extending across venter; micro-ornament of small concentric growth lines having a density of about eight per 1 mm.

Brachial valve subquadrate, geniculate; visceral disc moderately to gently concave, concavity extending on to umbo; posterolateral margins flat; trail projecting at about right angles from visceral disc, 7mm high at front of a valve 17 mm wide and 13 mm long; ornament reticulate on visceral disc, costellate on trail; costellae having same density as those on pedicle valve, and bearing minute spines 0.15 mm in diameter at base; up to twenty rugae, highest on posterolateral margins and at base of trail, and extending across visceral disc; micro-ornament as on pedicle valve.

Internal. Pedicle valve. Diductor muscle scars flabellate in outline, slightly impressed, smooth and pointed posteriorly, broader and coarsely striated anteriorly; adductor muscle scars not observed.

Brachial valve. Median septum extending two-thirds length of valve, arising from swollen region in front of cardinal process either as narrow ridge or as two low rounded ridges divided by median furrow extending one-third the length of septum; septum narrow and higher anteriorly, ending in high blade-like extremity; cardinal process usually bilobed internally and trilobed externally; inner lobes highly convex, separated by narrow furrow, extending posterodorsally into a median lobe; median lobe wedge-shaped externally, flanked by convex lateral lobes, external faces of all three lobes having lamellose myophores; cardinal process in some individuals bilobate (Fig. 24a), lacking median lobe, having divergent convex inner lobes, and deeply concave outer lobes bearing lamellose myophores; other variations having reduced lobes (Pl. 24, figs 6-8), or fused inner lobes; lateral ridges support sides of cardinal process, extending up to two-thirds along width of hinge (some specimens slightly divergent from hinge); adductor muscle scars in two pairs; inner scars subovate to triangular, tapering posteriorly with broader elevated anterior margins; outer scars low and triangular; front of visceral disc and trail bearing rows of fine endospines.

Measurements

Pedicle valve	Length	Length (of curvature)	Width (of hinge)	Width (of body)	Height
CPC 8645 Holotype	24.5	45 est.	20 est.	25	13.5

Brachial valve	Length	Width (of hinge)	Width (of body)	Length (median septum)
CPC 8526 Paratype	13 est.	14	18 est.	9
CPC 8528 Paratype	18.5	20	27	12
CPC 8529 Paratype	18.5	22	23	12.5

Remarks: Productus vaughni Muir-Wood (1928, pp. 65-70, pl. 2, figs 11-13) from the Avon Gorge section of Britain (K_2 to C_1 ? Zones) is comparable in shape with this species. *P. vaughni* is distinguished primarily by its lack of numerous spines arising from the costellae. It has fewer and larger erect spines over the body of the pedicle valve, rows of spines along the hinge of the pedicle valve, weaker rugae, and slightly denser costellae; the lateral ridges in the brachial valve are divergent from the hinge.

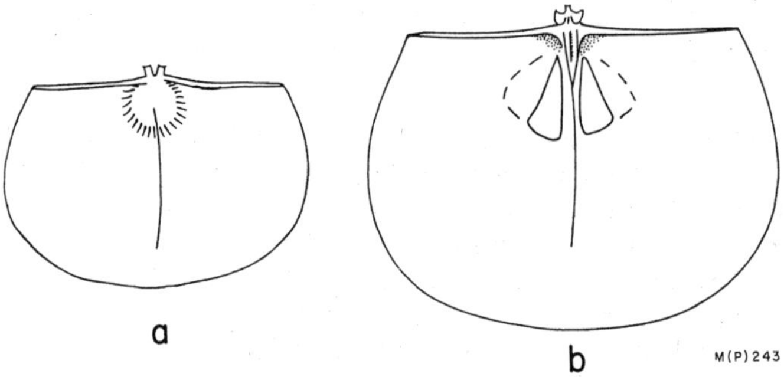

Figure 24. Brachial valve interiors of two specimens of *Acanthocosta teicherti* gen. et sp. nov. a is reconstructed from CPC 8526 and has the median septum arising from a swollen region in front of a bilobed cardinal process; b is reconstructed from CPC 8528, and has a median septum arising as two low ridges separated by a median furrow from near the front of a trilobed cardinal process. Both specimens are from locality 100/19B, Burt Range Formation. × 2.

A similar absence of small spines arising from the costae also distinguishes *Productus sedaliensis* Weller (1914, pp. 108-10, pl. 14, figs 1-7) from the Chouteau Formation of the Mississippi Valley, USA. *P. sedaliensis* has a wider hinge, broader posterolateral extremities, higher and coarser costae, and weaker rugae on the pedicle valve; the interior of *P. sedaliensis* has not been described.

Productus kinlingensis Chu (1933, pp. 21-24, pl. 3, figs 4-16) from the Kinling Limestone (Tournaisian) of China is comparable in size and shape to *A. teicherti*. It differs in having fewer spines arising from the costellae on the pedicle valve, and in the possession of a diaphragm and dendritic muscle scars in the brachial valve.

Specimens of *A. teicherti* from locality 100/24 in the Burt Range Formation, stratigraphically the highest locality from which the species is known, are invariably larger than those from other horizons (*see* Pl. 23, figs 9, 10, 18-21). The pedicle valve is higher, has widely flaring flanks, and a less convex trail. The brachial valve and external ornament are the same in specimens from all localities.

The specific name is in honour of Dr Curt Teichert, one of the first geologists to collect fossils from the Burt Range area.

Occurrence: Localities 100/17-100/23, locality 100/19B (the type locality), 7 miles northwest of Mount Septimus; locality 101/7C, 5.5 miles west of Mount Septimus; locality 117/1, Spirit Hill; locality 421/5, southern end of Enga Ridge; all in the Burt Range Formation. Locality 7/1, Ningbing Limestone (Tournaisian part), Ningbing Range.

Material: CPC 8523-8534, CPC 8645. Holotype CPC 9645; paratypes CPC 8524, 8526, 8528, and 8529. CPC 8523-8529, 8645 from locality 100/19B; CPC 8530, 8534 from locality 100/24; CPC 8531 from locality 100/18; CPC 8532 from locality 100/20; CPC 8533 from locality 100/22. More than 100 specimens.

Age: Tournaisian.

Genus Spinauris nov.

Type species: Spinauris cristata gen. et sp. nov. from the Burt Range Formation of the Bonaparte Gulf Basin.

Description: Pedicle valve subquadrate to subcircular, moderately to strongly convex, with spreading flanks; ornament of fine quincuncially arranged spine bases bearing fine sessile spines on body of valve, and rows or a group of larger curving erect spines on posterolateral extremities; rugae usually weak on pedicle valve; brachial valve moderately concave, geniculate, with a short trail, and also having quincuncially arranged spine bases; rugae moderately well defined over most of visceral disc; brachial valve interior with two pairs of adductor muscle scars; posterior adductor scars dendritic, anterior scars smooth and elevated; median septum extends two-thirds length of visceral disc, and usually notched posteriorly by an antron of varying strength; cardinal process leioproductid, bilobate internally, trilobate externally; lateral ridges extend along half the length of hinge, then curve across ears, and join a marginal ridge; brachial ridges extend horizontally from muscle field and enclose smooth ovoid brachial discs; front of visceral disc and posterior part of trail ornamented with small endospines.

Remarks: Spinauris gen. nov. is referred to the Overtoniidae because it has a leioproductid type of cardinal process and a fine quincuncially arranged spinose external ornament. It is well to note, however, that the posterior part of the median septum usually has an alveolus, which is characteristic of the Buxtoniidae, and that the dorsal muscle scars resemble those of genera in the Echinoconchidae. In addition, instead of the oblique brachial ridges typical of the Overtoniidae (Muir-Wood & Cooper, 1960, p. 183) *Spinauris* is characterized by brachial ridges given off horizontally from the adductor muscle field.

Externally *Spinauris* closely resembles *Waagenoconcha* Chao, 1927, which is common in the late Carboniferous and Permian (Muir-Wood & Cooper, 1960) and has been recorded as early as the Visean from Australia by Campbell (1956) and Roberts (1964). *Waagenoconcha,* as redefined by Muir-Wood & Cooper (1960), is distinguished from *Spinauris* by the possession of a non-geniculate brachial valve, a long-shafted and recurved cardinal process which is fully supported by the median septum, and shorter lateral ridges. On the pedicle valve exterior the gradation from coarse to fine spines anteriorly is much more marked, and the spines on the ears are of smaller diameter than those in *Spinauris.*

Pustula Thomas, 1914, also redefined by Muir-Wood & Cooper (1960), lacks a geniculate brachial valve, and has a large dorsally recurved cardinal process supported by a massive median septum, and shorter lateral ridges. *Pustula* has stronger rugae on both valves, and lacks strong spines on the ears of the pedicle valve.

Levipustula Maxwell, 1951, has much coarser quincuncially arranged spine ridges over the body of the shell, a single pair of smooth trigonal muscle scars in the brachial valve, weaker rugae on the pedicle valve, and lacks rugae on the brachial valve.

Krotovia Fredericks, 1928, as redefined by Muir-Wood & Cooper (1960), has a coarser external spinose ornament, a more concave brachial valve, a strongly shafted median septum supporting the cardinal process, and a single pair of smooth dorsal adductor muscle scars.

The genus is named after the dense brush of spines on the ears. The name is derived from the Latin *spina* — thorn, and *auris* — ear.

SPINAURIS CRISTATA gen. et sp. nov.

(Pl. 20, figs 1-8; Pl. 21, figs 1-16)

Diagnosis: Pedicle valve subquadrate to subcircular, strongly convex, bearing up to twelve curving spines in each row along the hinge, and five quincuncially arranged spines per 5 mm at the anterior margin; sinus absent; brachial valve gently concave, having a strong marginal geniculation, and a short but well-developed trail; fold absent; rugae moderately strong, covering the visceral disc; antron of variable strength, but usually present; front of the visceral disc bears fine endospines.

Description: External. Pedicle valve subquadrate to subcircular, moderately to strongly convex, with an evenly rounded venter; greatest width at or slightly in front of hinge; umbo pointed, incurved, extending a short distance behind hinge, and having concave umbonal shoulders; flanks steep and flaring laterally, postero-lateral portions extending behind hinge; ears wide; quincuncially arranged spine bases approximately 1.5 mm in length, bearing fine sessile spines, and having a density (measured longitudinally) of five per 5 mm near front of valve; three or four rows or a group of long erect to sub-erect spines on posterior and postero-lateral margins; spine rows diverging from hinge, extending on to lateral margins; spines long and curved, up to twelve in each row, and 1 mm in diameter at base; rugae weak, most strongly defined on umbonal shoulders, weak to absent on venter, but at times more common on trail; trail marked by fine concentric growth lines.

Brachial valve rounded-quadrate to rectangular, usually slightly wider than long, with a well-developed geniculation extending from hinge; trail highest at front of valve (5 mm on a valve 38 mm wide and 28 mm long), becoming lower towards hinge; visceral disc gently concave, deepest towards front of valve, and with a small triangular depression at umbo; posterolateral margins flat and elevated above visceral region; rugae low, numerous, closely spaced, moderately well defined over most of visceral disc, having a density of about ten per 10 mm near front of valve; rugae ornamented with closely spaced concentric growth lines; spine bases quincuncially arranged, bearing fine sessile spines, situated on or between rugae, largest (up to 1 mm long) at anterior margins of visceral disc; density of spines approximately five per 5 mm along one ruga 20 mm from umbo.

Internal. Pedicle valve. Diductor muscle scars moderately long, flabellate, marked with subparallel longitudinal striations, situated just in front of tip of umbo; adductor muscle scars not observed; low marginal ridge around lateral and anterior margins on larger valves.

Brachial valve. Adductor muscle scars in two pairs; posterior scars large, dendritic, irregularly rectangular to triangular, anteriorly divergent from median septum; posterior scars slightly impressed or level with surface of valve, or in large forms elevated on low platforms; anterior scars smooth, forming trigonal to subovate platforms, abutting against or partially enclosed by posterior adductor muscle scars; median septum extending two-thirds length of visceral disc; septum notched posteriorly by an antron of varying strength (Pl. 20, figs 1-7); posterior of most septa having a shallow antron and base of cardinal process supported by two rounded septal ridges or a thickened platform; in other (often smaller) specimens posterior part of septum thinner, apparently lacking an antron, arising from a platform in front of cardinal process; septum usually rounded and tapering to midlength of anterior adductor muscle scars, and then forming a narrow blade-like ridge rising to a high anterior termination (septum 1.5 mm high at front in a valve 33 mm long and 37 mm wide); cardinal process leioproductid, bilobate internally and trilobate externally; inner face usually on a short straight shaft, having small convex lateral lobes usually fusing medially at a furrow, but sometimes projecting posteriorly into a median lobe; outer face of process with a triangular median lobe surrounded by incurved extremities of lateral lobes; lateral ridges extending along half the length of hinge, then cutting across ears, and in larger specimens joining with a marginal ridge extending around front of visceral disc; lateral ridges curving posteriorly at base of cardinal process, slightly notched on either side of cardinal process, and highest behind muscle field; brachial ridges extending horizontally from midlength of anterior adductor muscle scars, becoming stronger around brachial discs, and reaching a maximum height on anteromedian parts of disc; a faint trace of brachial ridge curving from end of hooked portion to median septum (visible on left side of specimen in Pl. 21, fig. 1); brachial discs ovoid and slightly raised above floor of valve; front of visceral disc and posterior of trail ornamented with fine pustules; front of visceral disc (Pl. 21, fig. 1) with about fourteen short irregular and possibly branching mantle canals? extending to marginal ridge; trail occasionally with short wrinkles.

Remarks: An interesting feature of *Spinauris cristata* is the variation in the posterior of the median septum, and the base of the cardinal process. The range of variation is shown in Figures 1-7 of Plate 20. Most forms have an antron or its remnants (Pl. 20, figs 4-7), and the cardinal process is supported at the base by two ridges or a thickened platform behind the two ridges. The strength of the antron varies from deep (Pl. 20, fig. 6) to very shallow (Pl. 20, fig. 5). In other forms the posterior part of the median septum is thinner and the septum arises from a broad platform in front of the cardinal process (Pl. 21, figs 1, 2); the antron is usually absent in these forms, but in some specimens (Pl. 20, fig. 1; Pl. 21, fig. 3) there is a small elongate depression between the posterior of the septum and the cardinal process.

Measurements

Pedicle valve exterior	Length	Length (of curvature)	Width (of hinge)	Width (at midlength)	Height
CPC 8549 Paratype	33.5	50	35 est.	31	19

Brachial valve interior	Length	Width (at midlength)	Length (median septum)	Height (of trail)
CPC 8543 Holotype	33	37	20	4
CPC 8542 Paratype	34	39	21	4
CPC 8552 Paratype	20	28 est.	13	—
CPC 8553 Paratype	26.5	31 est.	16.5	—
CPC 8554 Paratype	28	37	16.5	6
CPC 8556 Paratype	31.5	39	20	4 est.
CPC 8557 Paratype	27	30	18	3

The irregular canals at the front of the visceral disc have been observed in one specimen only, and hence they are doubtfully referred to as mantle canals; convincing canals have been found in specimens of *Lomatiphora aquila* gen. et sp. nov. (Fig. 20) from the Burt Range Formation.

A comparison with *Spinauris sulcata* sp. nov., from the Septimus Limestone, Bonaparte Gulf Basin, is given on page 112. No other comparable species have, to my knowledge, been described.

The specific name (from the Latin *crista* — ridge) refers to Enga Ridge, where most specimens were collected.

Occurrence: Locality 100/29, 7 miles northwest of Mount Septimus; localities 101/13-101/16, 101/18-101/21, 4.5 miles west of Mount Septimus; localities 103/3, 103/6 (the type locality), 103/7, 109/2, 109/3, 109/5, 109/7, 128/1, 128/2, 128/6, middle part of Enga Ridge; locality 129/4, southern end of Enga Ridge; all in the Burt Range Formation. Locality 102/4, northern end of Enga Ridge; localities 103/15, 103/17B, 103/19, 109/8, middle part of Enga Ridge; in the Enga Sandstone. Localities 104/3, 104/6, 104/11, 104/11A, 104/13, 104/15, 104/17, 150/4, 150/6, Mount Septimus; localities 110/3, 110/4, Mount Zimmermann; localities 112/2, 112/3, Langfield Point; in the Septimus Limestone. Possibly in locality 210/6, an unnamed Tournaisian breccia in the Waggon Creek valley.

Material: CPC 8542-8558. Holotype CPC 8543; paratypes CPC 8542, 8549, 8556, and 8558. CPC 8542-8543, 8548-8549, 8556, 8558 from locality 103/6; CPC 8544-8546, 8552-8555, 8557 from locality 103/7; CPC 8547 from locality 101/18; CPC 8550 from locality 109/3; CPC 8551 from locality 104/11A. Several hundred specimens.

Age: Tournaisian.

Spinauris sulcata sp. nov.

(Pl. 22, figs 1-23)

Diagnosis: Pedicle valve subovate to subrectangular, moderately convex, with a shallow sinus on anterior half of valve; fine quincuncially arranged spine bases over visceral disc becoming finer and arranged in rows on lamellae on trail; coarser spines 0.3 mm in diameter, along hinge, rugae low, widely spaced, crossing venter, and becoming lamellose and more numerous on trail; brachial valve subrectangular, with a small anterior geniculation; visceral disc moderately concave, crossed by a median fold; rugae absent from juvenile visceral disc; valve usually lamellose anteriorly; median septum massive posteriorly, supporting base of cardinal process, and only occasionally bearing a trace of an antron; posteromedian parts of inner lobes of cardinal process forming a posteriorly or ventrally projecting median lobe; moderately coarse endospines on front of visceral disc.

Description: External. Pedicle valve subovate to subrectangular; hinge two-thirds width of valve, greatest width at about two-thirds length of valve; valve most convex at umbo, flatter towards front of trail; flanks moderately steep, projecting dorsally behind hinge; umbo narrow, strongly incurved, extending behind hinge; umbonal shoulders concave and sloping on to flattened region along hinge and at posterolateral margins; median sinus shallow, rounded, usually narrow, extending from near midpoint to anterior margin; spine bases small, elongate, quincuncially arranged on visceral disc, and in rows on lamellose trail; spines on visceral disc and trail prostrate, 0.1 to 0.2 mm in diameter at base, and up to 2.5 mm long; four irregular rows of curving sub-erect spines, 0.3 mm in diameter at base and more than 3 mm long, extending along hinge; rugae low, widely spaced, strongest on flanks, and generally crossing venter; in many specimens rugae become lamellose and more numerous towards front of trail, each lamella bearing one or more rows of spine bases.

Brachial valve subrectangular, with a small anterior geniculation; visceral disc moderately concave, deepest near midpoint of valve on either side of a gentle median fold, with a small V-shaped concavity at umbo; fold low, commencing at about one-third length of valve, broadening towards front; posterolateral margins gently concave, separated from visceral disc by two ridges running anterolaterally from umbo; ornament of fine quincuncially arranged spine bases over most of visceral disc; in most specimens, front of visceral disc and trail lamellose, each lamella bearing irregular rows of spines; spines similar in size to those on body of pedicle valve; rugae low, extending across valve, few in number, most prominent towards front of visceral disc, and absent from juvenile parts of valve.

Internal. Pedicle valve. Diductor muscle scars broad, flabellate, and striated; adductor muscle scars poorly defined but also striated.

Brachial valve. Adductor muscle scars in two pairs; posterior adductor scars deeply dendritic, ovoid in outline, with posterior parts abutting against median septum, and front diverging laterally and enclosing posterolateral margins of anterior adductor scars; anterior adductor scars trigonal or quadrangular, pointed or bluntly rounded posteriorly, with inflated subrounded anterior margins; median septum supports base of cardinal process and extends two-thirds length of valve; septum broad and rounded near cardinal process, tapering towards midlength, becoming high and blade-like anteriorly (1.5 mm high at tip on a valve 28 mm wide and 21 mm est. long); cardinal process on a short straight shaft, bilobed or trilobed internally, trilobed externally; two lateral lobes on inner face small, gently to moderately convex, at times slightly divergent from midline of valve, separated by a median furrow of varying strength; on some specimens posteromedian parts of lateral lobes project posteriorly into narrow, deeply concave median lobe, forming an internally trilobed process; on other specimens posteromedian parts of lateral lobes project posteriorly into narrow, deeply concave median lobe, forming an internally trilobed process; on other specimens posteromedian parts of lateral lobes project dorsally to trilobed outer face, resulting in an internally bilobed cardinal process; on external face, high lamellose lateral lobes enclose cone-like depressions on either side of a lamellose wedge-shaped median lobe; lateral ridges extend two-thirds length of hinge, becoming broader, lower, and slightly divergent from hinge at lateral extremities, and curving posteriorly towards base of cardinal process; brachial ridges extend horizontally from midpart of anterior adductor muscle scars, highest around margins of brachial discs; brachial discs ovoid, smooth, usually slightly elevated, but sometimes high medially; endospines form a dense mat of coarse spines around front of visceral disc, becoming finer on trail.

Measurements

Pedicle valve exterior	Length	Length (of curvature)	Width (at hinge)	Greatest width	Height
CPC 8538 Holotype	24 est.	40 est.	25 est.	29	13.5
CPC 8540 Paratype	25 est.	—	29	36 est.	14

Brachial valve interior	Length	Greatest width	Length (median septum)
CPC 8535	26	30.5	16

Remarks: In comparison with *Spinauris cristata* sp. nov., also from the Bonaparte Gulf Basin, *S. sulcata* is smaller, has a finer spinose ornament, is relatively more transverse, less convex on the pedicle valve, and has a weak fold and sinus. The brachial valve has fewer rugae, a juvenile visceral disc lacking in concentric ornament, and usually has a lamellose margin. In addition, the brachial valve has a weaker anterior geniculation, a shorter trail, and in the interior a massive posterior shaft on the median septum.

Spinauris sulcata is named from the Latin *sulcus* — a furrow, and refers to the sinus on the pedicle valve.

Occurrence: Locality 104/17, Septimus Limestone, Mount Septimus.

Material: CPC 8535-8541. Holotype CPC 8538; paratypes CPC 8535 and 8540. All from locality 104/17. Ten specimens.

Age: Tournaisian.

Genus MAGNUMBONELLA Carter, 1968

Type species: Magnumbonella macrura Carter, 1968, from the lower Burlington Limestone of Missouri, USA, by original designation of Carter, 1968.

Diagnosis: See Carter (1968, p. 1145).

MAGNUMBONELLA PROLATA sp. nov.

(Pl. 25, figs 1-21)

1959b cf. *Avonia proteus* Veevers, pp. 15-16, pl. 4, figs 35, 36.
1959b Gen. et sp. indet. Veevers, pp. 16-17, pl. 4, figs 32-34.

Diagnosis: Trail elongate, especially on pedicle valve, and bearing coarse low ribs at front; pedicle valve having a broad umbo, a narrow, usually weak median sinus, and a single row of hinge spines; ventral muscle field with impressed diductor muscle scars, and adductor muscle scars on a platform-like ridge; brachial valve with large lobes on the cardinal process, and a median septum with a robust posterior shaft and a high anterior blade; dorsal adductor muscle scars usually in two pairs.

Description: External. Pedicle valve elongate, subrectangular, having a long curved non-geniculate trail; large specimens broadening anteriorly; greatest width in front of hinge, and in large specimens near anterior margin; greatest convexity slightly in front of umbo; umbo tapering and projecting 1 to 2 mm behind hinge; lateral slopes steep; posterior margin flattened on either side of umbo and laterally bears small convex auricles; posterolateral margins flaring and projecting behind hinge; a narrow median sinus of more or less constant width commences 5 to 10 mm in front of umbo and extends to anterior margin of valve; ornament rugose and spinose; rugae confined to posterior part of valve, strongest on posterolateral margins, grading anteriorly into closely spaced wavy growth lines; growth lines cover much of visceral disc and all of trail, having a density of about twenty lines per 5 mm on median part of valve; trail of large specimens bears low coarse ribs; about six spines in a row along hinge, others scattered over body of valve, frequently most numerous towards posterior of valve; spine bases on body of valve about 0.5 mm in diameter, sometimes producing elongate longitudinal spine ridges.

Brachial valve geniculate, with steep spreading flanks, an extensive trail, and a moderately concave visceral disc; visceral disc rises gently to a flattened or slightly concave posterior margin having a small V-shaped depression at umbo; ears flat or slightly concave; fold absent except at anterior margin; valve aspinose; visceral disc ornamented by rugae (about fifteen on visceral disc of a valve 16 mm wide at hinge) and faint pits, and trail and flanks by concentric growth lines.

Internal. Pedicle valve thickened at umbo; diductor muscle scars tapering posteriorly and broadly flabellate anteriorly; posterior parts smooth and on thickened umbonal region; flabellate anterior portions coarsely striate and deeply impressed; adductor muscle scars on posterior part of a broad platform-like median ridge deeply impressed posteriorly, and in specimens with a high median ridge also impressed at front of muscle field; adductor muscle field irregularly pitted; median ridge broad, extending a short distance in front of adductor field, up to 3 mm high near anterior end, but usually lower and sloping gently to floor of valve; strong ridges on posterolateral margins of diductor scars across ears; ginglymus up to 0.4 mm long, striated, and divided by an extensive umbonal cavity.

Brachial valve. Median septum extends about two-thirds length of visceral disc, originating from a thickened platform in front of cardinal process; septum broad and sometimes low to midlength of muscle field, becoming blade-like and increasing in height (2 mm high at front in a valve 22 mm wide) towards front; cardinal process sessile, straight and trilobate; ventral face having two swollen rounded lateral lobes separated by a depression, and a wedge-shaped median lobe bearing a median groove; both median and lateral lobes swollen dorsally and having lamellose myophores; median lobe dorsally inclined over median groove separating two rounded lateral lobes; lateral ridges smooth, rounded, divergent from hinge and possibly cutting across ears; adductor muscle scars in two pairs, having high rounded elevations at front; posterior pair round or triangular, slightly impressed posteriorly, extending anterolaterally; anterior adductor scars set into antero-median part of posterior scars, separated by a rounded trough containing median septum; brachial ridges low, arising at right angles from junction of anterior and posterior adductor scars, curving anterolaterally a short distance from their origin, and enclosing club-shaped brachial discs; smooth areas extend obliquely from discs towards front of anterior adductor platforms; brachial ridges reach almost to margins of visceral disc; floor of valve near geniculation ornamented with rows of fine elongate endospines.

Measurements

Pedicle valve	Length	Length (of curvature)	Width (at hinge)	Width (of body)	Height
CPC 8500 Holotype	26	52	22	22.5	16
CPC 8499 Paratype	28 est.	60 est.	20.5	24	19
CPC 8510 Paratype	33.5	65	22 est.	28	22 est.

Brachial valve	Length	Width (at hinge)	Width (of body)	Height (of trail)	Length (median septum)
CPC 8506 Paratype	19	18	22.5	10	11.5

Remarks: Apart from the narrow median sinus and a single row of hinge spines on the pedicle valve, *Magnumbonella prolata* sp. nov. is externally close to *M. macrura* Carter (1968, p. 1146, pl. 146, figs 1-28), the type species of *Magnumbonella*. Internally, both forms have a similar morphology in the brachial valve, but most structures in *prolata* are larger than those in *macrura*. For example, the cardinal process has larger lobes, the median septum has a more robust posterior shaft and a higher anterior blade, and in some specimens the muscle field and brachial markings are better defined. The inner adductor muscle scars in the brachial valve have a similar morphology in both species. There may be variation in the strength of the lateral ridges in *prolata;* these extend to the margins in the best preserved specimen (Pl. 25, fig. 19), but are weaker in another (Pl. 25, fig. 17). In *macrura* the lateral ridges consistently extend to the lateral margins. *M. prolata* also appears to have a more deeply impressed ventral muscle field than *macrura.*

A comparison with *Leioproductus buttonensis* sp. nov., an external homeomorph from the Buttons Beds, is given on page ??.

The name of the species (Latin for elongate) refers to the length of the trail.

Occurrence: Locality 100/25, 6 miles northwest of Mount Septimus; localities 101/15, 101/18, 101/19, 4.5 miles west of Mount Septimus; localities 103/12, 109/5, 109/7, 128/2, middle part of Enga Ridge; locality 129/4, southern part of Enga Ridge; all in the Burt Range Formation. Locality 102/4, northern end of Enga Ridge (the type locality); locality 103/17B, 103/19, middle part of Enga Ridge; Enga Sandstone. Locality 210/6, unnamed Tournaisian breccia, Waggon Creek valley.

Material: CPC 8499-8508. Holotype CPC 8500, paratypes CPC 8499, 8501, 8502, 8503, 8506, 8507, and 8508. All from locality 102/4. Approximately 200 specimens.

Age: Tournaisian.

Genus PROTONIELLA Bell, 1929

Type species: Protoniella beedi Bell (1929) from the upper Windsor (Lower Carboniferous) of Nova Scotia, Canada, by original designation of Bell, 1929.

Remarks: Internal features of the brachial valve of *Protoniella* have previously been poorly known. Muir-Wood & Cooper (1960), working from Bell's original illustrations of the brachial valve, placed *Protoniella* in the Buxtoniidae (subfamily Buxtoniinae) probably because the cardinal process appeared to be dorsally recurved as in the Buxtoniidae.

A re-examination of the paratype specimen No. 7954a in the collection of the Geological Survey of Canada, the brachial valve of *Protoniella beedi* illustrated by Bell (1929, pl. 15, fig. 12), shows that mud adhering to the external face gave a false impression of the curvature and morphology of the cardinal process, and that the right-hand side of the valve is broken and at an angle to the remainder. The mud over the cardinal process was removed, revealing a trilobed external face with a partly corroded median lobe. The internal face of the cardinal process has a circular alveolus at the front, and two lateral lobes which posteriorly project dorsally into a median lobe. The external face has two deeply concave lateral lobes joined medially into a wedge-shaped median lobe which has a very deep median furrow. These features are typically found in the family Overtoniidae, and I have accordingly removed the genus *Protoniella* from the Buxtoniidae.

A description of the paratype brachial valve interior follows:

Cardinal process sessile and slightly ventrally recurved; internal face with two concave lobes separated anteriorly by a deep round alveolus; posteromedian part of inner lobes projecting dorsally into a median lobe; external face with two deeply excavate lateral lobes joining medially into a wedge-shaped median lobe; ventral tip of median lobe having a bridge-like connection with lateral lobes, and remainder of lobe with a very deep median furrow; lateral margins of lateral lobes thickened and lamellose; a flat heart-shaped platform occurring in front of alveolus and cardinal process; median septum arising from front of platform as a low rounded ridge; septum higher between muscle scars, blade-like towards front, and extending two-thirds length of visceral disc; adductor muscle scars in a single pair, elongate, sinuously curved, pointed posteriorly, and wider and rounded anteriorly; muscle scars separated by deep furrows from median septum; lateral ridges moderately developed, extending along half the width of hinge, then cutting across ears and joining a low ridge extending around margin of valve; front and lateral margins of visceral disc with two to three rows of short endospines arising from front of internal traces of costellae; area between spines and marginal ridge smooth; brachial markings obscure.

The specimens from the Tournaisian breccia in Waggon Creek valley, described below, are tentatively referred to *Protoniella* Bell, 1929. They differ from the type species in having a pedicle valve with a shorter hinge, steep posterior margins lacking flattened ears, steeper flanks, more spines on the body of the valve, and three to four rows of larger erect spines on the posterolateral extremities. The brachial valve has a small anterior geniculation, and does not appear to be lamellose anteriorly, as in the type species. Costae on both valves are weaker and fewer in number, and the brachial valve has a stronger reticulate ornament on the visceral disc. In the brachial valve interior, the Waggon Creek specimens have a stronger posterior shaft on the median septum, lack an alveolus, and do not have a well-defined marginal ridge, although one specimen (Pl. 26, fig. 14) has a rim-like structure which may be comparable with that in the paratype brachial valve.

PROTONIELLA? WAGGONENSIS sp. nov.

(Pl. 26, figs 1-24)

Diagnosis: Pedicle valve subquadrate, having a steep trail and flanks; posterior margin steeply inclined or vertical to hinge, and lacking flattened ears; costae low, having a density of about five per 5 mm at front of valve (in the adult), and bearing longitudinal rows of numerous sub-erect to erect spines; larger erect spines in a row along hinge, and in three or four longitudinal rows down the lateral extremities; brachial valve with a small anterior geniculation, and a low median fold; visceral disc coarsely reticulate, trail costate, and both bearing small spines; posterior shaft of median septum usually supports cardinal process.

Description: External. Pedicle valve subquadrate, having rounded anterolateral margins; widest and highest at or in front of midlength; umbo strongly incurved, very broad anteriorly, tapering posteriorly, projecting a short distance behind hinge; umbonal shoulders strongly concave; posterolateral extremities very steeply inclined or vertical to hinge; venter flattened at apex and lacking a median sinus; flanks and trail steep; ornament of spine ridges on posterior of valve giving rise to broad rounded costae bearing longitudinal rows of spines on median and anterior parts; spines numerous on body of valve, sub-erect posteriorly to erect anteriorly, usually 0.3 mm in diameter at base; larger erect spines, 0.5 mm in diameter at base, in three or four longitudinal rows down lateral extremities, in a single row along hinge, and occasionally scattered on trail (Pl. 26, fig. 1); usually six spines in each row on lateral extremities, and three in row along hinge; costae low and rounded, separated by shallow furrows of similar width, and sometimes irregular at spine bases and broader and higher towards front of trail; costae increasing mainly by bifurcation, but occasionally joining together into a single broad costa; five costae per 5 mm at front of a valve 19.5 mm long and 19 mm wide; rugae highest at posterolateral extremities, sometimes extending across venter; micro-ornament of fine concentric growth lines.

Brachial valve with a small anterior geniculation; visceral disc moderately concave, having a broad low fold extending from in front of a subcircular concavity at umbo on to posterior of trail; posterolateral margins flat; rugae extend across visceral disc, strongest at posterolateral extremities, and form a coarse reticulate pattern with wide costae; trail coarsely costate; costae increasing by bifurcation, bearing spine bases 0.3 mm in diameter along their length; spines most numerous on trail; micro-ornament as on pedicle valve.

Internal. Pedicle valve. Diductor muscle scars smooth posteriorly and striated anteriorly.

Brachial valve (Fig. 25). Median septum broad posteriorly, tapering to a blade-like structure anteriorly, and extending two-thirds length of valve; posterior part of septum supports base of cardinal process, or arises from a thickened area in front of process, and in some specimens possesses a small antron (Pl. 26, fig. 21); two low supporting ridges extend anterolaterally from platform at posterior of septum on one specimen (Pl. 26, fig. 17); cardinal process sessile, with two rounded inner lobes separated medially by a deep furrow; process trilobed externally, with a peg-like median lobe having a small concave muscle attachment? at base surrounded laterally by two curving outer lobes; lateral ridges support sides of cardinal process, extend two-thirds along hinge and then cut across ears; adductor muscle scars in a single pair, smooth, rounded-triangular in outline; scars deeply impressed, sometimes elevated anteriorly, surrounded anteriorly and laterally by rounded ridges; brachial ridges arise from lateral margins of muscle field, extend obliquely and enclose smooth ovoid brachial discs; visceral disc ornamented with irregular radial ridges, and trail by rows of fine pustules.

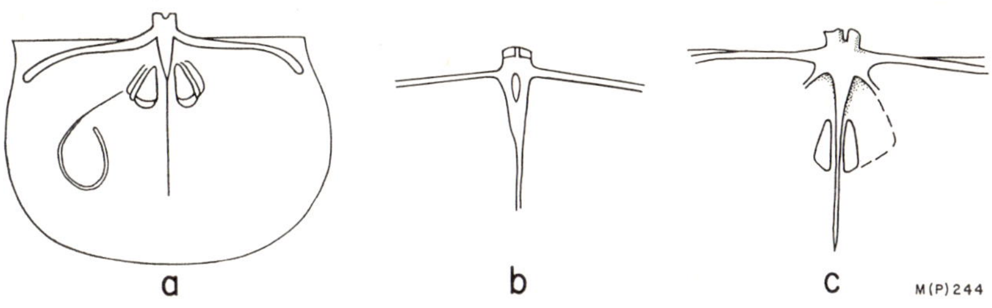

Figure 25. Brachial valve interiors of three specimens of *Protoniella? waggonensis* sp. nov. a is reconstructed from CPC 8591, b is the posteromedian part of CPC 8594 showing the antron, and c is the posteromedian part of CPC 8593 showing buttressing ridges extending anterolaterally from the platform in front of the cardinal process. The single pair of muscle scars is bordered by depressed areas which could be a second pair of scars. Specimens from locality 210/6, unnamed Tournaisian breccia in the Waggon Creek valley.

Measurements

Pedicle valve	Length	Length (of curvature)	Width (of hinge)	Width (of body)	Height
CPC 8586 Paratype	23	40	21	23	15
CPC 8588 Holotype	20	35	19.5	21.5	13
CPC 8589 Paratype	20	32	18.5	19.5	11

Brachial valve	Length	Width (at hinge)	Width (of body)	Length (median septum)
CPC 8591 Paratype	—	14	—	9.5
CPC 8595 Paratype	20.5	19	21	12

Remarks: Except for the possession of fewer costae, *Protoniella? waggonensis* sp. nov. is externally close to specimens of *P? parva* (Meek & Worthen) from the Mississippian Pella Formation of Iowa, figured by Muir-Wood & Cooper (1960, pl. 91, figs 8-13). The interior of *P? parva* has not been described.

The spinose ornament on the pedicle valve is also close to that on specimens of *Productus muricatus* Phillips (now *Antiquatonia muricatus)* from the Lower Carboniferous at Harelow, Northumberland, and Campsie, Stirlingshire, Great Britain, figured by Muir-Wood (1928, pl. 3, figs 5-7). *A. muricatus* is distinguished by the possession of higher and more regular costae, and a single row of spines down the posterolateral margins.

Spinocarinifera adunata gen. et sp. nov. from the lower part of the Burt Range Formation has a pedicle valve with better defined and more regular costae bearing a smaller number of body spines; the umbo on the pedicle valve is also narrower, and there are larger flattened auricles. In the brachial valve of *adunata* the median septum arises from in front of an alveolus, and the lateral ridges are parallel with the hinge.

The specific name is taken from Waggon Creek which runs close by the type locality in the unnamed Tournaisian breccia.

Occurrence: Locality 210/6 in an unnamed Tournaisian breccia in the Waggon Creek valley, and localities 103/15 and 103/17B, Enga Sandstone, middle part of Enga Ridge.

Material: CPC 8586-8595. Holotype CPC 8588, paratypes CPC 8586, 8589, 8591, 8594, and 8595. All from locality 210/6. One hundred and seventy specimens.

Age: Tournaisian.

Genus STEGACANTHIA Muir-Wood & Cooper, 1960

Type species: Stegacanthia bowsheri Muir-Wood & Cooper (1960, pp. 199-200, pl. 48, figs 1-12) from the Lake Valley Formation, New Mexico, by original designation by Muir-Wood & Cooper, 1960.

STEGACANTHIA STRIGIS sp. nov.

(Pl. 27, figs 1-9)

Diagnosis: Pedicle valve subrectangular, moderately convex, with a well-defined median sinus, especially at front of valve; concentric ornament with a density of three lamellae per 10 mm at 25 to 35 mm from the umbo; spines on the lamellae have a density of ten per 10 mm at 25 mm from the umbo; four to six concentric rows of erect spines near ears; brachial valve gently concave posteriorly, becoming deeper towards front; median fold on anterior two-thirds of valve; concentric lamellae more closely spaced than on pedicle valve, having a density of six per 10 mm at 40 mm from umbo; brachial valve interior with lateral ridges slightly divergent from hinge and cutting across ears; adductor muscle scars in two pairs; posterior scars dendritic, and round to ovoid; anterior scars smooth and triangular.

Description: External. Pedicle valve subrectangular, with rounded lateral margins, and moderately steep spreading flanks; ears flat and well differentiated; valve moderately convex, greatest convexity on umbo and visceral disc, becoming less convex on trail; umbo small, incurved, projecting over hinge, having concave umbonal shoulders; greatest width at midlength or on spreading posterolateral flanks; hinge with a flat articulatory surface; median sinus commencing slightly in front of umbo, U-shaped in section, slightly broader anteriorly, and usually best defined at front of valve; ornament of small spine ridges and weak rugae on juvenile parts of valve replaced anteriorly by overlapping concentric lamellae bearing elongate spine ridges; lamellae broad posteriorly, marked with concentric growth lines, and bearing usually a single, occasionally two, rows of spine ridges; spine ridges frequently extend between successive lamellae; density of lamellae three per 10 mm at 25 to 35 mm from umbo; spines prostrate, averaging 0.4 mm in diameter at base, originating from spine ridges up to 4 mm long; spines have a density of ten per 10 mm at 25 mm from umbo; four to six poorly defined concentric rows of erect spines on ears, each row having as many as six spines, and with spine bases averaging 0.5 mm in diameter.

Brachial valve transverse, with rounded lateral margins; gently concave posteriorly, deeper around midregion or towards anterior margin; a small V-shaped depression at umbo isolated from other concave areas by commencement of median fold at one-third length of valve; fold rounded and well defined anteriorly; posterior surface rugose, with concentric rows of spine bases, passing anteriorly into a lamellose and spinose ornament; concentric lamellae more closely spaced than on pedicle valve, having a density of six per 10 mm between 20 and 30 mm from umbo; lamellae ornamented with fine concentric growth lines, bearing a single row (occasionally two rows) of small elongate spine bases; spine ridges much shorter than those on pedicle valve, having a density of ten per 10 mm at 40 mm from umbo; spines on mature parts of shell more than 6 mm long, and averaging 0.3 mm in diameter at base; spines finer posteriorly and poorly defined on juvenile parts of shell.

Internal. Pedicle valve. Adductor muscle scars elongate, subrectangular, and separated by a shallow median furrow; diductor muscle scars broadly triangular, smooth and pointed in posterior third, and broader and coarsely striated anteriorly.

Brachial valve. Adductor muscle scars in two pairs; posterior scars round to ovoid, faintly dendritic, and flanking posterior margins of anterior scars; anterior adductor muscle scars smooth, triangular, and having pointed posterior and broad rounded anterior extremities; median septum buttressing base of cardinal process, rounded posteriorly, becoming blade-like towards front, and extending half the length of valve; cardinal process gently recurved dorsally, trilobate internally; lateral lobes arise from a thickened platform, convex, slightly divergent, and separated by a median furrow; inner margins of lateral lobes unite medially and form a posterodorsally projecting median lobe; external face not observed; lateral ridges slightly divergent from hinge, extending almost entire width of valve, and curving anteriorly across ears; ridges strong and well rounded in midportion of valve, but broader and lower on ears; brachial ridges not observed; endospines around margins of visceral disc coarsest on anterior and anterolateral margins.

Remarks: Stegacanthia strigis differs from the type species, *S. bowsheri* Muir-Wood & Cooper, in its larger size, and possession of a well-defined fold and sinus; in the brachial valve interior, *S. strigis* has longer lateral ridges which cut across the posterolateral margins, the cardinal process has a well-developed median lobe on the inner face, and the adductor muscle scars are differentiated into two pairs.

Similar adductor muscle scars are found in *S. sibirica* (Sarycheva) from the Upper Tournaisian and Lower Visean of the Kuznetsk Basin, Siberia (Sarycheva et al., 1963, p. 152, fig. 59). Sarycheva's illustration (fig. 59), which appears to be taken from *S. sibirica artyshtensis* (Sarycheva) (pl. 17, fig. 4), shows more inflated anterior adductor muscle scars than those in *S. strigis,* and lateral ridges extending parallel with the hinge. *S. sibirica sibirica* (Sarycheva) (Sarycheva et al., 1963, pl. 17, figs 1-3), which is externally closer to *S. strigis,* has smaller and more numerous spine bases on the pedicle valve.

Stegacanthia abbotti larga (Cvancara) (1958, pp. 864-65, pl. 110, figs 14-19) from the *Rhipidomella fortimuscula* Zone at Barrington, New South Wales, is smaller, less transverse, has smaller and more numerous spine bases on both valves, and a greater number of concentric lamellae on the pedicle valve. In the brachial valve interior, *S. abbotti larga* is distinguished by its longer median septum, bilobed inner face on the cardinal process, lateral ridges extending along the hinge line, and posterior adductor muscle scars with deeper dendritic markings.

The specific name *strigis* (Latin for furrow) refers to the median sinus on the pedicle valve.

Occurrence: Localities 107/1 and 108/4A (the type locality), Utting Calcarenite, Utting Gap; locality 305/1, 1.5 miles northeast of Ningbing Homestead; locality 306, Burvill Point; locality 435/0, 5 miles northeast of Point Spring in the Weaber Range; Burvill Beds.

Material: CPC 8481-8486. Holotype CPC 8482. CPC 8481, 8483, 8485 from locality 306; CPC 8482 from locality 108/4A; CPC 8484 from locality 107/1; CPC 8486 from locality 305/1. Twenty-five specimens.

Family BUXTONIIDAE Muir-Wood & Cooper, 1960

Subfamily BUXTONIINAE Muir-Wood & Cooper, 1960

Genus MARGINATIA Muir-Wood & Cooper, 1960

Type species: Productus fernglenensis Weller, 1909, from the Fern Glen Formation, Missouri, USA, by original designation of Muir-Wood & Cooper, 1960.

MARGINATIA MIMICA sp. nov.

(Pl. 28, figs 1-20)

Diagnosis: Pedicle valve subquadrate; sixteen to twenty costellae per 10 mm at 10 mm from umbo; rugae almost obsolete over venter, and ornament non-reticulate; spines in poorly defined concentric rows on the visceral disc, long halteroid spines on the trail, and a single row of approximately seven erect spines along either side of the hinge; brachial valve gently concave, with a long trail; reticulate ornament confined to the posterior of valve; adductor muscle scars in the brachial valve in two pairs in large specimens; posterior adductor scars dendritic, and round to ovoid; anterior adductor scars triangular, smooth and swollen; median septum lacking an antron in large specimens; lateral ridges long, and continuous with the marginal rim of the valve.

Description: External. Pedicle valve subquadrate, moderately to strongly convex, with a slightly flattened venter; greatest convexity at umbo; trail long and incurved; greatest width slightly in front of midlength of valve; umbo pointed, slightly incurved, and projecting behind hinge; umbonal shoulders steep; hinge approximately two-thirds width of valve, having small flattened auricles at almost right angles to trail; flanks steep, subparallel, projecting well behind hinge; costellae narrow, usually evenly developed, becoming coarser on extremities of flanks and trail, and occasionally wrinkled at front of visceral disc; costellae increase mainly by bifurcation, usually from immediately in front of spine bases, with a density of sixteen to twenty costellae per 10 mm at 10 mm from umbo; spines on body of valve arise from costellae; spines arranged (1) in poorly defined concentric rows with erect to sub-erect spines, 0.3 mm in diameter, on visceral disc, and slightly larger halteroid spines, 0.5 mm in diameter at base, on trail; (2) in a row of approximately seven erect spines on either side of hinge; and (3) in some specimens, twenty to thirty coarse spines on front of flanks; rugae restricted to visceral disc, well defined on posterolateral margins, becoming nearly obsolete across venter; twelve rugae on a valve 31 mm wide and 30 mm long; concentric ornament of fine growth lines over most of valve.

Brachial valve gently concave, geniculate, with a long steep trail extending around flanks to hinge; visceral disc deepest and most concave at front, with a small triangular concavity at umbo; posterolateral margins flat to slightly concave, and just below the level of hinge; external face of cardinal process supported on a small lophidium; costellae sometimes twisted towards front of visceral disc, parallel on trail, increasing by bifurcation and intercalation, with same density as on pedicle valve; costellae on trail crossed by fine concentric growth lines; rugae low, strongest towards hinge and on posterolateral margins, crossing visceral disc, and forming a reticulate ornament; more than twenty rugae on visceral disc of a valve approximately 30 mm wide; valve aspinose.

Internal. Pedicle valve. Adductor muscle scars narrow, elongate, pointed posteriorly, wider and less distinct anteriorly, and separated by a narrow furrow; diductor muscle scars flabellate, smooth posteriorly and coarsely striated anteriorly; marginal rim narrow, diverging slightly from hinge, and extending a short way down ears.

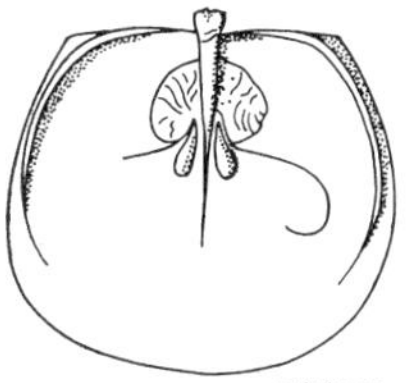
M(P)245

Figure 26. Brachial valve interior of a large specimen of *Marginatia mimica* sp. nov. showing the lateral ridges extending into the marginal rim, two pairs of adductor muscle scars, and the brachial ridges. Reconstructed mainly from CPC 8596.

Brachial valve (Fig. 26). Adductor muscle scars in two pairs; posterior scars large, coarsely dendritic, impressed into floor of valve, and round to ovoid in shape; at front, posterior adductor muscle scars diverge from median septum and flank the pointed posterior extremities of anterior adductor muscle scars; anterior adductor scars smooth, triangular to crescent-shaped, swollen medially, and separated from median septum by a wide trough; in younger specimens adductor scars trigonal, not differentiated into two pairs, non-dendritic posteriorly, and only slightly elevated anteriorly; median septum extends approximately two-thirds length of visceral disc; septum broad and flattened to slightly rounded posteriorly, and, sometimes in association with a broad platform, buttresses base of cardinal process; septum high and well rounded between posterior adductor scars, tapering and lower between anterior adductor scars, becoming higher and forming a blade-like anterior termination; in younger specimens, septum more delicate, and base of cardinal process supported by two ridges enclosing an antron; lateral ridges diverge from hinge, cut across posterolateral margins, and extend around edge of visceral disc on to thickened marginal rim at front of valve; lateral ridges low on either side of cardinal process, having shallow socket-like excavations; ridges high and well rounded posteriorly, especially on posterolateral margins, and low towards front; lateral ridges shorter and confined to posterior of valve in small specimens; cardinal process straight, sessile, bilobate internally, and trilobate externally;

anterior face with two swollen lateral lobes separated by a median furrow; on external face, posteromedian parts of lateral lobes produced posterodorsally into short lamellose wedge-shaped median lobe; lateral lobes convex dorsally, bear a series of lamellae, and enclose deep pits on either side of median lobe; brachial ridges arise at a low angle to horizontal from anterolateral margins of posterior adductor muscle scars, curve anterolaterally, and surround slightly elevated ovoid brachial discs; front of visceral disc and trail with irregular rows of elongate spine ridges; posterolateral regions pitted.

Measurements

Pedicle valve	Length	Length (of curvature)	Width (of hinge)	Width (of body)	Height
CPC 8603 Paratype	—	—	23.5	30 est.	—
CPC 8605 Paratype	27	45	22.5	28	15 est.

Brachial valve	Length	Width (of hinge)	Width (of body)	Length (median septum)
CPC 8596 Holotype	33.5	24	32	—

Remarks: Marginatia fernglenensis (Weller), the type species, has a shorter hinge and a slightly coarser costellate ornament. The body of the pedicle valve has fewer spines, and lacks the row along the hinge and the group of spines on the flanks. In the brachial valve, the adductor muscle scars are not readily divisible into two pairs, the septum has a deep pit-like antron, and the lateral ridges are shorter.

M. deruptoides Sarycheva, from the Tournaisian of the Kuznetsk Basin, Siberia (Sarycheva et al., 1963, pp. 188-91, pl. 26, figs 1-3; pl. 27, figs 1-4; pl. 28, figs 1-4), is morphologically extremely close to this species, differing only in its larger size, the possession of a greater number of spines on the trail of the pedicle valve, and the lack of a row of spines along the hinge of the pedicle valve.

M. burlingtonensis (Hall) from the Burlington Limestone of USA (Weller, 1914, pp. 104-6, pl. 9, figs 1-10) has more pronounced auricles, stronger rugae, and a sinus; it lacks the row of spines along the hinge of the pedicle valve. The Siberian specimens figured by Sarycheva et al. (1963, pl. 28, figs 5-8), from the Tournaisian of the Kuznetsk Basin, have a similar costate ornament to that on *mimica.*

M. vaughni (Muir-Wood) (1928, pp. 65-70, pl. 2, figs 12, 13), which ranges from the K_2 to the C_1? Zones of Great Britain, is distinguished by its non-dendritic triangular adductor muscle scars and more divergent lateral ridges in the brachial valve; it has fewer body spines over the pedicle valve.

The specific name (from the Latin *mimicus*-imitative) refers to the close resemblance to the Siberian species *M. deruptoides* Sarycheva.

Occurrence: Locality 102/4, Enga Sandstone, northern end of Enga Ridge. Localities 104/2, 104/6, 104/11, 104/11A, 104/12, 104/13, 104/15, 104/17, Mount Septimus; localities 110/1, 110/3, 110/4, 110/6, 110/7, Mount Zimmermann; locality 112/2, Langfield Point; Septimus Limestone. Locality 110/11 and possibly locality 110/10, Zimmermann Sandstone at Mount Zimmermann.

Material: CPC 8596-8608. Holotype CPC 8596; paratypes CPC 8597, 8598, 8603, 8604, and 8605. CPC 8596-8599, 8601-8606, 8608 from locality 104/17; CPC8600, 8607 from locality 112/2. Fifty silicified specimens.

Age: Tournaisian.

Family ECHINOCONCHIDAE Stehli, 1954

Subfamily ECHINOCONCHINAE Stehli, 1954

Genus ECHINOCONCHUS Weller, 1914

Type species: Productus punctatus J. Sowerby, 1822, by subsequent designation of Chao, 1927.

Diagnosis: A revised diagnosis of the genus is given by Muir-Wood & Cooper (1960, pp. 243-44).

Remarks: A group of species typified by *Echinoconchus elegans* (M'Coy), and containing *E. elegans, E. venustus* (I. Thomas), *E. subelegans* (I. Thomas), *E. eximius* (I. Thomas), and *E. gradatus* Campbell, has long been distinguished from the group similar to the type species (Campbell, 1956, p. 474). Campbell considered that because there were only minor internal differences between the two groups there was no reason to separate them generically. The precise morphology of the cardinal process of the type species is still obscure, but a specimen of *E. punctatus* from Castleton on the Isle of Man (British Museum B 10531) has a bilobate internal face, and appears to be dorsally recurved as in the specimens of *E. alternatus* (Norwood & Pratten) illustrated by Muir-Wood & Cooper (1960, pl. 83, figs 5-9). If subsequent work confirms that the process of *punctatus* is dorsally recurved, then there is a significant difference between the *elegans* and *punctatus* groups: the cardinal process of specimens of *E. elegans* (M'Coy) from Settle, Yorkshire, England (British Museum B 20459), is short bilobed internally, trilobed externally, and is not recurved. The processes of *E. biseriatus* (Hall) and *E. gradatus* Campbell have a morphology similar to that of *elegans*. Future work may distinguish the *elegans* group as a subgenus of *Echinoconchus* typified by its small size, lack of a fold and sinus, a small sessile cardinal process, and a dorsal median septum which frequently, although not invariably, arises a short distance in front of the cardinal process.

The *elegans* group of species is distinguished from *Karavankina* Ramovs, 1966, which at present appears to lack a valid type species, by the absence of two diverging buttress plates supporting the cardinal process, more irregular and larger spines, the absence of a median sinus, smaller ears, and a smaller ventral umbo (Ramovs, 1966; Winkler Prins, 1968). Sarycheva (1968, p. 94), on the other hand, includes many *elegans*-group species in *Echinoconchus (Karavankina)*. Further comment is not warranted until the type species of *Karavankina* is described.

ECHINOCONCHUS GRADATUS Campbell

(Pl. 24, figs 16-20; Pl. 29, figs 1-14)

1956 *Echinoconchus gradatus* Campbell, p. 474, pl. 49, figs 14-18.

1964 *Echinoconchus gradatus* Campbell, Roberts, pl. 5, figs 6-8.

Description: External. Pedicle valve elongate-oval, moderately to strongly convex; relatively evenly convex in lateral profile, greatest convexity usually at umbo; valve longer than wide, widest and highest at midlength, with hinge approximately two-thirds width of valve; umbo narrow, tapering posteriorly, projecting behind hinge (2 mm on a valve 13.5 mm wide and 15 mm long) and moderately to strongly incurved; umbonal shoulders concave, passing laterally into small flat ears; venter evenly rounded; flanks and trail moderately to steeply inclined (becoming flared on larger specimens); ornament of successively raised concentric bands bearing a row of elongate spine bases, five to six per 10 mm on midpart of valve; concentric bands becoming crowded on posterolateral margins; smaller spine bases not observed; density of spine bases not measurable.

Brachial valve rounded-quadrate, moderately concave, particularly at umbo; visceral disc deepest around midpart, gradually rising on posterolateral margins to small flat ears; valve non-geniculate; concentric ribs concave, having raised overlapping anterior margins, and bearing both fine and coarse spines; ribs more closely spaced than on pedicle valve, having a density of five to six per 5 mm on midanterior part of valve.

Internal. Pedicle valve with a pair of raised linear adductor muscle scars extending 5.5 mm along midlength of a valve 11 (est.) mm long and 12.5 mm wide; internal surface ornamented with impressions of external concentric bands.

Brachial valve. Cardinal process sessile, bilobed internally, with two short convex lobes separated by a deep median furrow; external face trilobed, with long narrow dorsally facing lobes; lateral ridges curving posteriorly at base of cardinal process, extending parallel with hinge, and then appearing to curve anteriorly across posterolateral margins; inner lobes of cardinal process supported by two ridges extending to posterior of adductor muscle scars; adductor scars elongate, trigonal, and elevated above floor of valve; median septum narrow, arising from in front of cardinal process and extending a short distance in front of muscle field; internal surface ornamented with strong concentric ribs.

Measurements	Length (pedicle valve)	Length (brachial valve)	Width	Height
CPC 10927	18	15	17	4 est.
CPC 10928	15	—	15	6
CPC 10929	12.5	—	13	5
CPC 10930	—	18	23	—
CPC 10931	—	12	12	—
CPC 10934	20	—	18.5	8
CPC 10926	20	15	17	13

Remarks: The specimens described above are from a friable medium-grained quartz sandstone, and a ferruginized limestone; they have poorly preserved spine bases on the exterior of the shell. From the two well-preserved brachial valve interiors available for study, this material differs from the eastern Australian specimens described by Campbell (1956) and Roberts (1964) in having the extremities of the lateral ridges cutting across the ears. I do not consider this to be of specific significance.

A comprehensive comparison of *E. gradatus* with members of the *elegans* group was given by Campbell (1956, p. 476). *E. gradatus* was compared most closely with *E. biseriatus* (Hall) and the specimens of *E. elegans* (M'Coy) described by Paeckelmann (1931, pp. 161-65, pl. 16, figs 6-8).

Specimens of *E. biseriatus* (Hall) figured by Muir-Wood & Cooper (1960, pl. 52, figs 4-6) as *Stegacanthia biseriata,* and others in the U.S. National Museum, have a cardinal process in which the inner face is inflated anteriorly, and has a weak posteromedian furrow supporting two dorsally inclined lateral lobes. In the specimens of *E. gradatus* from the Bonaparte Gulf Basin, the process is not swollen anteriorly, and the inner lobes are separated by a strong furrow extending posteriorly from the end of the median septum. *E. biseriatus* ranges from the

Keokuk Limestone to the St. Louis Limestone, and has been recorded from the Salem Limestone (Weller, 1914) and members ?2 and 3 of the Paradise Formation of Arizona (Hernon, 1935). Specimens in the collection of the U.S. Geological Survey and the U.S. National Museum are from the Keokuk Limestone, Iowa, the Salem Limestone, Indiana, the Warsaw Limestone, Missouri and Alabama, and the lower St. Louis Limestone, Illinois. Paeckelmann's specimens of *elegans* came from the Visean of Germany.

The morphology of the cardinal process, the posterior part of the median septum, the ridges supporting the lateral lobes on the inner face of the cardinal process, and the extension of the lateral ridges around the ears in specimens of *E. elegans* (M'Coy) figured by Chao (1927, pl. 6, figs 1-3) from the Visean Choniukou Formation of Kansu, China, is closely comparable with that in this material. The Chinese specimens are distinguished by a more circular outline and fewer concentric bands on the exterior of both valves.

E. gradatus Campbell is a junior secondary homonym of *Productus gradatus* Swallow, 1863 (M. Gordon Jr, pers. comm.). I propose to continue using *E. gradatus* Campbell because: (1) *Productus gradatus* Swallow is a junior synonym of *E. vittatus* (Hall) (Weller, 1914, p. 138); (2) *Productus gradatus* was not figured; and (3) Swallow's name can be classed as a forgotten name (International Code of Zoological Nomenclature Article 23b).

Occurrence: Locality 306, Burvill Beds, Burvill Point. Locality 458/12, Point Spring Sandstone, 740 feet above the base of section 458, 4.25 miles northeast of Point Spring near the top of the Weaber Range. In cores 23 (6697-6718 feet) and 27 (7752-7763 feet) in the Tanmurra Formation in Kulshill No. 1 Well near Port Keats.

Material: CPC 10926-10934. CPC 10926 from locality 306; CPC 10927-10934 from locality 458/12. Twenty-five specimens, most preserved in friable quartz sandstone.

Age: Visean to Namurian.

Family DICTYOCLOSTIDAE Stehli, 1954

Subfamily DICTYOCLOSTINAE Stehli, 1954

Genus DICTYOCLOSTUS Muir-Wood, 1930, emend. Muir-Wood & Cooper, 1960

Type species: Productus semireticulatus (Martin), 1809, from the Visean of Great Britain, by original designation of Muir-Wood, 1930.

Remarks: The material described below, which has close affinities with members of the Subfamily Dictyoclostinae Stehli, is tentatively referred to *Dictyoclostus* sensu stricto. When compared with the type species *D. semireticulatus,* this form has a pedicle valve with a longer and more convex trail, fewer spines on the flanks, and an extremely weak reticulate ornament; the valve lacks the diagnostic row of hinge spines. The brachial valve is geniculate, has two pairs of adductor muscle scars, and is always more reticulate than the pedicle valve.

Tolmachoffia Fredericks, revised by Sarycheva et al. (1963, pp. 169-70), has a greater number of randomly placed body spines on the pedicle valve, and spinose auricles and a finely reticulate ornament on the brachial valve; the cardinal process is also dissimilar (*see* Sarycheva et al., 1963, fig. 72).

Dictyoclostus? funiferus sp. nov.
(Pl. 30, figs 1-16)

Diagnosis: Pedicle valve convex, rounded-quadrate, with a long strongly incurved coarsely costate trail; sinus shallow; costae with a density of eight to twelve per 10 mm at 20 mm from the umbo; rugae weak, rarely extending across the umbo, and producing a weak reticulate ornament; five to ten low rugae at the posterior of the valve; two to three sub-erect spines on the posterolateral margins, and a band of up to four spines near the anterior margin; occasional smaller prostrate body spines scattered over the remainder of the valve; hinge spines absent; brachial valve moderately concave, geniculate, with a low median fold; rugae generally more prominent than on pedicle valve, especially at the front of the visceral disc; valve aspinose; median septum extending two-thirds the length of visceral disc; adductor muscle scars in two pairs: a round to ovoid dendritic posterior pair, and a smooth swollen anterior pair.

Description: External. Pedicle valve convex, rounded-quadrate (viewed ventrally), flattened across venter, with a long strongly incurved, often wrinkled trail; greatest convexity near umbo; umbo pointed in small specimens to swollen in large individuals, barely incurved, projecting 2 mm over hinge on a valve approximately 35 mm wide; umbonal shoulders concave, produced into small but well-defined convex auricles; flanks steep, extending dorsally behind hinge; greatest width towards front of valve; median sinus narrow, shallow, commencing just in front of umbo, and extending to front of trail; valve thickened posteriorly; ginglymus absent; costae rounded, becoming extremely coarse, more quadrate in profile, and at times irregular on trail; density of costae eight to twelve per 10 mm at 20 mm from umbo; costae increasing by bifurcation and intercalation, bifurcation usually occurring on anterior, and intercalation on posterior of valve; intercostal furrows rounded and always less than half as wide as costae, valve ornamented with fine concentric growth lines; rugae few in number, confined to posterior of valve, generally weak, and rarely extending completely across umbo; five to ten low rounded rugae present on most individuals; two to three sub-erect spines on the flanks usually pointing parallel with hinge, and up to four larger spines, 1.5 mm in diameter at base, in a band towards anterior margin; a few smaller prostrate body spines, arising from costae, scattered over valve.

Brachial valve quadrate, moderately concave, geniculate, with a low median fold commencing at 5 mm from umbo on a valve approximately 35 mm wide; valve deepest towards posterior of visceral disc, with a small rounded concavity at umbo; posterolateral regions sloping gradually upwards into well-defined flat to slightly concave auricles; trail short but well differentiated from visceral disc; small lophidium at base of cardinal process; ornament reticulate; costae with same density as those on pedicle valve, increasing by intercalation and bifurcation; rugae extending across visceral disc, strongest at margins of disc and towards hinge; ten rugae per 25 mm on a valve approximately 35 mm wide; valve aspinose.

Internal. Pedicle valve. Adductor muscle scars narrow, elongate-ovoid, faintly dendritic, and on a short platform; diductor muscle scars flabellate, smooth posteriorly, coarsely striated and frequently deeply impressed into shell anteriorly; posterior of valve thickened and bearing a marginal ridge extending along hinge and across posterolateral extremities; internal surface of visceral disc striated, and inner surface of trail bearing impression of external ornament.

Brachial valve. Median septum buttresses the base of cardinal process, and extends about two-thirds length of valve; septum broadly rounded posteriorly, lower and narrower between posterior adductor muscle scars, and higher and blade-like between anterior adductor muscle scars; adductor muscle scars in two pairs; posterior scars round to ovoid, slightly impressed, or on well-elevated platforms, and strongly dendritic; anterior scars smooth, ovoid, highly convex, and enclosed posteriorly by posterior adductor scars; cardinal process straight, trilobed internally and externally, and with a short internal shaft; on internal face, lateral lobes swollen, well rounded, separated ventrally by a median furrow, and with their inner margins projecting into a short posteriorly sulcate median lobe; external face with a high lamellose wedge-shaped median lobe surrounded by convex lamellose lateral lobes; lateral ridges strong, rounded, curving posteriorly towards base of cardinal process, and extending along hinge; brachial ridges arise at a low angle to horizontal from lateral margins of anterior adductor muscle scars, and enclose relatively smooth ovoid brachial discs.

Measurements

Pedicle valves	Length	Length (of curvature)	Width (of hinge)	Width (of body)	Height
CPC 8644 Holotype	52	93	40 est.	48.5	28
CPC 8643 Paratype	33	75 est.	30	32	20
CPC 8641 Paratype	—	—	30	34 est.	15

Brachial valve	Length	Width	Length (median septum)
CPC 8638	20	25	15

Remarks: The description of a marginal ridge in the pedicle valve and the brachial markings is based on small specimens from locality 306 in the Burvill Beds (Pl. 30, figs 6, 7). There are no specimens from the type locality in the Utting Calcarenite in which these features are visible. Material from the Burvill Beds and the Point Spring Sandstone in the Weaber Range, near Point Spring, is externally close to the small and medium-sized specimens from the type locality, and has been identified as *Dictyoclostus? funiferus* sp. nov. The internal features of these specimens, especially those from the Point Spring Sandstone, are obscure.

No species closely comparable with *D? funiferus* are known to me. A comparison with the type species, *D. semireticulatus* (Martin), is given on page 125.

The specific name is derived from the Latin *funis* — a rope, and *ferro* — carry, and refers to the rope-like costae.

Occurrence: Localities 107/1, 107/6A (the type localities), 107/7, 108/0, 108/3, 108/4A, 143/1, Utting Calcarenite, Utting Gap. Locality 305/1, 1.5 miles northeast of Ningbing Homestead; localities 306, 453/5, Burvill Point; locality 25, 2.5 miles east-northeast of Point Spring in the Weaber Range; locality 435/0, 5 miles northeast of Point Spring, in the Weaber Range; locality 301, Milligans Hills; Burvill Beds. In Point Spring Sandstone 1 mile east of Point Spring (no locality number). Locality 73/2, Milligans Beds, Spirit Hill.

Material: CPC 8636-8644. Holotype CPC 8644; paratypes CPC 8636, 8638, 8641, and 8643. CPC 8636, 8641 from locality 107/6A; CPC 8637 from locality 305/1; CPC 8638, 8642-8644 from locality 107/1; CPC 8639-8640 from locality 306. More than 100 specimens.

Age: Visean to Namurian.

Family LINOPRODUCTIDAE Stehli, 1954
Subfamily LINOPRODUCTINAE Stehli, 1954
Genus OVATIA Muir-Wood & Cooper, 1960

Type species: Ovatia elongata Muir-Wood & Cooper, 1960, from the late Mississippian (Chester-Lower Fayettville Shale) of Oklahoma, by original designation of Muir-Wood & Cooper, 1960.

OVATIA sp. A
(Pl. 29, figs 21-28)

Description: External. Pedicle valve ovoid, having a long, slightly medianly arched trail, and flaring lateral margins; visceral disc strongly convex and almost at right angles to trail; flanks steep and bearing well-defined rugae near hinge; umbo blunt, broadly rounded, projecting slightly behind hinge, and moderately incurved; umbonal shoulders concave; ears not preserved; costellae high, rounded, separated by grooves of about the same width as each costella, and crossed by fine concentric growth lines; costellae sinuous at times, having markedly angular changes in direction, frequently splitting and rejoining along their length, and increasing mainly by intercalation; costellae giving rise to scattered node-like spine bases mainly on trail, and sometimes bifurcating in front of spine bases; density of costellae sixteen per 5 mm on midpart of trail of a valve 32 mm wide and 33.5 mm long; hinge margin obscure and no hinge spines observed; five rounded rugae extending from hinge on to umbonal shoulders, but becoming completely obsolete on visceral disc.

Brachial valve subquadrate, moderately concave; deepest at about midlength, with a small rounded concavity at umbo; lateral shoulders bearing two divergent rows of small round pits possibly opposite spine bases on pedicle valve; rugae more numerous than on pedicle valve, extending from hinge across visceral disc; rugae strong on lateral margins, but weakening across visceral disc; costellae similar to those on opposite valve, but increasing mainly by bifurcation; valve aspinose.

No internal features have been observed.

Remarks: Ovatia sp. A differs from the type species, *O. elongata* Muir-Wood & Cooper (1960, pp. 312-14, pl. 114, figs 1-4, 7, 11, 12), in having a wider shell with flaring extremities on the flanks and trail, and slightly finer costellae. On the pedicle valve, the trail is shorter, and the umbo is smaller and less incurved. The brachial valve of *O.* sp. A is less convex.

O. laevicosta (White) from the Kinderhook of Iowa is in need of revision because White (1861) did not figure the original specimens. Muir-Wood & Cooper (1960, p. 313) suggest that the specimens figured by Weller (1914, pl. 16, figs 12-15) as *Productus ovatus* Hall may be *Ovatia laevicosta* (White). When compared with *Ovatia* sp. A, Weller's specimens have coarser costae, lack node-like spine bases on the front of the trail, have narrower flanks and a smaller umbo on the pedicle valve, and have a more concave brachial valve.

Productus tenuicostus Hall (1858, pp. 675-76, pl. 24, fig. 2), from the St. Louis Limestone of Illinois, has flaring lateral margins on the pedicle valve comparable with those in *O.* sp. A. Hall's figures will not permit a closer comparison, but specimens from the St. Louis Limestone figured by Weller (1914, pl. 11, figs 1-5) have much coarser costae, and rugae extending across the venter on the pedicle valve.

Occurrence: Locality 7/1, Ningbing Limestone (Tournaisian part), Ningbing Range.

Material: CPC 10940-10942. All from locality 7/1. Five specimens.

Age: Tournaisian.

OVATIA sp. B

(Pl. 29, figs 16-20)

Description: External. Pedicle valve elongate-oval, evenly convex in profile, with steep flanks and a long trail; umbo small, pointed, and very slightly incurved; greatest width towards front of valve; hinge slightly shorter than greatest width; umbonal shoulders moderately concave, bearing three rugae, and produced into short convex rectangular ears; rugae entirely obsolete on venter; costellae regular, rounded, approximately as wide as intercostal furrows, and having a density of fourteen to sixteen per 5 mm at 10 mm from umbo; costellae increasing mainly by intercalation; no spines observed.

Brachial valve quadrate, with a small but abrupt anterior and lateral geniculation forming a short trail; trail 4.5 mm long on a valve 20 mm long and 19.5 mm wide; visceral disc gently concave, linked with a broad concavity at umbo; posterolateral margins flat to slightly concave; rugae well defined, crossing visceral disc, and strongest on posterolateral margins and around margins of visceral disc.

Internal. Pedicle valve interior not observed. Brachial valve. External face of cardinal process with two posteriorly convergent lateral lobes.

Remarks. The few specimens of *Ovatia* sp. B available for study are preserved in medium-grained quartz sandstone; the sandstone is friable, and the fossils are poorly preserved. They warrant description because they are from the youngest outcropping Carboniferous fossil horizon (top of the Point Spring Sandstone) in the Bonaparte Gulf Basin.

Ovatia sp. B differs from the Tournaisian species *Ovatia* sp. A, from locality 7/1 in the Ningbing Limestone, in having coarser and more regular costellae, a more slender and less incurved umbo, a more evenly curved profile, less flaring flanks, and in the apparent absence of spines on the body of the pedicle valve.

Occurrence: Locality 458/12, Point Spring Sandstone, 740 feet above the base of section 458, 4.25 miles northeast of Point Spring in the Weaber Range. Locality 301, Burvill Beds, Milligans Hills.

Material: CPC 10936-10939 from locality 458/12. Ten specimens.

Age: Visean to Namurian.

Family GIGANTOPRODUCTIDAE Muir-Wood & Cooper, 1960

Subfamily GIGANTOPRODUCTINAE Muir-Wood & Cooper, 1960

Gigantoproductinid gen. et sp. indet.

(Pl. 31, figs 1-11)

Description: External. Pedicle valve subovate to subquadrate, moderately convex posteriorly, with a long flat trail, and wide, flat to convex posterolateral extremities; valve highest near umbo, widest at or near hinge; umbo tapered at tip, but widening abruptly anteriorly; umbo large, usually strongly incurved, and flanked by deeply concave umbonal shoulders; venter broad, and rounded to medially flattened; sinus absent; flanks sloping moderately to steeply to flat lateral margins; spines (1) in a single row along the hinge, each spine 0.5 mm in diameter at base, and (2) arising from costellae, infrequently scattered over body of valve, and up to 1 mm in diameter at base; costellae rounded, frequently irregular, increasing by bifurcation and intercalation, sometimes decreasing in number by joining with one another in areas of irregularity; density of costellae ten to fourteen, usually twelve, per 5 mm at 20 mm from umbo; visceral disc ornamented with closely spaced growth lines giving a faintly reticulate ornament; posterior surface of some valves bearing rhomboidal markings produced by two sets of parallel ridges.

Internal. Brachial valve interior with a short median septum; internal surface of valve bearing many radially arranged papillae.

Remarks: This form is referred to the Gigantoproductinae on the basis of its size and shape, costellate ornament, row of spines along the hinge, scattered body spines, and lack of a prominent ventral interarea. There is insufficient internal detail available to refer it to a genus; both the pedicle valve interior and brachial valve exterior have not been observed. Comparison with other species is hampered by the apparent wide range of variation in the shape of this form, and the small quantity of material available for description.

Occurrence: Locality 143/1, Utting Calcarenite at Utting Gap: locality 108/7, Burvill Beds at Utting Gap; locality 435/0, Burvill Beds, 5 miles northeast of Point Spring at the foot of the Weaber Range; and locality 435/400, Point Spring Sandstone, 5 miles northeast of Point Spring in the Weaber Range.

Material: CPC 10943-10949. CPC 10943 from locality 108/7; CPC 10944-10949 from locality 435/0. Ten specimens.

Age: Visean to Namurian.

Productacean gen. et sp. A

(Pl. 18, figs 14-18)

Remarks: The generic affinities of the species described below are difficult to determine because of the lack of specimens showing internal structures. The gross form of the median septum and brachial markings are described from a partially decorticated brachial valve, and one specimen displays the inner face of the cardinal process. The distinctive fine concentric ribs on the rugae separate this form from other productid genera.

Description: External. Pedicle valve subovate, with well-rounded lateral and anterior margins; evenly convex in lateral profile, except for a slight increase in convexity at umbo; greatest width at about midlength; umbo small, rounded, poorly differentiated from body of valve, and barely projecting behind hinge; umbonal shoulders short, gently concave; posterolateral margins rounded, non-auriculate; visceral disc moderately convex, venter evenly rounded, and flanks sloping gently to margins; both trail and sinus absent; rugae small, rounded, crowded on umbo, commencing at hinge, and extending across venter; rugae bearing small scattered spine bases 0.25 mm long; occasional spine bases in troughs between rugae; both rugae and intervening troughs ornamented with narrow concentric ribs having a density of six per 5 mm near front of valve.

Brachial valve with very evenly rounded lateral and anterior margins; visceral disc moderately concave and extending into a rounded depression at umbo; posterolateral margins very broad and flat; trail absent; ornament strongly rugose and bearing regularly developed pits in furrows between rugae; spines and micro-ornament the same as those on pedicle valve.

Internal. Pedicle valve interior not observed.

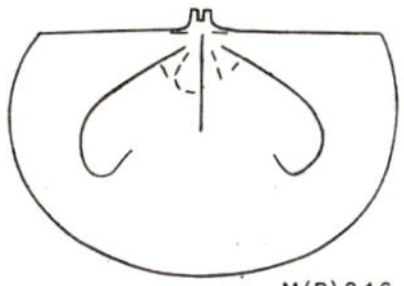

Figure 27. Brachial valve interior of Productacean gen. et sp. A. Details of the muscle field, brachial ridges, and median septum from CPC 8577, a partially decorticated valve, and of the cardinal process from CPC 8576.

Brachial valve (Fig. 27). Median septum extending almost one-third length of valve; brachial ridges originating at about 2 mm from umbo on a valve 25 mm wide and 16 mm long, given off at 35° to horizontal, curving towards anterolateral margins, and enclosing ovoid brachial discs at their extremities; cardinal process sessile, having two small convex lobes separated by a small median furrow on its internal face; lateral ridges slender, and parallel with hinge.

Occurrence: Locality 7/1, Ningbing Limestone (Tournaisian part), Ningbing Range. Locality 101/7B, Burt Range Formation, 5.5 miles west of Mount Septimus.

Material: CPC 8575-8578. All from locality 7/1. Four specimens.

Age: Tournaisian.

Productacean gen. et sp. B

(Pl. 17, figs 1-10)

Description: External. Pedicle valve rounded-quadrate, strongly convex, with a moderately long incurved trail and steep subparallel flanks; greatest convexity towards front of visceral disc, greatest width at about four-fifths length of valve; umbo broad, moderately incurved, extending only a short distance over hinge; venter slightly flattened; umbonal shoulders straight, and normal to hinge; hinge shorter than greatest width of valve, terminating in small flat ears; spine ridges on umbo and posterior of visceral disc elongate and well spaced, passing anteriorly rather abruptly into coarse costae; many costae terminate at front of visceral disc, resulting in trail being ornamented with a few widely spaced costae as well as strong concentric growth lines; costae on visceral disc broad, rounded, and separated by intercostal furrows of equal width; costae increasing by intercalation and bifurcation, having a density of four to five per 5 mm at 15 mm from umbo; about twelve rugae, low and poorly differentiated from one another; rugae strongest near hinge, and extending across umbo and posterior of visceral disc; spines (1) arising from spine ridges at posterior of valve, (2) arising from costae and forming two poorly defined concentric rows (with spines up to 0.5 mm in diameter at base), one across visceral disc, and the other across trail, and (3) in three or four rows, each with about six spines, 0.5 mm in diameter at base, extending from umbonal shoulders down flanks; concentric growth lines coarse, cord-like, extending over entire valve, but most prominent on trail.

Brachial valve quadrate, with a pronounced anterior geniculation; visceral disc strongly concave, the concavity extending evenly from umbo to trail, and deepest near midpoint of valve; posterolateral shoulders steep, passing posterolaterally into flat ears; trail steep, almost normal to visceral disc, and more than 5 mm high on a valve 20 mm wide and 17 mm long; visceral disc ornamented by small quincuncially arranged pits and very weak rugae posteriorly, and by poorly defined costae anteriorly; trail coarsely costate; valve aspinose; three to four rows of pits, opposite the longitudinal rows of spines on pedicle valve, extending down flanks from hinge; about fifteen rugae, strongest at hinge and on the flanks, and extending across umbo only; micro-ornament of cord-like concentric growth lines over entire valve.

Internal. Pedicle valve interior not observed. Brachial valve. Cardinal process bilobed internally and externally; median septum extending to midlength of visceral disc; brachial ridges given off at about 45° to horizontal at one-third length of median septum, and enclosing subquadrate brachial discs.

Remarks: Four specimens of Productacean gen. et sp. B were available for description. The description of the brachial valve interior was made mainly from structures visible through the exterior of the valve.

The specimens referred to *Productus meisteri* Peetz by Nalivkin (1937, pl. 5, figs 12-17), from the Meister Beds (Famennian) of northeastern Kazakhstan, resemble this species in the size and shape of the pedicle valve, and in possessing well-spaced costae on the pedicle valve. The costae on *P. meisteri,* however, do not become obsolete on the trail, and are even more widely spaced than in this species. Deep pits on the brachial valve are more abundant than in Productacean gen. et sp. B, and the valve has stronger rugae, narrower posterolateral shoulders, a less concave visceral disc, and lacks an abrupt anterior geniculation.

Semiproductus irregularicostatus (Krestovnikov & Karpyshev), from the lowest part of the Tournaisian in the Kuznetsk Basin (Sarycheva et al., 1963, pl. 13, figs 8-12), has a similar style of ornament on the pedicle valve, with spine ridges passing anteriorly into costae which in turn become obsolete on the trail. *S. irregularis* has stronger rugae on the visceral discs of both valves.

Occurrence: Locality 146/11, Buttons Beds, 7 miles northwest of Mount Septimus.

Material: CPC 8653-8655. All from locality 146/11. Four specimens.

Age: Famennian.

Order RHYNCHONELLIDA Kuhn, 1949
Superfamily RHYNCHONELLACEA Gray, 1848
Family TRIGONIRHYNCHIIDAE McLaren, 1965
Genus SEPTEMIROSTELLUM nov.

Type species: Camarotoechia septima Veevers, 1959b, pp. 12-13, pl. 4, figs 11-15, 19, and 21, from the Septimus Limestone at Mount Septimus.

Description: Shell small to medium-sized, rounded- or oval-triangular, with a low apical angle; fold and sinus developed anteriorly; commissure uniplicate and serrated; shell material thin; costae few in number, covering entire shell, narrow and having rounded crests posteriorly, higher and more angular anteriorly; pedicle valve gently convex to nearly flat, with a well-defined umbo; brachial valve more strongly convex but not inflated; dental plates short, with concave inner surfaces in transverse section; median septum extending about one-half the length of valve; inner hinge plate thin, covering anterior half of septalium, extending along inner margins of outer hinge plates, and having a long narrow apical foramen; outer hinge plates concave, and flanked by high inner socket ridges.

Remarks: Septemirostellum gen. nov., from present work, is known definitely from the Tournaisian, but may range throughout the Lower Carboniferous. The early Carboniferous genus *Rossirhynchus* Gaetani (1964), from the early or middle Tournaisian of Iran, resembles *Septemirostellum* in having a relatively small number of angular costae. *Rossirhynchus,* however, is much larger, more inflated, has a thicker shell, and costae which are much wider and are separated by narrow intercostal furrows. When compared with *Septemirostellum* internally, it has a shorter septalium supported by the posteriormost part of the median septum only, a thicker inner hinge plate which always seems to be preserved, straight dental plates (in transverse section), and deeply impressed muscle scars. Gaetani (1964, p. 638) drew attention to the swollen preumbonal region of *Rossirhynchus* and used it as one of the main distinguishing features of the genus. I have observed similar preumbonal swellings in some specimens of *Septemirostellum amnicum* (Veevers) and *Grammorhynchus eganensis* (Veevers), and suggest that this is essentially a growth retardation caused by a change in the environment; hence I would hesitate to use it as a generic character. This does not mean that the inflation of *Rossirhynchus* lacks significance; even neglecting the preumbonal swelling, *Rossirhynchus* is still much thicker than *Septemirostellum.* Two small specimens of *Rossirhynchus adamantinus* (Gaetani, 1964, pl. 47, figs 5, 6) appear to have a larger number of costae on the apical parts than are present on the other illustrated specimens. However, because the larger individuals all appear to be decorticated, it is difficult to determine whether the small specimens are juveniles of *R. adamantinus* or belong to another species. Of the species referred to *Septemirostellum,* the closest to *Rossirhynchus* is *S. amnicum* (Veevers). This species is moderately large in size, and has wider costae than the type species, both characters being reminiscent of *Rossirhynchus. S. amnicum* has a thinner shell, and is smaller and much less inflated than *R. adamantinus.* Internally, *amnicum* has a more delicate inner hinge plate covering a relatively longer septalium, and dental plates which are strongly concave internally in transverse section.

Sinotectirostrum Sartenaer (1961a), so far known only from the Famennian of North America and the Soviet Arctic, resembles *Septemirostellum* in having a delicate inner hinge plate covering the front of the septalium, and a similar convexity in the pedicle valve. It is distinguished from *Septemirostellum* by its smaller beak, a larger number of lateral and median costae, a wider sinus, larger straight (in transverse section) dental plates, and a much higher median septum in the brachial valve.

Cupularostrum Sartenaer (1961b), from the Givetian, is smaller, has a more convex pedicle valve, a less transverse outline, a relatively thicker shell with deeply impressed muscle scars, a high median septum, and a thick inner hinge plate which projects in front of the septalium.

Apart from the type species, the only others that can be assigned with certainty to *Septemirostellum* are *S. simplex* sp. nov. from the middle and upper part of the Burt Range Formation, *S. amnicum* (Veevers) from the Enga Sandstone, and *Camarotoechia mitcheldeanensis* Vaughan (1905, pp. 302-3, pl. 26, fig. 7) from the Tournaisian of the Avon Gorge section in Britain. *S? tereticostum* sp. nov. from the lower part of the Burt Range Formation is tentatively referred to *Septemirostellum.* Other species which have yet to be described in detail and which may belong to *Septemirostellum* include *Camarotoechia biplex* (Tolmachoff) described by Sarycheva et al. (1963) from the early and middle Tournaisian of the Kuznetsk Basin, USSR; *C. acutirugata* (de Koninck, 1887) from the Tournaisian of Belgium and the late Tournaisian of the Moscow Basin (Sarycheva & Sokolskaya, 1952); *C. chouteauensis* Weller (1914) from the Chouteau Formation of Missouri, USA; and possibly *C. metallica* (White), *C. tuta* (Miller), *C. inequa* Shaw, and *C. tidwellae* Shaw from the Madison Group of Montana and Wyoming, USA. *C. tuta* (Miller) has previously been described from the Chouteau Formation, Missouri, by Weller (1914) and Branson (1938), and from the Lake Valley Formation, New Mexico, and the lower Burlington Formation, Missouri, by Weller (1914).

SEPTEMIROSTELLUM SEPTIMUM (Veevers)
(Pl. 32, figs 1-20)

1959b *Camarotoechia septima* Veevers, pp. 12-13, pl. 4, figs 11-15, 19, 21.
Not *Camarotoechia septima* Veevers, 1959b, pl. 4, figs 1-10.

Diagnosis: Shell rounded to oval-triangular, commissure moderately uniplicate in the adult; ventral umbo forming an apical angle of 90°; ventral median sinus shallow anteriorly, bearing two to five, usually three, costae; each flank of the pedicle valve with from seven to nine costae.

Description: External. Shell of small to medium size, rounded- or oval-triangular, unequally biconvex, and swollen towards front in older specimens; commissure serrated and uniplicate; greatest width at midlength; costae commencing at umbo, narrow and having rounded crests posteriorly, higher and more angular anteriorly, and separated by broad rounded intercostal furrows.

Pedicle valve gently convex, greatest convexity at umbo or at front of median sinus; umbo narrow, slightly to moderately incurved, having a round to ovoid apical foramen, and subtending an apical angle of 90°; delthyrium open; umbonal shoulders steep, and abruptly separated from small concave palintropes; flanks gently convex, each bearing seven to nine costae; median sinus commencing at midlength of valve, projecting into a broad dorsally curved lingual extension; sinus bearing two to five, usually three, costae.

Brachial valve transverse, more strongly convex than pedicle valve, especially on anterior flanks of larger individuals; umbo flat or with a small median depression, elevated on shallowly concave umbonal shoulders; flanks moderately convex, steeper at front of large valves, each bearing six to eight costae; fold commencing at midpoint of valve, strongest and curving ventrally near anterior margin, and bearing three to six, usually four, costae.

Internal (Fig. 28). Pedicle valve. Dental plates short, inner surfaces slightly concave in transverse section, and flanking lateral margins of pedicle cavity; teeth small, rounded, and expanding anterolaterally; pedicle cavity short, having a rounded floor; adductor muscle scars subrectangular, pointed and impressed posteriorly, and broader anteriorly.

Brachial valve. Median septum triangular in cross section posteriorly, slender anteriorly, extending to midlength of valve; septalium amphora-shaped, covered at anterior and along lateral margins by inner hinge plate; inner hinge plate thin, having an elongate apical foramen; outer hinge plates concave, flanked postero-

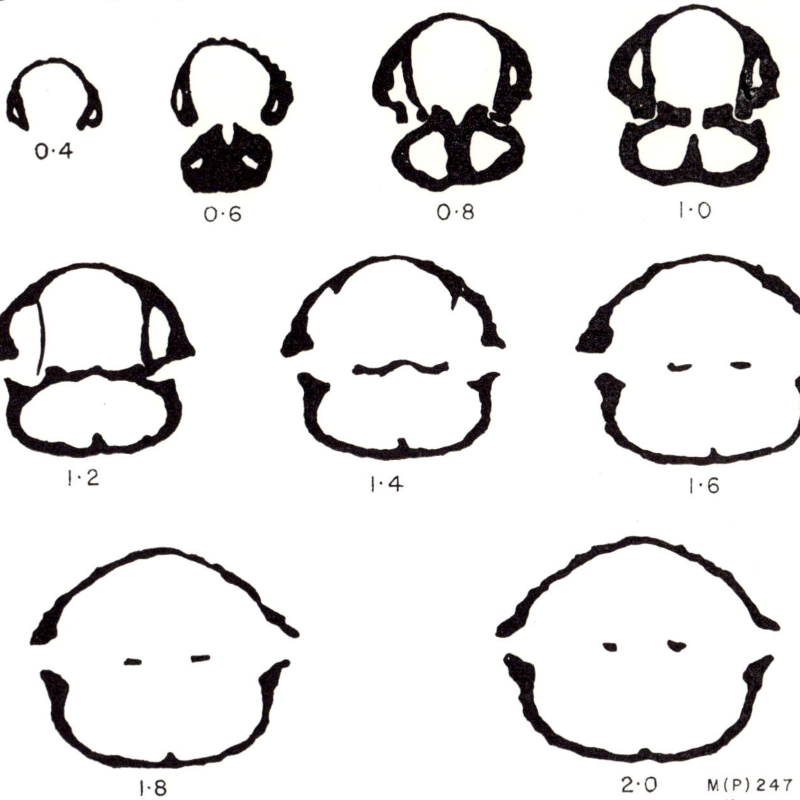

Figure 28. Transverse serial sections of a topotype specimen of *Septemirostellum septimum* (Veevers). CPC 11087, locality 104/11A, Septimus Limestone, Mount Septimus. × 3.

laterally by high rounded inner socket ridges; sockets deep, expanding antero-laterally, with rounded floors and robust inner socket ridges; crural bases extending along inner margin of outer hinge plate, then curving ventrally into pedicle valve interior, adductor muscle scars in long slightly impressed pits on either side of median septum.

Measurements	Length (pedicle valve)	Length (brachial valve)	Width	Height
CPC 8487	9.5	8	10.5	6.5
CPC 8488	8	7	8.5	6
CPC 8489	6.5	5.5	7	4
CPC 8490	—	5	5.5	—
CPC 8494	8.5	—	9.5	—

Remarks: Septemirostellum septimum (Veevers) is morphologically closest to *S. simplex* sp. nov. from the upper part of the Burt Range Formation, Bonaparte Gulf Basin. A comparison with *S. simplex* is given on page 140. Specimens from

J

Sandy Creek, originally assigned to *septimum* by Veevers (1959b, pl. 4, figs 1-10), are now tentatively regarded as *S. simplex* sp. nov. They differ from *septimum* in having fewer costae on the flanks (five to seven compared with seven to nine in *septimum*), a stronger median dorsal deflection of the anterior commissure, and a deeper median sinus.

S. mitcheldeanensis (Vaughan) from the Tournaisian (K, Z, and basal C Zones) of the Avon Gorge Section in Britain (Vaughan, 1905, pp. 302-3, pl. 26, fig. 7) resembles *septimum* in having a comparable shape, costae, and an anteriorly covered septalium, but has fewer costae on the flanks (five to six as against seven to nine in *septimum*), and two instead of the usual three costae in the sinus.

S? chouteauensis (Weller), from the Chouteau Formation of Missouri (Weller, 1914, pp. 176-77, pl. 24, figs 34-40), differs from *S. septimum* in having a more inflated brachial valve, and a more strongly defined fold and sinus. Weller's serial section (1914, fig. 5) shows that *S? chouteauensis* lacks an inner hinge plate over the septalium, but this could have been broken away.

S? tidwellae (Shaw) (1962, p. 633, pl. 97, figs 32-36) from the Madison Group in the Rocky Mountains, USA, is distinguished from *septimum* by its smaller size, narrower costae, more inflated brachial valve, and deeper sinus. *S? inaequum* (Shaw), from the same location (Shaw, 1962, p. 632, pl. 97, figs 1-9), has an outline resembling that of the topotype specimens described in this Bulletin; *inaequum* is characterized by having two small costae on either side of two large costae in the sinus, and a higher brachial valve. The internal morphology of both *inaequum* and *tidwellae* remains undescribed.

Veevers (1959b) compares *septimum* with *S? acutirugatum* (de Koninck) (1887, p. 141, pl. 16, figs 1-14) from the Tournaisian of Belgium, and a species identified as *acutirugatum* by Sarycheva & Sokolskaya (1952) from the late Tournaisian of the Moscow Basin. He considered the Russian form to be closer to *septimum* because there were two to three costae in the sinus compared with three to four in the Belgian forms. De Koninck's figures show a variation of two to four as opposed to three to four in his description, and hence the difference may not be significant. Additional differences are the deeper median sinus, and larger size of the Belgian specimens.

S? biplex (Tolmachoff), described by Maximova in Sarycheva et al. (1963, pl. 39, figs 5-10) from the early and middle Tournaisian of the Kuznetsk Basin, USSR, resembles *septimum* in outline, and in having seven to nine costae on each lateral slope of the pedicle valve. It is distinguished from *septimum* by a deeper median sinus which bears two instead of the usual three costae, and laterally curving costae adjacent to the front of the fold and sinus. The interior of *biplex* has not been described.

Occurrence: Localities 104/11, 104/11A, 104/12, 104/17, 150/4, 150/5, 150/6, Mount Septimus; localities 110/4-110/7, Mount Zimmermann; localities 112/1, 112/2, Langfield Point; localities 138/1, 138/2, Burt Range Amphitheatre; Septimus Limestone. Locality 110/11, Zimmermann Sandstone, Mount Zimmermann.

Material: CPC 8487-8494, CPC 11087. CPC 8487-8491, 8494 from locality 104/11; CPC 8492-8493 from locality 138/1; CPC 11087 from locality 104/11A. Several hundred silicified specimens.

Age: Tournaisian.

SEPTEMIROSTELLUM AMNICUM (Veevers)

(Pl. 34, figs 1-20)

1959b *Camarotoechia amnica* Veevers, pp. 11-22, pl. 4, figs 18, 20, 22-31.
1959b *Camarotoechia* sp. cf. *C. lucida* Veevers, Veevers, pl. 4, fig. 17.

Description: External. Shell rounded-triangular, dorsally biconvex, wider than long, with greatest width just in front of midlength; greatest thickness at midlength; commissure serrated and uniplicate; costae rounded posteriorly, high, broad and angular anteriorly, and separated by narrow intercostal furrows; furrows rounded posteriorly, deeper and more angular anteriorly.

Pedicle valve most convex behind midlength; preumbonal region swollen in some specimens; flanks gently convex; umbo long, narrow, erect to sub-erect, forming an apical angle of 90-95°; palintropes concave, differentiated from umbonal shoulders by prominent beak ridges; delthyrium open; pedicle foramen small, ovoid, and submesothyrid to mesothyrid; sinus commencing at about one-quarter length of valve, narrow, becoming deep and rectangular in profile anteriorly, and produced into a prominent tongue; sinus containing two, less commonly three, strong angular costae; bordering costae well defined, but always slightly lower than those on their lateral margins; costae numbering six to nine on each flank.

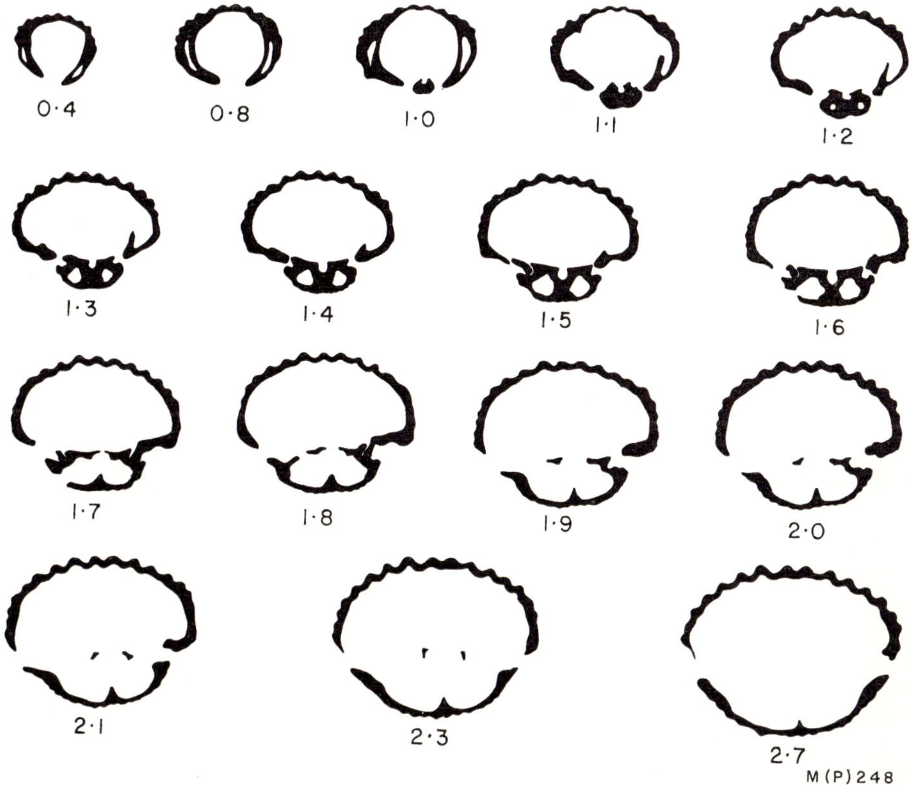

Figure 29. Transverse serial sections of *Septemirostellum amnicum* (Veevers). The inner hinge plate, visible in many silicified specimens, is missing. CPC 11088, locality 102/4, Enga Sandstone, northern end of Enga Ridge. × 4.

Brachial valve subelliptical and moderately convex; greatest convexity at umbo or on anterolateral margins, greatest height at or near front of fold; flanks steep; fold narrow, commencing at one-quarter length of valve, becoming higher anteriorly; fold bearing three, less commonly four, strong angular costae; each flank with five to eight costae.

Internal (Fig. 29). Pedicle valve. Dental plates short, inner surfaces concave in transverse section; plates extending along margins of pedicle cavity; teeth small and rod-like; muscle scars not observed.

Brachial valve. Median septum V-shaped in transverse section, usually broad and low, occasionally narrow and higher, extending one-third to one-half length of valve; septalium amphora-shaped, concave, covered anteriorly; inner hinge plate thin, usually flat, concave over front of septalium, and extending posteriorly along margins of outer hinge plates leaving a narrow elongate apical foramen; outer hinge plates flat to gently concave, bounded by prominent inner socket ridges; outer hinge plates curving ventrally towards front and extending a short distance along crural bases; crural bases forming low ridges along inner margins of outer hinge plates, curving ventrally from distal parts of outer hinge plates; sockets long, narrow, crenulated, expanding anterolaterally, and having a rounded floor; sometimes blocked at front by a thickening extending from lateral parts of outer hinge plates; muscle scars not observed.

Measurements	Length (pedicle valve)	Length (brachial valve)	Width	Height
CPC 8269	14	13	15.5	9.5
CPC 8270	13.5	11	14	7.5
CPC 8271	12	10.5	14	7
CPC 8272	16	13.5 est.	18	—

Remarks: Septemirostellum amnicum (Veevers) was first described from the Moogooree Limestone in the Carnarvon Basin, Western Australia, by Veevers (1959b). Thomas (1962b) subsequently included the species in faunal lists from 140 feet, and from the interval between 430 feet and 500 feet above the base of the Moogooree Limestone. The type specimens came from beds equivalent to those in the interval 478 feet to 488 feet in the type section (Veevers, 1959b, p. 12). Specimens from the Enga Sandstone in the Bonaparte Gulf Basin are larger than those from the type locality. Internally they have a stronger median septum, and broader hinge plates, presumably because of their larger size, but in all other respects they are closely comparable with the type specimens. The specimens referred to *Camarotoechia* sp. cf. *C. lucida* Veevers by Veevers (1959b, p. 17) are now referred to *S. amnicum*. The locality from which they were collected, locality E of Utting (1957), is the same as locality 102/4, and is regarded as being within the Enga Sandstone, not the Burt Range Formation.

S. amnicum differs from *S. septimum* (Veevers), the type species, in its large size, less inflated brachial valve, wider costae, and narrower intercostal furrows. The fold and sinus of *S. amnicum* are narrower and more strongly defined, commence at about one-quarter the length of the shell compared with about half the length in *S. septimum,* and the sinus usually has two instead of three costae.

The internal structures of most species comparable with *S. amnicum* have not been described, so comparisons are based wholly on external features. Externally, one of the closest species is *'Camarotoechia' acutirugata* (de Koninck), which I have tentatively referred to *Septemirostellum* (p. 134).

Specimens referred to this species by Sarycheva & Sokolskaya (1952, pl. 46, fig. 252) from the C_1tsch (late Tournaisian) of the Moscow Basin are distinguished by the possession of fewer costae on the flanks, and a deeper median sinus. Those from the Tournaisian of Belgium (de Koninck, 1887, p. 141, pl. 16, figs 1-14) have a more inflated pedicle valve, and a wider and deeper median sinus.

'Camarotoechia' kinlingensis Grabau, described by Chu (1933, pp. 25-28, pl. 4, figs 21, 22) from the Kinling Limestone of China, has a comparable external appearance except for a raised median costa in the sinus, a more inflated pedicle valve, and a wider fold and sinus.

The species referred to *Camarophoria? moshihlingensis* Ozaki (1939, pp. 248-49, pl. 40, fig. 2) from the Shi-teng-tzu Limestone, Hunan, China, has a comparable exterior except for the greater number of costae in the sinus.

Occurrence: Locality 102/4, northern end of Enga Ridge; localities 103/17B, 103/19, 109/8, middle part of Enga Ridge; localities 139/1, 139/2, eastern dip slope of Enga Ridge, 1.3 miles northwest of Mount Zimmermann; Enga Sandstone. Locality 210/6, unnamed Tournaisian breccia, Waggon Creek valley.

Material: CPC 8269-8275, CPC 11088. All from locality 102/4. More than 100 free specimens, and some casts and moulds preserved in fine-grained sandstone.

Age: Tournaisian.

SEPTEMIROSTELLUM SIMPLEX sp. nov.

(Pl. 33, figs 21-35)

1959b *?Camarotoechia septima* Veevers, pl. 4, figs 1-10.

Diagnosis: Shell small, changing from elongate in the juvenile to transverse in the adult; commissure nearly straight in small specimens to strongly uniplicate in large individuals; costae high anteriorly, and those immediately adjacent to the front of the fold and sinus curving laterally; one to three, nearly always two, costae in the sinus, and six to seven costae on each lateral slope of the pedicle valve; ventral beak elongate, forming an apical angle of 70-80°, and having long palintropes; sinus deep anteriorly.

Description: External. Shell small, rounded-triangular, changing from elongate in juvenile to transverse in adult; shell almost equally convex in juveniles and specimens of moderate size (5 mm long and 5 mm wide), to dorsibiconvex in large individuals (8.5 mm long and 10 mm wide); greatest width at midlength, greatest height at midpoint; commissure serrated, nearly straight in small specimens, to strongly uniplicate in adult; costae commencing at apex of shell, narrow and rounded posteriorly, becoming more angular and higher but still retaining rounded crests anteriorly, and separated by rounded intercostal furrows; costae immediately adjacent to front of fold and sinus usually curving laterally.

Pedicle valve having a narrow elongate beak subtending an apical angle of 70-80°; umbo moderately incurved, with a round to ovoid apical foramen; umbonal shoulders separated by prominent beak ridges from narrow elongate concave palintropes; valve slightly convex, most convex at front of sinus; greatest height at midpoint; each flank bearing six to seven costae; sinus commencing at or just behind midlength of valve, deep anteriorly, forming a prominent lingual extension; floor of sinus bearing from one to three, nearly always two, rounded to angular costae.

Brachial valve more strongly convex than pedicle valve, greatest convexity at anterolateral margins; umbo broadly rounded, flanked by high umbonal shoulders, frequently with a shallow median concavity behind fold; flanks gently convex medially, steeper towards margins, each bearing five to six costae; fold commencing at one-third to one-half length of valve, bearing two to four, usually three, costae, and highest near two-thirds length of valve.

Internal. Pedicle valve. Dental plates short, slightly divergent and concave outwards, extending along pedicle cavity to immediately beneath teeth; teeth rounded, extending anterolaterally, bordered posterolaterally by an articulatory furrow; muscle field usually not observed; in a thicker-shelled specimen from locality 133/2, muscle field rounded and impressed between dental plates, becoming wider, less impressed, and extending to near midlength of valve.

Brachial valve. Median septum thickened near front of septalium, but slender along remainder of length to just short of midpoint of valve; septalium amphora-shaped, covered anteriorly by a fragile inner hinge plate; hinge plate having an apical foramen, usually not preserved; outer hinge plates concave, flanked laterally by high rounded inner socket ridges; crural bases, along inner margins of outer hinge plates, extending a short distance anteriorly, and then curving ventrally; sockets elongate, with rounded or flat grooved floors; muscle scars not observed.

Measurements	Length (pedicle valve)	Length (brachial valve)	Width	Height
CPC 10950 Paratype	3.5	3	3	2
CPC 10951 Paratype	5	4.5	5	3
CPC 10952 Holotype	7	6	7.5	4.5
CPC 10953 Paratype	8.5	7.5	10	5.5
CPC 10956 Paratype	5	—	5.5	—

Remarks: Septemirostellum simplex sp. nov. from the upper part of the Burt Range Formation is morphologically close to, and was probably the fore-runner of, *S. septimum* (Veevers) from the Septimus Limestone (Veevers, 1959b, pp. 12-13, pl. 4, figs 11-15, 19, 21). *S. simplex* is distinguished from *septimum* by its longer ventral beak and palintropes, narrower apical angle, deeper median sinus, and fewer costae in the sinus and on the flanks.

In the sinus there are usually two costae in *simplex* compared with usually three in *septimum,* and on each flank of the pedicle valve there are six to seven costae in *simplex* compared with seven to nine costae in *septimum.* The costae in *simplex* are slightly more angular and higher anteriorly than those in *septimum,* and at the front of the shell, costae on either side of the fold and sinus curve laterally in *simplex,* but are straight in *septimum.*

Specimens from Sandy Creek in the Bonaparte Gulf Basin, referred by Veevers (1959b) to *Camarotoechia septima* and now tentatively placed in *S. simplex* sp. nov., have five to seven costae on each flank of the pedicle valve, and three to four, usually three, costae in the sinus. They differ from the type specimens of *simplex,* and material of this species from elsewhere in the Burt Range Formation, in having a larger number of costae in the sinus, a slightly larger size, a more swollen anterior, and costae adjacent to the fold and sinus which do not curve laterally.

Of overseas species, *S. simplex* is closest to *S. mitcheldeanensis* (Vaughan) from the Tournaisian (K, Z, and basal C Zones) of the Avon Gorge section in Britain (Vaughan, 1905, pp. 302-3, pl. 26, fig. 7). The number of costae in the sinus is the same in both species, and on the flanks of the pedicle valve there are five to six in *mitcheldeanensis* and six to seven in *simplex*. *S. mitcheldeanensis* differs in having costae which are straight rather than curving laterally near the front of the valve, a narrower median sinus, and a shorter beak.

S? acutirugatum (de Koninck) from the Tournaisian of Belgium (de Koninck, 1887, p. 141, pl. 16, figs 1-14) is larger, has a more inflated brachial valve, usually a shorter beak, and in some specimens a greater number (four) of costae in the sinus than *S. simplex*. Comparable features in both species include a deep median sinus, and costae which curve laterally at the front of the shell. The specimens from the late Tournaisian of the Russian Platform referred to *acutirugatum* by Sarycheva & Sokolskaya (1952, pl. 46, fig. 252) are closer to *simplex;* they are distinguished by a shorter beak and a higher fold.

S? biplex (Tolmachoff), described by Maximova in Sarycheva et al. (1963, pl. 39, figs 5-10) from the early and middle Tournaisian of the Kuznetsk Basin, USSR, is externally close to *S. simplex,* being similar in shape, convexity, and in the depth and number of costae in the sinus. There are from seven to nine costae on each flank of the pedicle valve (*simplex* has six to seven), and the costae adjacent to the front of the fold and sinus curve laterally as in *simplex*. The interior of *S? biplex* is not figured, and a close comparison cannot be made.

The specific name refers to the arrangement of the costae, which is simpler than in the type species.

Occurrence: Localities 101/13, 101/15, 101/18-101/21, approximately 4.5 miles west of Mount Septimus; localities 128/1, 128/6 (type locality), middle part of Enga Ridge; locality 133/2, Alpha Hill; and possibly from Sandy Creek (Veevers, 1959b).

Material: CPC 10950-10956. Holotype CPC 10952; paratypes CPC 10950-10951, 10953-10956. All from locality 128/6. More than 100 silicified specimens, and approximately 40 specimens preserved in limestone.

Age: Tournaisian.

SEPTEMIROSTELLUM? TERETICOSTUM sp. nov.

(Pl. 33, figs 1-20)

Diagnosis: Shell transversely subovate with a small beak, and straight inclined cardinal margins; apical angle 105-110°; costae with rounded apices, high and broad anteriorly, flaring ventrally below the general level of the valve on the anterolateral extremities of large specimens, and numbering three to six, usually four or five on each flank of the pedicle valve; one to three, usually two, costae in the sinus; sinus commencing at one-third the length of the valve, becoming deep anteriorly; fold highest at or just behind the anterior extremity; commissure strongly uniplicate.

Description: External. Shell transversely subovate, subequally biconvex in juveniles to dorsibiconvex in adults; commissure serrated, nearly straight in juveniles to strongly uniplicate in adults; greatest width near midlength; costae commencing at umbo, low and rounded posteriorly, much higher and more angular towards front but still retaining rounded crests; intercostal furrows rounded; micro-ornament of closely spaced concentric growth lines.

Pedicle valve. Beak short, poorly differentiated from remainder of valve; cardinal margins nearly straight, almost level with beak, inclined, with beak forming an apical angle of 105-110°; umbo nearly straight, with a small round apical foramen; palintropes concave, bounded by sharp beak ridges; posterior part of valve gently convex; anterolateral margins nearly flat (but strongly ribbed) in specimens of moderate size (8 mm long and 9 mm wide), flaring ventrally in large individuals; valve highest at midpoint or on anterolateral margins; sinus commencing at one-third length of valve, deep anteriorly, forming a high dorsally projecting tongue; floor of sinus with one to three, usually two, low rounded costae; each flank bearing three to six, usually four or five, costae.

Brachial valve. Umbo low, broad, poorly differentiated from posterior of valve; valve highest at or a short distance behind front of fold; valve moderately convex medially, with steep flanks, especially on anterolateral margins; fold commencing at or slightly in front of midlength, bearing two to four, usually three, costae; each flank bearing two to five, usually three or four, costae.

Internal (Fig. 30). Pedicle valve. Dental plates slender, inner surfaces straight or concave, extending along margins of pedicle cavity; teeth club-shaped in transverse section, having a flat articulating surface.

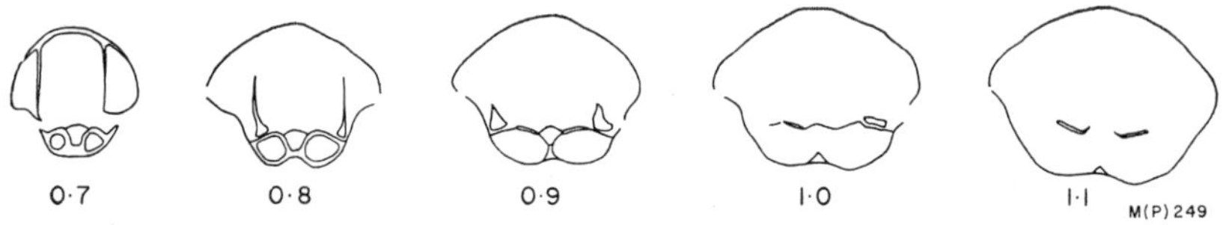

Figure 30. Transverse serial sections of *Septemirostellum? tereticostum* sp. nov. The sectioned specimen, CPC 11089, from locality 100/14, Burt Range Formation, is decorticated externally, and has a coarsely crystalline matrix internally. × 5.

Brachial valve. Septalium short, U-shaped in transverse section, supported on thickened posterior end of median septum; septalium covered by a ventrally arched inner hinge plate; apical foramen not observed in transverse section; outer hinge plates convex posteriorly, becoming concave towards front of septalium; inner socket ridges and sockets poorly defined.

Measurements	Length (pedicle valve)	Length (brachial valve)	Width	Height
CPC 8310 Holotype	8	7	9.5	5
CPC 8311 Paratype	8	7	9	5
CPC 8312 Paratype	8	7	9	4.5
CPC 8313 Paratype	7	6	7.5	3

Remarks: The available specimens of *Septemirostellum? tereticostum* sp. nov. are poorly preserved, being partially decorticated and usually recrystallized internally. It is particularly difficult to determine the morphology of the sockets because of decortication. Serial sections using acetate peels and photographs of calcined specimens (Fig. 30) indicate a possible relationship with the genus *Septemirostellum,* but they should not be relied on for comparisons with other species.

S? tereticostum is recorded from the lower part of the Burt Range Formation only. It is readily distinguished from *Septemirostellum simplex* sp. nov., from the middle and upper parts of the same formation, by its shorter beak, much wider apical angle, and the morphology and number of costae on the flanks. The costae on *S? tereticostum* are wider and more rounded apically, lower posteriorly, higher and flaring ventrally on the anterolateral margins, and are fewer in number on the flanks; on the flank of a pedicle valve there are usually four or five costae in *tereticostum,* compared with six to seven in *simplex.* Partial decortication of some of the illustrated specimens may accentuate the roundness of the costal crests, but specimens with the original thin shell material still preserved have rounded crests on the costae also. These same distinctions apply to *S. septimum* (Veevers) (Pl. 32, figs 1-20), from the Septimus Limestone, which has seven to nine costae on each flank of the pedicle valve, and usually three costae in the median sinus. *S. septimum* is also distinguished by its shallower sinus, and smaller median dorsal deflection of the anterior commissure.

The specific name (derived from the Latin *teres* — smooth or rounded off, and *costa* — a rib) refers to the rounded crests on the costae.

Occurrence: Localities 100/13C and 100/14, Burt Range Formation, 7 miles northwest of Mount Septimus; localities 101/1, 101/1A, and 101/4, Burt Range Formation, 6 miles west of Mount Septimus.

Material: CPC 8310-8313, CPC 11089. Holotype CPC 8310, paratypes CPC 8311-8313. All from locality 100/14. More than 50 specimens.

Age: Tournaisian.

SEPTEMIROSTELLUM? sp.

(Pl. 37, figs 1-8)

Description: External. Shell transversely subovate, biconvex, with greatest width at midlength; commissure uniplicate; costae commencing at umbo, rounded posteriorly, higher and angular anteriorly, and separated by trough-like furrows.

Pedicle valve with a long narrow beak, curved cardinal margins, and an apical angle of 70-80°; umbo slightly incurved; palintropes concave, elongate; delthyrium probably closed by deltidial plates; valve most strongly convex at front of sinus, highest at one-third length, or at ventrally flaring anterolateral margins; sinus beginning at between one-half and two-thirds length of valve, wide, moderately deep anteriorly, with usually four strong angular costae; flanks steep medially, flatter laterally, and flaring ventrally at anterolateral margins; each flank bearing eight costae.

Brachial valve nearly flat posteriorly, but in large specimens swollen anteriorly, and more convex than pedicle valve; greatest height at midlength or near front of fold, greatest convexity at anterolateral margins; umbo low, with a median longitudinal furrow; fold commencing at about half the length of valve, bearing three to five strong costae; flanks steep laterally and anteriorly, each bearing nine costae.

Internal. Pedicle valve. Dental plates extending along margins of pedicle cavity, diverging laterally beneath hinge; teeth not observed.

Brachial valve. Septalium long, narrow, supported by thick posterior end of median septum; septum with a sharp keel, a wide base, and extending to midlength of valve; outer hinge plates large, concave, curved ventrally at distal extremities, bounded posterolaterally by high inner socket ridges; sockets deep and very broad; inner hinge plate either absent or not preserved; adductor muscle field subquadrate, divided by anterior half of median septum, impressed into floor of valve; mantle canals originating from lateral margins of median septum immediately in front of septalium, passing into pitted areas of vascula genitalia in lateral parts of valve.

Remarks: The few specimens of this species collected from the Utting Calcarenite are tentatively referred to *Septemirostellum* gen. nov. on the basis of their angular costae, dorsibiconvex profile, and short dental plates. Some specimens are larger than usual for *Septemirostellum,* the shell is thicker and bears impressed muscle scars and mantle canals, both of which have not been observed in the other species of *Septemirostellum,* and the inner hinge plate is either missing or absent. *Septemirostellum?* sp. is distinguished from *S. simplex* sp. nov., *S. amnicum* (Veevers), and *S. septimum* (Veevers) by the possession of a larger number of costae in the sinus (usually four), ventrally flared anterolateral margins on the pedicle valve, and a thickened shell in large individuals.

The closest species is *S? metallicum* (White) described by Shaw (1962, pp. 631-32, pl. 97, figs 10-26) from the Madison Group of Montana, USA, which has a comparable arrangement of costae on the flanks and sinus, similar ventrally flared anterolateral margins on the pedicle valve, an anteriorly inflated brachial valve in the adult, and a narrowly triangular juvenile stage of growth. Shaw does not record a covering plate on the septalium in *S? metallicum.*

Some specimens of *'Rhynchonella' multirugata* de Koninck (1887, pp. 54-55, pl. 15, figs 68-87, particularly fig. 72) are externally comparable in shape and ribbing with *Septimirostellum?* sp.; others, however, have a greater number of costae in the sinus. *'Rhynchonella' multirugata* is recorded from the Visean (stage III) of Belgium.

Occurrence: Localities 108/0 and 108/3, Utting Calcarenite, Utting Gap.

Material: CPC 10957-10962. All from locality 108/0. Ten silicified specimens, most of which are broken.

Age: Visean.

Genus GRAMMORHYNCHUS nov.

Type species: Camarotoechia eganensis Veevers, 1959b, from the Laurel Formation, Canning Basin, Western Australia.

Description: Shell trigonal and slightly biconvex in juveniles, to elongate, subovate or subrectangular, inflated, and dorsibiconvex in adults; commissure serrated and straight in juveniles, strongly uniplicate in adults; costae coarse, few in number on both flanks and sinus, rounded posteriorly, more angular and higher anteriorly, but still retaining rounded apices; shell thick; ventral umbo narrow, and incurved; palintropes small; pedicle cavity short; dental plates nearly straight in transverse section; brachial valve with short septalium supported on posteriormost part of median septum; septalium covered anteriorly by thin inner hinge plate having elongate apical foramen; outer hinge plates concave; median septum extending to about half length of valve; muscle scars impressed in both valves.

Remarks: Grammorhynchus gen. nov. is characterized by its elongate subovate to subrectangular outline, gibbous profile, and a small number of apically rounded costae on the flanks and in the sinus. Internally, the dental plates are straight in

Figure 31. Transverse serial sections of a topotype specimen of *Grammorhynchus eganensis* (Veevers). CPC 11090, locality KCH 18 of Veevers (1959b), Canning Basin, Western Australia. × 5.3.

transverse section, the short septalium is covered anteriorly by a delicate inner hinge plate, and the muscle scars are impressed into the thick shell. Serial sections of topotype specimens of *G. eganensis* illustrating these features are given in Figure 31.

Rossirhynchus Gaetani (1964), from the early or middle Tournaisian of Iran, has a small number of costae, a thick shell, straight dental plates, and a short septalium. It is distinguished by its more transverse growth form, higher and more angular costae, and a thick inner hinge plate that always seems to be preserved.

Pleuropugnoides Ferguson (1966), which has *Terebratula pleurodon* Phillips from the Visean of England as type species, resembles *Grammorhynchus* in having rounded costae and a gibbous brachial valve. The growth form of *Pleuropugnoides,* however, is transverse rather than elongate, the pedicle valve is flatter and in some cases concave (*see P. greenleightonensis* Ferguson, 1966, pl. 23, figs 7-13, from the Namurian of England), and the outer hinge plates are poorly differentiated from the septalium. Ferguson regards the septalium as being uncovered, but this is difficult to confirm from serial sections alone.

Septemirostellum gen. nov. from the Tournaisian of the Bonaparte Gulf Basin differs from *Grammorhynchus* in its smaller size, different growth form, more angular costae, thinner shell, dental plates which have concave inner surfaces in transverse section, and relatively larger septalium.

The generic name, derived from the Greek *grammodes* — linear, and *rhynchos* — nose, refers to the elongate growth form of adult specimens.

GRAMMORHYNCHUS EGANENSIS (Veevers)

(Pl. 34, figs 21-41)

1935 *Camarotoechia pleurodon* Phillips var. *tripla* Prendergast, p. 20, pl. 2, figs 1-6.
1959b *Camarotoechia eganensis* Veevers, pp. 9-11, pl. 3, figs 4-7.

Description: External. Shell subtrigonal and slightly biconvex in small individuals; elongate, subovate or subrectangular, extremely inflated, and dorsibiconvex in large specimens; greatest width towards anterior margin, greatest height at about midlength; commissure serrated, straight in juveniles, strongly uniplicate in adults; costae coarse, rounded at posterior, higher and subangular anteriorly, but always with rounded crests; intercostal furrows narrow; shell material thick, especially at apex of large specimens.

Pedicle valve elongate, usually inflated around umbo and preumbonal regions, and moderately convex, greatest convexity at umbo or at front of sinus; umbo narrow, incurved over brachial valve, bearing a small ovoid foramen, and subtending an apical angle of 90-100°; umbonal shoulders steep, separated by prominent beak ridges from narrow concave palintropes; delthyrium open; flanks moderately convex in smaller specimens, higher and extremely steep on larger valves; five to nine costae on each flank; sinus broad, moderately deep, commencing at one-quarter to one-third length of large valve, projecting anteriorly into dorsally curved tongue; floor of sinus flat, bearing two to four, usually two or three, high costae; sinus obsolete in small specimens.

Brachial valve subcircular to subovate, more inflated than pedicle valve, with greatest convexity at umbo or at anterior; umbo broad, moderately well defined; umbonal shoulders steep and slightly concave; fold commencing at one-quarter to one-third length of valve, low posteriorly, higher towards front; fold bearing three to five, usually three or four, high costae; flanks moderately steep on small specimens to very steep on adults, each ornamented with from four to eight costae.

Internal (Fig. 32). Pedicle valve. Pedicle cavity short, narrow apically, broadening anteriorly, having an evenly rounded floor, and separated from muscle field by a distinct ridge; dental plates slender, almost straight in transverse section, bordering lateral margins of pedicle cavity, and extending a short distance anteriorly on either side of muscle field; in large specimens, anterior extension of dental plates obscured by callus; teeth small, short, becoming slightly broader anterolaterally; muscle field subovate to subrectangular, expanding anteriorly, and moderately to deeply impressed posteriorly, especially at front of dental plates; diductor muscle scars elongate-oval, separated by narrow adductor muscle scars.

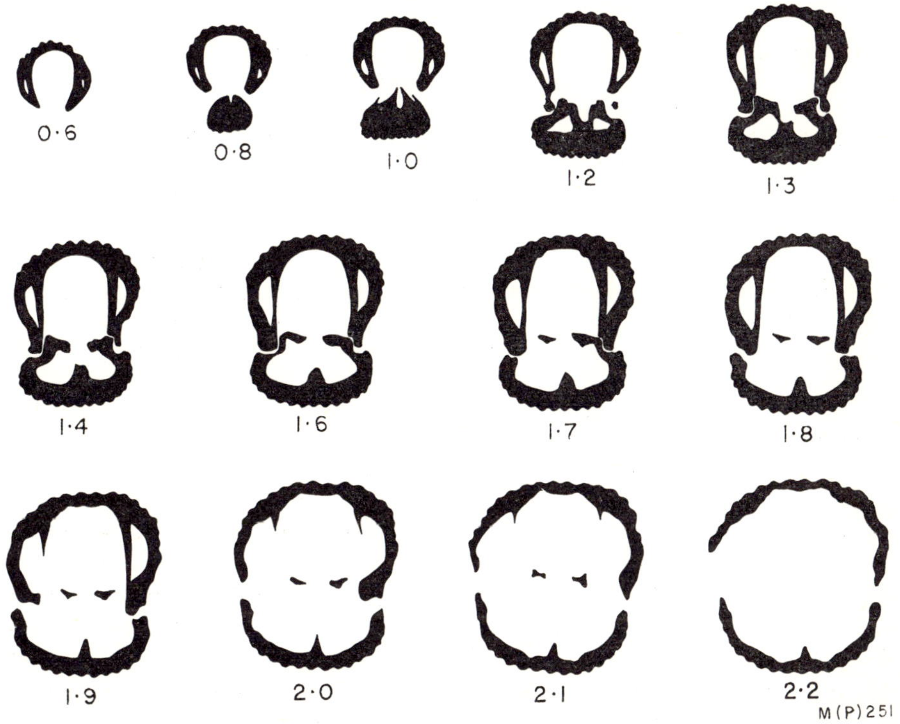

Figure 32. Transverse serial sections of *Grammorhynchus eganensis* (Veevers). The inner hinge plate is missing. CPC 11091, locality 103/12, Burt Range Formation, middle part of Enga Ridge. × 2.

Brachial valve. Median septum triangular in cross section, extending to near midlength of valve, and usually broadest at front of amphora-shaped septalium; septalium deep, U-shaped in cross section, supported by posteriormost part of septum; posterior part of septum and umbonal cavities obscured by apical thickening in large individuals; in these, septalium rests on floor of valve, bears a low keel-like median ridge, and gives off median septum from its anterior margin; inner hinge plate convex, wedge-shaped, covering anterior part of septalium, extending posteriorly along margins of outer hinge plates, and having an elongate foramen over posterior part of septalium; inner hinge plate thin, frequently not preserved; outer hinge plates gently concave, bordered posterolaterally by high inner socket ridges, and distally curving ventrally with crural bases; crural bases robust, projecting anteroventrally; sockets expanding anterolaterally, with rounded floors, and high thick inner socket ridges; adductor muscle scars moderately to strongly impressed, elongate, expanding anteriorly, with narrow rounded posterior margins.

Measurements	Length (pedicle valve)	Length (brachial valve)	Width	Height
CPC 8280	10	—	11	—
CPC 8281	11.5	10	11	5.5
CPC 8282	20	—	16.5	—
CPC 8283	—	19.5	18	9
CPC 8284	13.5	12	11.5	9
CPC 8285	22	21	18	18.5

Remarks: Grammorhynchus eganensis (Veevers) is subtrigonal, moderately convex, and has a straight but serrated commissure in juvenile stages of growth, but becomes elongate, subovate, inflated, and has a strongly uniplicate commissure in large adult specimens. This style of growth differs markedly from that of other Carboniferous rhynchonellids, notably species of *Septemirostellum,* in the Bonaparte Gulf Basin. An analysis of the growth of *eganensis,* following Rudwick (1959), indicates that after the juvenile stage there was an increase in vertical growth, a decrease in radial growth, and increased serial and median vertical deflections; these are assumed to be relative to a constant anterior growth component. Growth series such as this are found in topotype specimens from the Laurel Formation near Egan's Bore in the Canning Basin, and in large collections from localities 103/12, 103/14, and 109/5 in the Burt Range Formation on Enga Ridge. The topotype specimens do not reach the size of some of the large specimens from locality 103/12.

Veevers (1959b, pp. 9-10) contended that the name *Camarotoechia pleurodon* Phillips var. *tripla* of Prendergast (1935) was invalid under Section 114 (2) of the Copenhagen decisions on zoological nomenclature. His reason for invalidating the name was that Prendergast did not intend it to be used in zoological nomenclature, and designated a 'variety' and termed the figured specimens 'morphotypes'; two new species described elsewhere in her paper were based on either a holotype or a group of syntypes, satisfying the rules of zoological nomenclature.

Veevers (1959b, p. 11) compared *eganensis* with specimens of *Camarophoria? isorhyncha* M'Coy, 1844, illustrated by Davidson (1860, pl. 25, figs 1, 2) and de Koninck (1887, pl. 17, figs 1-8). Davidson's specimens, from the Carboniferous of Ireland and England, are closer to *eganensis* than those of de Koninck, but are larger, even more inflated, and have a sinus which commences near the umbo; the three costae in the sinus mentioned by Veevers as an external difference are within the range of variations of *eganensis*. The Belgian specimens are more transverse, particularly at the posterior, have a lower pedicle valve, and a greater number of costae.

Camarotoechia subglobosa Weller (1914, pl. 24, figs 29-33) from the Burlington Limestone of the Mississippi Valley, also compared with *eganensis* by Veevers, resembles small specimens of *eganensis* in outline, but not in convexity; it has lower costae, and is much smaller.

Occurrence: Localities 103/8, 103/9, 103/9A, 103/11, 103/12, 103/14, 109/3, 109/5, 109/6, 109/7, and possibly 128/6, middle part of Enga Ridge; locality 129/4, southern part of Enga Ridge; Burt Range Formation.

Material: CPC 8276-8285, CPC 11090, CPC 11091. CPC 8276-8280 from locality 109/5; CPC 8281-8285, 11091 from locality 103/12; CPC 11091, a topotype from locality KCH 18, Canning Basin. Several hundred specimens, silicified, in limestone, and as casts and moulds in fine-grained sandstone.

Age: Tournaisian.

Genus PTYCHOMALETOECHIA Sartenaer, 1961a

Type species: Rhynchonella omaliusi Gosselet, 1877, by original designation of Sartenaer, 1961a.

Diagnosis: As given by Sartenaer (1961a, p. 7), except that a delicate inner hinge plate covers the septalium.

Remarks: Sartenaer (1961a) originally described *Ptychomaletoechia* as lacking a cover on the septalium, but subsequent sections of the type species (Gaetani, 1965, p. 710; Sartenaer, 1966, p. 7) have revealed a septalial cover. Gaetani described the septalium as 'completely covered', and made no mention of an apical foramen; the exact morphology of the inner hinge plate will remain unknown until Sartenaer publishes the relevant serial sections.

Gaetani (1963) assigned to *Ptychomaletoechia Rhynchonella turanica* Romanowski, 1880, from Kazakhstan, *Camarotoechia baitalensis* Reed, 1922, from Pamir, *P. elburzensis* Gaetani, and possibly *P? deltidialis deltidialis* Gaetani and *P? deltidialis transversaria* Gaetani; all of Gaetani's species are from Iran. Sartenaer (1966) placed *elburzensis* Gaetani in the genus *Ripidorhynchus* Sartenaer, 1966, and in a later publication (Sartenaer, 1967) suggested that the other species may belong to a new genus or subgenus.

PTYCHOMALETOECHIA LUCIDA (Veevers)

(Pl. 36, figs 1-12)

1935 *Camarotoechia pleurodon* (Phillips), Prendergast, pp. 19-20, fig. 4.
1959a *Camarotoechia lucida* Veevers, pl. 10, figs 1-11.
Not *Camarotoechia* sp. cf. *C. lucida* Veevers, Veevers, 1959b, p. 17.

Description: External. Shell rounded-triangular, biconvex in small specimens to dorsibiconvex in large individuals; commissure serrated and nearly straight in juveniles, to strongly uniplicate in adults; greatest width at two-thirds length of shell; costae commencing at umbo; rounded posteriorly, and very broad and angular and with a sharp median keel anteriorly; intercostal furrows narrow.

Pedicle valve. Umbo prominent, slightly incurved, narrow, with steep umbonal shoulders; cardinal margins inclined, slightly curved, and, with umbo, subtending an apical angle of 85-90°; palintropes long, flat, and separated from flanks by strong cardinal ridges; valve highest slightly in front of umbo, most convex at front of sinus; sinus commencing at between one-third and one-half length of valve, wide, deep anteriorly, and in large specimens having strong dorsal median deflection; sinus bearing two to three, usually three, broad angular costae; flanks moderately convex, becoming flat at anterior, and bearing eight to eleven, usually eight, costae.

Brachial valve transversely subovate, moderately convex posteriorly, and in large individuals high and swollen anteriorly; greatest convexity at front of flanks, greatest height at crest of fold, at two-thirds length of valve; umbo broad, with shallow median longitudinal depression, and concave umbonal shoulders; fold commencing at one-third to one-half length of valve, highest at two-thirds length of valve, and then abruptly curving ventrally; costae on fold high and keel-like to highest point, becoming rounded on ventral deflection, and numbering three to four, usually four; flanks moderately convex posteriorly, and in large individuals extremely high and steep anteriorly; costae on flanks numbering six to ten, usually seven.

Internal. Pedicle valve with short divergent dental plates.

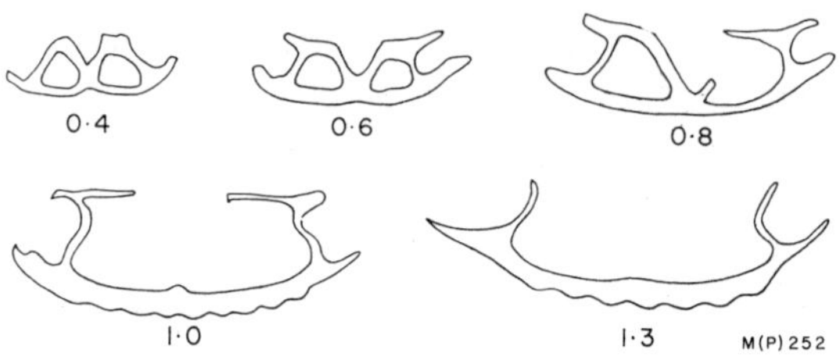

Figure 33. Transverse serial sections of a brachial valve of *Ptychomaletoechia lucida* (Veevers). The inner hinge plate is missing. CPC 11092, locality 140/3, Ningbing Limestone, southern end of Ningbing Range. × 5.

Brachial valve (Fig. 33). Septalium U-shaped in transverse section, uncovered, and supported by posterior end of median septum; septum low and triangular in transverse section in front of septalium, extending one-third length of valve; outer hinge plates flat to slightly concave, bordered posterolaterally by high inner socket ridges; sockets rounded in transverse section.

Measurements	Length (pedicle valve)	Length (brachial valve)	Width	Height
CPC 8646	9.5	9	12.5	5
CPC 8647	8.5 est.	7.5	9.5	4
CPC 8648	—	12	16	11

Remarks: Material from the Ningbing Limestone referred to *Ptychomaletoechia lucida* (Veevers) was obtained by calcining blocks of tough back-reef limestone. The specimens are sufficiently well preserved to be referred to *lucida* on the basis of shape and external ornament, but there are insufficient numbers for an adequate comparison of the interior, especially that of the brachial valve. The costae are broad and angular, and on each flank of the pedicle valve there are from eight to eleven, usually eight, in this material, and from five to eight in specimens from the Canning Basin; the type and figured specimens (Veevers, 1959a, pl. 10, figs 1-11) have eight costae on each ventral flank. Veevers (1959a, figs 52-53) recorded two types of apical structures in the brachial valves of specimens from the Fairfield Formation in the Canning Basin: most specimens had a short median septum supporting a short but well-defined septalium; in others there was an incipient septalium at the posterior extremity of a very short median septum. Sectioned specimens from the Ningbing Limestone resemble the majority group with a well-defined septalium. In the Canning Basin *P. lucida* is recorded from the Fairfield, Napier, Geikie, and Pillara Formations, and from the Bugle Gap Limestone, and probably ranges from the toII-toVI zones of the Famennian; in the Bonaparte Gulf Basin, the species is found in rocks of toII-toIII age (Druce, pers. comm.), and may range as high as the toV zone in section 443 at Jeremiah Hills. The specimens referred to *Camarotoechia* sp. cf. *C. lucida* Veevers by Veevers (1959b, p. 17) are now referred to *Septemirostellum amnicum* (Veevers). The locality from which they were collected, locality E of Utting (1957), is the same as locality 102/4, and is regarded as being within the Enga Sandstone, not the Burt Range Formation.

When compared with the type species, *P. omaliusi* (Gosselet), figured by Sartenaer (1961a, pl. 1, fig. 6) from the Famennian of Belgium, *P. lucida* is slightly larger, has wider and more angular costae, and a more variable development of the septalium in the brachial valve. Both species have a low pedicle valve, and a brachial valve which is swollen anteriorly.

P? deltidialis deltidialis Gaetani (1965, pp. 716-25, pl. 69, figs 2-8; pl. 70, figs 1-4), from the Famennian of Iran, resembles *lucida* in having a similar internal structure (i.e. to most specimens of *lucida*), and a comparable external ornament. As already pointed out by Gaetani (1965, p. 724) *lucida* differs in its smaller size, and smaller deltial plates. The number of costae in specimens from the Ningbing Limestone is comparable to that in *deltidialis deltidialis.*

Further comparisons with overseas species are given by Veevers (1959a, p. 94). Of these, *P. lucida* is morphologically very close to *'Camarotoechia' sobrina* Stainbrook from the Famennian Percha Shale of New Mexico (Stainbrook, 1947, pl. 47, figs 19-22).

Occurrence: Locality 140/3, Ningbing Range; localities 443/22 and 443/23, Jeremiah Hills; Ningbing Limestone (Famennian part).

Material: CPC 8646-8648, CPC 11092. All from locality 140/3. Approximately 50 specimens, nearly all broken, preserved in limestone.

Age: Famennian.

Genus PLEUROPUGNOIDES Ferguson, 1966

Type species: Terebratula pleurodon Phillips, 1836, by original designation of Ferguson, 1966.

Diagnosis: See Ferguson (1966, p. 354).

PLEUROPUGNOIDES sp.

(Pl. 37, figs 9-11)

Description: External. Shell transversely subovate, and dorsibiconvex; commissure serrated and uniplicate; greatest width near posterolateral margins; costae broad and rounded on body of shell, but more angular at front of fold and sinus.

Pedicle valve with a short beak, and nearly horizontal cardinal margins; valve convex posteromedially; sinus broad, shallow, commencing in anterior half of valve, having a strong dorsal median deflection, and bearing three to four, usually four, high costae; flanks nearly flat, but flaring ventrally at their anterolateral extremities.

Brachial valve slightly convex posteriorly, strongly inflated anteriorly, with steep anterolateral margins; fold commencing in anterior half of valve, and bearing four or five costae.

Internal. Pedicle valve. Dental plates short, and divergent.

Brachial valve. Median septum extending nearly half the length of valve, supporting a short septalium; sockets with rounded grooved floors; details of hinge plates not observed.

Remarks: Despite poor preservation, the available specimens of *Pleuropugnoides* sp. are externally close to *P. pleurodon* (Phillips) from the Carboniferous Limestone at Bolland, Yorkshire, England (Ferguson, 1966, pl. 23, figs 1-3). A study on variation by Ferguson (1966) and later work by Parkinson (1969) indicate a considerable degree of variation within *pleurodon,* particularly in the number of costae on the fold, sinus, and flanks. I have been unable to ascertain the variation in *P.* sp. because of the lack of material. Some specimens from Bolland illustrated by Davidson (1860, pl. 23, figs 1, 2) are larger, more transverse, and have a fold and sinus which are stronger, and start nearer the umbo, but others (pl. 23, figs 4, 5) are morphologically closer to *P.* sp. A stronger fold and sinus is found also in material from the Visean (stage III) of Belgium illustrated by de Koninck (1887, pl. 15, figs 1-23). A more pronounced umbo and inclined cardinal margins are additional differences distinguishing the Belgian specimens from *P.* sp.

Occurrence: Locality 306, Burvill Beds, Burvill Point.

Material: CPC 11963. From locality 306. Nine specimens preserved as casts and moulds in ferruginized sandstone.

Age: Visean.

Family ?PUGNACIDAE Rzhonsnitskaya, 1965

Genus NINGBINGELLA nov.

Type species: Ningbingella flexuosa gen. et sp. nov. from reef limestone of Tournaisian age (locality 7/1) in the Ningbing Limestone, Ningbing Range.

Description: Shell large, crescent-shaped, with a flat to gently convex pedicle valve, a strongly convex brachial valve, and a very strong median dorsal deflection; commissure strongly uniplicate, and serrated; fold and sinus commencing at one-half to two-thirds length of shell; apical angle large; ventral umbo small and pointed, dorsal umbo broad and flat; median plicae few in number, commencing from near posterior of fold and sinus, stronger and more angular anteriorly; lateral plicae few in number, restricted to margins; micro-ornament of well-spaced radial lirae; dental plates divergent and concave in transverse section; brachial valve with a long median septum supporting an uncovered septalium; septalium broad and V-shaped in tranverse section; outer hinge plates flat medially, bordered by robust inner socket ridges; articulation strong; crural bases originating from anteroventral parts of septalium and in inner anterior parts of outer hinge plates; in transverse section, crurae crescent-shaped with a median projection immediately in front of septalium; crurae becoming higher, straighter, and having inflected tips anteriorly.

Remarks: Ningbingella is differentiated from *Pugnax* Hall & Clarke, 1893, by the possession of a well-defined median septum and septalium in the brachial valve (compare Figs 34 and 35 with Schmidt, 1965, fig. 6). The type species of *Pugnax, P. acuminatus* (Sowerby), usually lacks costae, and has a high keel-shaped brachial valve, whereas *P. pugnus* (Martin) is externally very close to *Ningbingella flexuosa* gen. et sp. nov.

Shumardella Weller, 1910, is smaller, has a triangular outline, a weaker dorsal median deflection, and a smaller septalium and median septum in the brachial valve when compared with *Ningbingella.* The Famennian genus *Porostica* Cooper, 1955, has a shape similar to *Ningbingella,* but differs in having a micro-ornament of fine capillae separated by rows of fine pits, a much smaller septalium supported by a more slender median septum, and longer dental plates. *Yunanella* Grabau, 1923, a genus with a wide septalium and a strong median septum in the brachial valve, has a smaller apical angle, a smaller dorsal median deflection, a more convex pedicle valve, and a capillate ornament over the whole shell. From *Pugnoides* Weller, 1910, *Ningbingella* is distinguished by its larger size, more transverse outline, larger apical angle, more inflated brachial valve, stronger dorsal median deflection, and the absence of an inner hinge plate over the front of the septalium. The septalium in *Ningbingella* is broader than that in *Pugnoides ottumwa* (White), the type species, figured by Sartenaer (1964, pl. 1) and Schmidt (1965, fig. 21).

Pugnoides boonensis (Shumard) from the Fern Glen Formation and the Burlington Limestone of the Mississippi Valley area, USA, is a possible species of *Ningbingella.*

NINGBINGELLA FLEXUOSA gen. et sp. nov.

(Pl. 35, figs 22-32)

Diagnosis: Pedicle valve with a broad sinus and a very high lingual extension; fold forms a short but pronounced anterior projection; two to three plicae in the sinus, and two to three on each lateral slope on the pedicle valve; radial lirae with a density of three per 1 mm on the lateral margins.

Description: External. Pedicle valve subovate in juvenile growth stages, becoming crescent-shaped in adults because of a strong dorsal median deflection; hinge curving anteriorly, forming rounded cardinal margins; lateral margins rounded; umbo small, very slightly incurved, pierced apically by a small foramen; umbonal shoulders short, gently convex, separated by beak ridges from small concave palintropes; flanks varying from flat or slightly concave to gently convex; sinus very broad, more than half as wide as valve, commencing at one-third to one-half length of valve, forming an extremely high dorsal lingual projection; lateral margins of sinus low and gently convex posteriorly, becoming higher, steeper, and concave towards front; floor of sinus moderately concave; two or three low rounded plicae commencing at posterior of sinus, becoming broader and more angular towards front; flanks mainly smooth, bearing two or three short rounded plicae at anterior extremities; mature specimens with a zig-zag commissure with a strong median dorsal deflection; micro-ornament of concentric growth lines, and fine widely spaced lirae having a density of three per 1 mm on lateral margins of a mature valve; lirae not observed to increase by either bifurcation or intercalation.

Brachial valve strongly convex, much more inflated than pedicle valve, with a broad smooth venter anteriorly, and steep ventrally curving flanks; umbo low, broadly rounded, with very short concave umbonal shoulders; fold commencing at midlength of valve, becoming broader and much higher anteriorly, and jutting out in a squarish projection from front of valve; median plicae low and broadly rounded at posterior of fold, higher and more angular at front; fold ornamented by three or four plicae, the median ones being higher than those on margins; lateral plicae low, broadly rounded, becoming slightly more angular towards commissure; anteriormost plicae long, extending from midpart of valve, and other plicae progressively shorter and less well developed posteriorly.

Internal (Figs 34, 35). Pedicle valve with short slender slightly divergent dental plates; dental plates concave in transverse section, supporting stout rounded teeth.

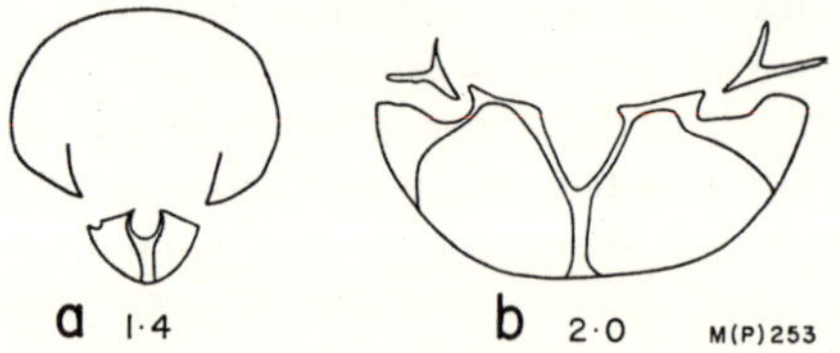

Figure 34. Transverse serial sections of a partially decorticated specimen of *Ningbingella flexuosa* gen. et sp. nov. showing the dorsal median septum and septalium. CPC 11093; a, × 2; b, × 7.5.

Brachial valve. Sockets rounded, tapering posteriorly, with thickened floors; inner socket ridges robust; outer hinge plates narrow apically, and mainly flat; median septum slender, extending one-third length of valve, supporting a short (1.2 mm in length in a sectioned specimen) wide uncovered septalium at posterior of valve; septalium deep and V-shaped in cross section, with slightly curved lateral margins; crural bases arising from inner margins of outer hinge plates and anteroventral parts of septalium; crurae crescent-shaped with a median inwardly directed projection immediately in front of septalium; crurae higher and straighter anteriorly, with gradual loss of median projection.

Ontogeny: The ontogeny of this species is closely comparable with that of its external homeomorph *Pugnax pugnus* (Martin) described by Rudwick (1964, p. 148). Its zig-zag commissure is assigned to subgroup B_1 of group B in the classification of zig-zag commissures proposed by Rudwick (1964).

In juvenile stages the shell is wider than long, and has well-rounded lateral margins, and a curving hinge. The pedicle valve is more arched than the brachial valve, and has a broad convex venter, and slightly convex flanks; the brachial valve has gently convex flanks, and in some cases a slight median depression. A very weak median ventral deflection may have been responsible for the arched nature of the venter on the pedicle valve.

In later stages of growth there is a weak dorsal median deflection which is affected by a gentle undulate serial deflection. The combination of these deflections gave rise to a shallow median sinus and a low fold bearing low rounded plicae. The flanks on the pedicle valve are flat to slightly concave, and those on the brachial valve moderately convex.

Adult growth stages are characterized by an increase in strength of the dorsal median and serial deflections, giving rise to a high dorsal lingual projection bearing high angular plicae. Serial deflections on the lateral parts of the shell form lateral plicae. In late adult stages of growth the serial deflections (median and lateral) underwent 'absolute angulation' (Rudwick, 1964), the plicae became increasingly angular towards the margins, and the flanks became straighter. The resulting shell has steep flanks and a short but protruding fold on the brachial valve, and flat flanks and a long deep sinus on the pedicle valve. Both the fold and sinus are plicate along their entire length, whereas the lateral plicae are mainly confined to the margins of the flanks.

Measurements

	Length	Width	Height
CPC 8307 Paratype	15	21	16
CPC 8308 Paratype	18.5	26.5	20.5
CPC 8309 Holotype	20	28.5	22

Remarks: Pugnoides boonensis (Shumard) described by Weller (1914, pp. 195-97, pl. 25, figs 22-26) from the Burlington Limestone and Fern Glen Formation of the Mississippi Valley area, USA, may belong to *Ningbingella.* It is distinguished from *N. flexuosa* by a more acute apical angle, a shorter fold, less angular

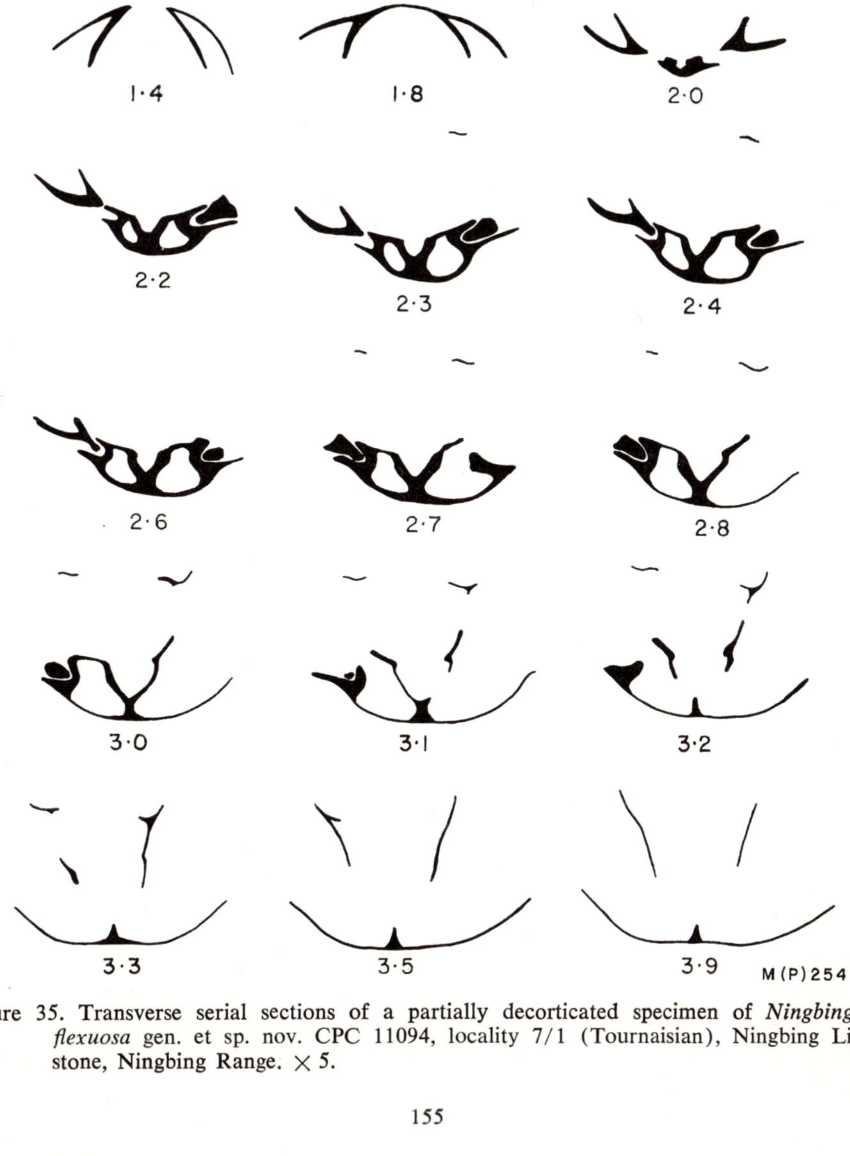

Figure 35. Transverse serial sections of a partially decorticated specimen of *Ningbingella flexuosa* gen. et sp. nov. CPC 11094, locality 7/1 (Tournaisian), Ningbing Limestone, Ningbing Range. × 5.

plications, and the apparent lack of radial micro-ornament. Serial sections (Weller, 1914, fig. 10) show a small median ridge in the pedicle valve which has not been observed in *N. flexuosa,* and a shorter septalium supported on a high slender median septum in the brachial valve.

The specimens of *Pugnax* sp. cf. *P. pugnus* (Martin) described by Veevers (1959a, pp. 110-13, pl. 11, figs 22-25) from the Frasnian Sadler Formation in the Canning Basin, Western Australia, are less transverse posteriorly, have a wider fold and sinus, and lack a median septum and septalium in the brachial valve (Veevers, 1959a, fig. 68).

The name *flexuosa* refers to the strong dorsal median deflection in the adult shell.

Occurrence: Locality 7/1 (Tournaisian part), Ningbing Limestone, Ningbing Range.

Material: CPC 8307-8309, CPC 11093, CPC 11094. Holotype CPC 8309, paratypes CPC 8307-8308. All from locality 7/1. Twenty individual specimens obtained by calcining dense reef limestone.

Age: Tournaisian.

Genus PUGNOIDES Weller, 1910

Type species: Rhynchonella ottumwa White, 1862, from the St. Louis Limestone, Ottumwa, Iowa, USA, by original designation of Weller, 1910.

PUGNOIDES ERUGATUS sp. nov.
(Pl. 36, figs 13-36)

Diagnosis: Shell rounded-pentagonal to subovate; apical angle 90-115°; ventral beak small, and poorly differentiated from the umbonal shoulders; costae broad, rounded; from three to four on each ventral flank, and one or two in the sinus; septalium small.

Description: External. Shell rounded-pentagonal to subovate, moderately dorsibiconvex, usually wider than long; greatest width at midlength; umbonal regions smooth; costae broadly rounded, up to twice as wide as intercostal furrows, and commencing at one-third to one-quarter length of shell; commissure serrated and uniplicate; large specimens lamellose at margins; micro-ornament not observed.

Pedicle valve most convex at umbo and towards front of median sinus; umbo small, moderately incurved, perforated by a small apical foramen, and poorly differentiated from umbonal shoulders; palintropes slightly concave, separated by strong beak ridges from umbonal shoulders; delthyrium triangular, and open; apical angle 90-115°; flanks moderately convex medially, flatter towards margins, each bearing three to four costae; sinus commencing at midlength of valve, deep anteriorly, and produced into a prominent dorsally directed tongue; floor of sinus flat, with one or two prominent rounded costae.

Brachial valve subovate, moderately convex, with a small median depression behind fold; umbo low and broad; flanks flattened at posterolateral margins, steep towards front, with slightly weaker costae than those on pedicle valve; fold commencing near midlength of valve, forming a prominent projection at front.

Internal. Pedicle valve. Pedicle cavity with a flat to slightly rounded floor; dental plates extending slightly in front of pedicle cavity, subparallel on floor of valve, but curving dorsolaterally beneath teeth; teeth and muscle scars not observed.

Brachial valve. Sockets narrow, elongate, having rounded floors, and flanked by low inner socket ridges; outer hinge plates triangular, with a flat to slightly concave ventral surface; crural bases flange-like, originating at anterolateral margins of septalium, projecting ventrally from mid-anterior region of outer hinge plates; broken plate-like tissue on inner margins of outer hinge plates suggests presence of an inner hinge plate; septalium small, concave, supported by posterior of median septum; septum low, narrow, extending approximately half the length of valve; muscle scars not observed.

Measurements	Length (pedicle valve)	Length (brachial valve)	Width	Height
CPC 8314 Holotype	7	6.5	7.5	4
CPC 8315 Paratype	10	9	12	5
CPC 8316 Paratype	6.5	6	6.5	3.5
CPC 8660 Paratype	4.5	4	4.5	2

Remarks: Pugnoides erugatus sp. nov. is characterized by wide rounded costae on the flanks. Those on the type species, *P. ottumwa* (White), described by Weller (1914, pp. 193-95, pl. 25, figs 7-17) from the St. Genevieve Limestone of Illinois, are higher and more angular. In addition, *P. ottumwa* has a smaller apical angle of approximately 90°, and a greater range of variability with from one to five costae in the sinus. The septalium in *erugatus* may be smaller than in *ottumwa,* but it is difficult to compare silicified specimens with serial sections of topotype specimens of *ottumwa* given by Sartenaer (1964) and Schmidt (1964); the septalium in *erugatus* is probably covered anteriorly by a delicate inner hinge plate as in *ottumwa*.

No other morphologically close species are known to the author. The specific name *erugatus* (Latin for clear of wrinkles or smooth) refers to the smooth posterior of the shell.

Occurrence: Localities 101/2 (the type locality), 101/1 and 101/1A, 6 miles west of Mount Septimus.

Material: CPC 8314-8320, CPC 8660. Holotype CPC 8314, paratypes CPC 8312, 8315-8320, and 8660. All from locality 101/2. Ninety silicified specimens.

Age: Tournaisian.

Family CAMAROTOECHIIDAE Schuchert & Le Vene, 1929

Subfamily CAMAROTOECHIINAE Schuchert & Le Vene, 1929

Genus RUGALTAROSTRUM Sartenaer, 1961

Type species: Leiorhynchus madisonense Haynes, 1916, from the Three Forks Formation, Montana, USA, by original designation of Sartenaer, 1961a.

Diagnosis: See Sartenaer (1961a, p. 6, pl. 1, fig. 5; pl. 2, fig. D).

Remarks: Rugaltarostrum australe sp. nov., from the Ningbing Limestone, has stronger lateral costae than any of the previously described species of *Rugaltarostrum*. According to Sartenaer (1961a) the genus either lacks costae, or has rare lateral costae which are mostly restricted to the anterior of the shell. McLaren (1965, p. H583) described the flanks of *Rugaltarostrum* as smooth, probably because costae are not evident on the poorly preserved holotype of *R. madisonense* Haynes illustrated by Sartenaer (1961a, pl. 1, fig. 5). The original figures of *madisonense* (Haynes, 1916, pl. 7, figs 11-13) show costae starting in front of the umbo and extending across most of the flanks. In *R. australe,* the lateral costae commence a short distance in front of the umbo, are rounded posteriorly, and towards the margins become higher, slightly angular, and in some cases increase by bifurcation or intercalation. Internally, *australe* is closer to the type species, having short dental plates and a wide septalium supported by a median septum.

The capillate micro-ornament and large V-shaped septalium exclude *australe* from *Calvinaria* Stainbrook. Sartenaer (1967, fig. 1) considers that *Calvinaria* is restricted to the Frasnian, whereas *Rugaltarostrum* ranges from the 'lower part (highest horizons) of the Lower Famennian to the middle (and perhaps the upper) part of the Upper Famennian' (Sartenaer, 1967, p. 1053).

The material described comes from one nest of rhynchonellids from reef rock in the Ningbing Limestone, and was extracted by calcination. Small specimens only were available for sectioning because large individuals were invariably filled with coarsely crystalline calcite.

RUGALTAROSTRUM AUSTRALE sp. nov.

(Pl. 35, figs 1-21)

Diagnosis: Shell transversely elliptical; costae on flanks well defined, commencing a short distance in front of umbo, rounded posteriorly, becoming higher, more angular, and at times increasing by bifurcation or intercalation anteriorly; three to six costae on each flank; three to four costae on floor of the sinus; accessory costae on flanks of the fold and sinus in some specimens; dental plates short, and divergent; septalium deep and V-shaped in transverse section; outer hinge plate mainly horizontal in transverse section, with dorsally curved inner margins in front of septalium passing anteriorly into crescent-shaped crurae.

Description: External. Shell dorsibiconvex, transversely elliptical, much wider than long, greatest width at midlength; lateral margins evenly rounded; commissure serrated, and uniplicate; costae on fold and sinus narrow, rounded, originating a short distance in front of tip of umbo, and numbering three to four in sinus; flanks of fold and sinus occasionally with accessory costae; costae on flanks usually slightly weaker and commencing a short distance farther towards front than those on fold and sinus, having a rounded profile, becoming higher and subangular at front of large individuals, and at times increasing by bifurcation or intercalation at margins; micro-ornament of intercalating capillae (so far observed between costae in sinus only), and regularly spaced concentric bands or groups of growth lines having a density of four to five per 1 mm near front of a mature shell.

Pedicle valve moderately convex, flattened on lateral and posterolateral extremities, with a prominent dorsally deflected tongue at front; umbo small, rounded, erect, and projecting 1 mm behind hinge on a valve 18 mm wide and 13.5 mm long; palintropes narrow, slightly concave, separated from umbonal shoulders by weak beak ridges; delthyrium and pedicle aperture not observed; sinus beginning just in front of umbo, broader and dorsally deflected anteriorly, having steep flared margins, a flat floor, and a rounded to square profile at front.

Brachial valve higher and more strongly convex than pedicle valve, greatest height at two-thirds length of valve; umbo small; flanks steep, bordering narrow flattened posterolateral extremities; fold beginning immediately in front of umbo, broader and higher towards front of valve, with steep angular lateral margins and a quadrangular profile.

Internal (Fig. 36). Pedicle valve with short slender divergent dental plates along sides of pedicle cavity; teeth narrow, pointed, extending well in front of dental plates.

Brachial valve. Septalium small, deep, narrowly V-shaped in tranverse section, supported on a short slender median septum; septum low in front of septalium; outer hinge plates mainly horizontal in transverse section, and at front of septalium having dorsally curved inner margins extending anteriorly into crescent-shaped crurae; sockets narrow and ovoid in transverse section, bounded by low inner socket ridges.

Remarks: As already indicated in the remarks on the genus (p. 157), *Rugaltarostrum australe* sp. nov. is characterized by strong lateral costae which commence a

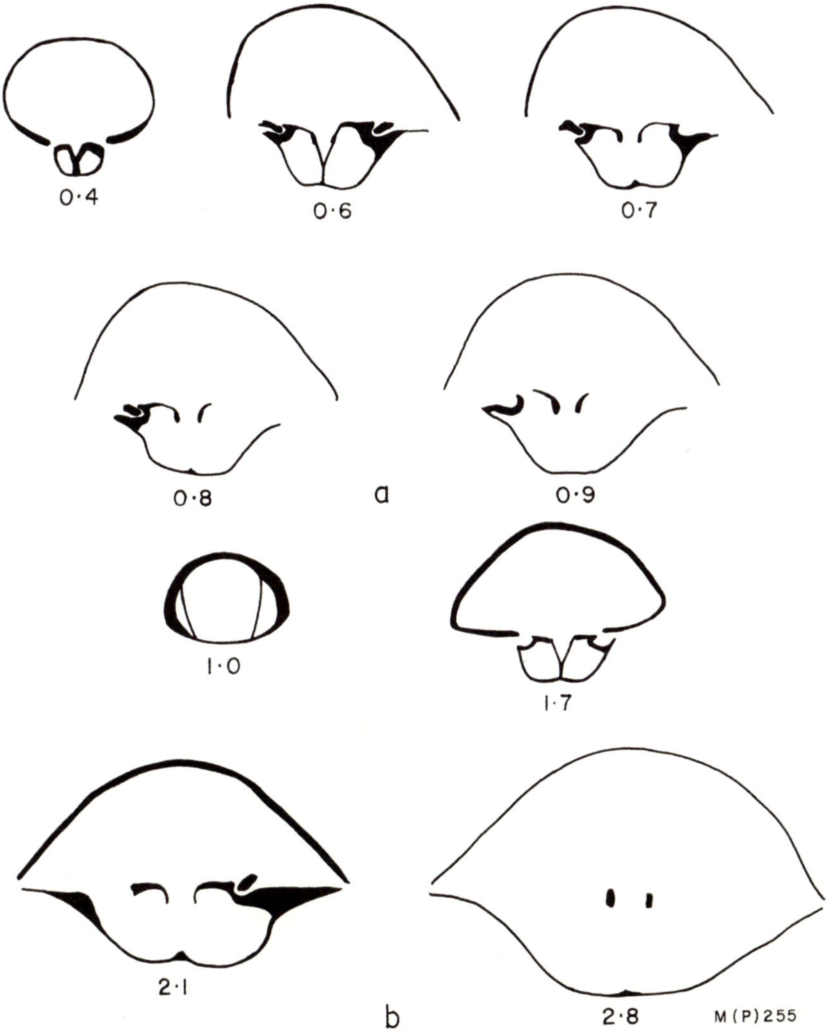

Figure 36. Transverse serial sections of two specimens of *Rugaltarostrum australe* sp. nov. Locality 455/2, Ningbing Limestone, Ningbing Range. a-CPC 11095, × 2; b-CPC 11096, × 4.

Measurements	Length (pedicle valve)	Length (brachial valve)	Width	Height
CPC 8302 Holotype	15	14	25	15
CPC 8303 Paratype	13.5	12.5	20.5	11
CPC 8304 Paratype	13	12	17.5	9.5
CPC 8305 Paratype	11.5	11	14.5	6
CPC 8306 Paratype	13.5	—	18.5	—

short distance in front of the umbo, and become slightly angular and in some instances increase by bifurcation or intercalation towards the margins. *R. gibbosum* (Haynes) from the Three Forks Formation, Montana, USA (Haynes, 1916, pp. 39-40, pl. 7, figs 14-16) is the most strongly costate species of *Rugaltarostrum* in North America. It differs from *australe* in having a smaller number (one to three) of lower lateral costae, usually a larger number of costae in the sinus, and a less transverse outline. *R. madisonense* (Haynes) and *R. jeffersonense* (Haynes), also from the Three Forks Formation (Haynes, 1916), either lack or have obsolete lateral costae.

Occurrence: Localities 455/2 (the type locality), in reef rock 2.5 miles south-southeast of the gap in which Tanmurra Creek breaches the Ningbing Range, and 9/4, in back-reef north of the Surprise Creek Gorge through the Ningbing Range; Ningbing Limestone (Famennian part).

Material: CPC 8302-8306, CPC 11095, CPC 11096. Holotype CPC 8302, paratypes CPC 8303-8306. All from locality 455/2. Seventy specimens.

Age: Famennian.

Family YUNNANELLIDAE Rzhonsnitskaya, 1959

Genus NAYUNNELLA Sartenaer, 1961

Type species: Yunnanella synplicata Grabau, 1931, by original designation of Grabau, 1931a.

Diagnosis: See McLaren (1965, p. H586).

Remarks: Sartenaer (1961c; 1962) designated *Nayunnella* to replace *Yunnanellina* Grabau, 1931, an objective synonym of *Yunnanella* Grabau, 1923. The internal structure of *Nayunnella* has not been fully described because of difficulty in obtaining Chinese material, but sections given by Grabau (1931b, fig. 30) show dental plates in the pedicle valve, and a septalium supported by a median septum in the brachial valve. Specimens from the Ningbing Limestone, referred to *Nayunnella turgida* sp. nov., have, in most instances, a septalium supported by a median septum; in some specimens these are obscured by callus deposits or recrystallization of the matrix (Figs 37, 38), and in others the septalium and median septum are absent (Fig. 39). The callus-like deposits in the brachial valve are reminiscent of those in sections of *Schnurella schnuri* (de Verneuil), 1840, given by McLaren (1965, fig. 462, 5d-i), in which the posterior part of the septum is poorly defined. Compared with the type species, *Nayunnella synplicata* (Grabau) (1931a, pp. 142-49, pl. 12, figs 1-7; pl. 13, figs 2, 3; pl. 14, figs 1-3), *N. turgida* sp. nov. has a more inflated brachial valve, a shorter ventral beak and shorter palintropes, slightly coarser costellae, three instead of two costae at the front of the sinus, and a larger number of costae at the margins of the flanks. *Schnurella* Schmidt (1964), typified by *S. schnuri,* is characterized by a convex brachial valve, but has a much lower and wider fold, a wider and more shallow sinus, and a more elongate outline.

Nayunnella turgida sp. nov.

(Pl. 32, figs 21-36)

Diagnosis: Shell dorsibiconvex, with an inflated brachial valve in adult specimens; greatest width at midlength; beak short, and subtending an apical angle of 100-120°; costellae having a density of nine to ten per 5 mm at 10 mm from umbo; three costae in the sinus, and up to seven costae on each flank; septalium and median septum usually present, frequently obscured by callus, but sometimes absent; outer hinge plates nearly horizontal in transverse section.

Description: External. Shell dorsibiconvex, subround to pentagonal, approximately as wide as long; greatest width at midlength; apical angle 100-120°; commissure weakly uniplicate in small individuals, strongly uniplicate and serrated in large specimens; shell costellate, becoming coarsely costate at margins of adults, up to three costellae coalescing to form each costa; costellae rounded, commencing at umbo, occasionally increasing by bifurcation on fold and sinus; density of costellae nine to ten per 5 mm at 10 mm from umbo; one shell 19.5 mm long and 20 mm wide has sixteen costellae on each flank, and thirteen costellae on fold and sinus; costae smooth, largest at front, V-shaped in section, usually with a slightly rounded apical keel; four costae on fold, three within sinus, and up to seven on each flank; micro-ornament of fine concentric growth lines.

Pedicle valve subpentagonal, lower and less convex than brachial valve; valve moderately convex posteriorly, but dorsally deflected, and more strongly convex at front; umbo narrow, erect, extending 3.5 mm behind hinge on a valve 19.5 mm long and 20 mm wide; pedicle foramen probably permesothyrid; delthyrium open; palintropes concave, especially posteroventrally, bordered by strong beak ridges, and ornamented with horizontal growth lines; sinus absent or weakly developed in small specimens, commencing at about midlength, usually shallow, with a wide flat floor produced into a dorsally deflected tongue; greatest width of sinus from one-half to two-thirds width of valve; in large specimens, front of sinus bordered by high angular costae; flanks moderately convex, becoming flatter on anterolateral margins.

Brachial valve subround, inflated, highest at midpoint; umbo broadly rounded, incurved, projecting a short distance behind hinge (1.5 mm on a valve 17 mm long and 19 mm wide); fold low, commencing in posterior third of larger valves, and truncated by dorsally deflected sinus of pedicle valve; flanks steep and flaring laterally.

Internal (Figs 37, 38). Pedicle valve. Dental plates extending along margins of pedicle cavity, usually free, but sometimes fused with walls; teeth broad, club-like in transverse section.

Brachial valve. Septalium and median septum variably developed, usually present, frequently partially obscured by apical thickening or recrystallization, but sometimes completely absent; septalium short, U-shaped in transverse section; septum short, triangular in transverse section; outer hinge plates straight or slightly flexed in transverse section, their inner margins giving rise to crural bases; in specimens without a septalium, outer hinge plates straight or with dorsally curved inner margins; outer hinge plates bordered laterally by concave inner socket ridges projecting below level of hinge plate.

Measurements	Length (pedicle valve)	Length (brachial valve)	Width	Height
CPC 8495 Paratype	14.5	13.5	15.5	10
CPC 8496 Paratype	18.5	17	19 est.	13.5
CPC 8498 Holotype	19.5	17.5	20	15

Remarks: Nayunnella turgida sp. nov. has a variably developed septalium and median septum in the brachial valve. From serial sections, I have determined that both structures are present in most valves, but they are frequently obscured by apical thickening or recrystallization of the matrix (Figs 37, 38); some valves, however, lack a septum and septalium (Fig. 39b). Variation in the septalium and median septum in rhynchonellids has previously been recorded by Veevers (1959a, p. 93, figs 52, 53) in *Ptychomaletoechia lucida* (Veevers) from the Famennian of the Canning Basin; most specimens of *lucida* have a well-defined septum and septalium, but in some, an incipient septalium is present at the apex of the valve only, and the septum is very short.

The species of *Nayunnella* described from the Famennian of the Mugodzhar region, USSR, by Rozman (1962) — *N. ericsoni* (Grabau), *N. zuleika* (Nalivkin), *N. nalivkini* (Rozman), and *N. acutiplicata* (Rozman) — all have costae which develop at an earlier stage of growth, and are higher and more angular than those in *N. turgida;* in *acutiplicata,* the most extreme case (Rozman, 1962, pl. 20, figs 1-5), the costae commence a short distance in front of the umbo. The brachial valves of the Soviet species are less inflated than in *N. turgida.*

A comparison with the type species, *N. synplicata* (Grabau), is given in the remarks on the genus (p. 160).

The specific name, Latin for inflated, refers to the strongly convex brachial valve.

Occurrence: Localities 8/1 (fore-reef) and 8/6 (back-reef) of the Ningbing Limestone, Surprise Creek, Ningbing Range.

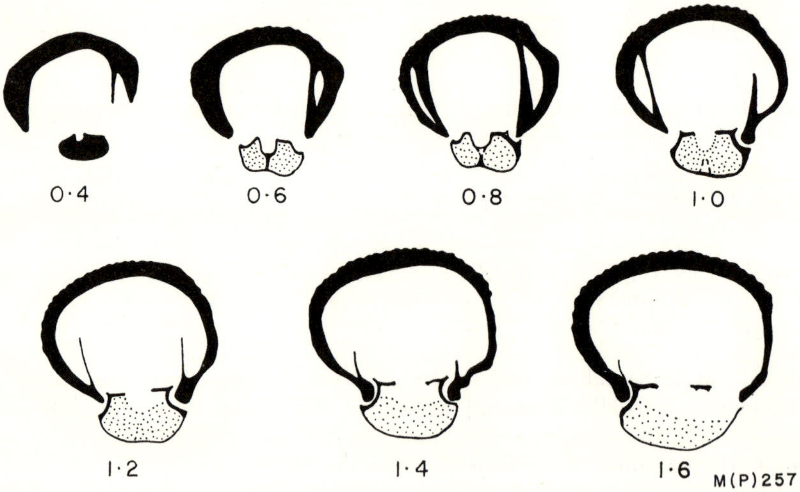

Figure 37. Transverse serial sections of *Nayunnella turgida* sp. nov. showing the septalium and median septum. The stippled region in the brachial valve represents apical thickening, or recrystallization of the matrix. CPC 11098, locality 8/1, Ningbing Limestone, Surprise Creek, Ningbing Range. × 4.

Material: CPC 8495-8498, CPC 11097, CPC 11098. Holotype CPC 8498, and paratypes CPC 8495-8497. All from locality 8/1. Approximately 60 specimens.

Age: Famennian.

Figure 38. Transverse serial sections of *Nayunnella turgida* sp. nov. showing the manner in which apical thickening or recrystallization of matrix obscures the septalium and median septum. CPC 11097, locality 8/1, Ningbing Limestone, Surprise Creek, Ningbing Range. × 4.

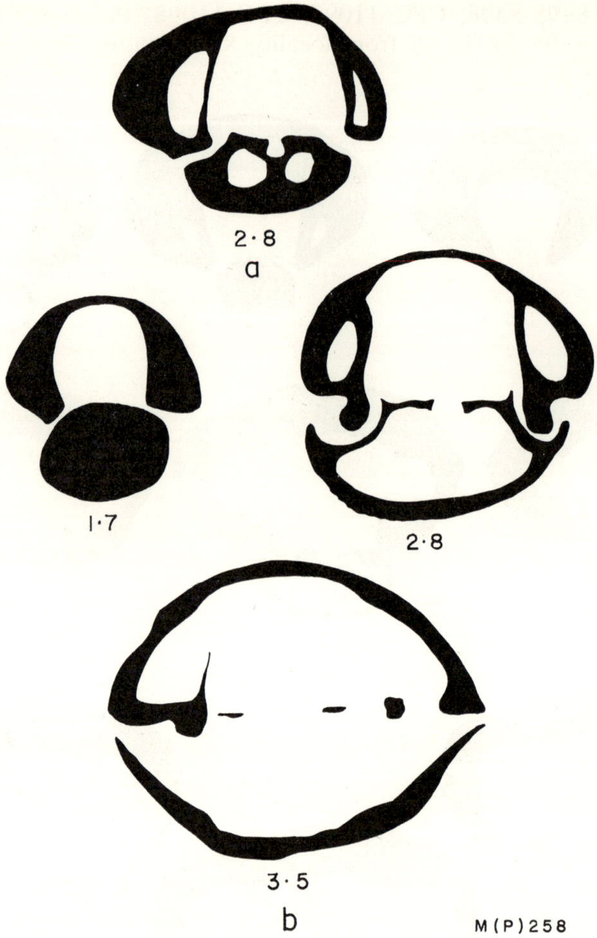

Figure 39. Transverse serial sections comparing two specimens of *Nayunnella turgida* sp. nov. a has a well-developed septum and septalium at 2.8 mm from the umbo, and b is without a septum or septalium at the same distance from the umbo. Locality 8/1, Ningbing Limestone, Surprise Creek, Ningbing Range.

Superfamily STENOSCISMATACEA Oehlert, 1887
Family STENOSCISMATIDAE Oehlert, 1887
Subfamily STENOSCISMATINAE Oehlert, 1887
Genus Coledium Grant, 1965

Type species: Coledium erugatum Grant, 1965, from the Moorefield Formation of Oklahoma, USA, by original designation of Grant, 1965.

Coledium? sp.
(Pl. 36, figs 37-40)

Description: External. Pedicle valve pentagonal with a long narrow beak; greatest width in front of midlength; anterior and lateral margins rounded; apical angle 90°; palintropes short and flat; delthyrium open; posterior regions of valve smooth, evenly convex, separated from remainder by a prominent growth halt; flanks less convex, on posterolateral margins bearing up to three short prominent rounded plicae; commissure uniplicate in adult individuals; sinus commencing in anterior third of valve, deep, with steep sides, and a dorsally projecting tongue; floor of sinus wide, concave, bearing a single strong rounded plication, and in one case two smaller accessory costae.

Internal. Pedicle valve. Spondylium probably sessile, bearing a low median ridge in one individual, extending one-third length of valve, broad, and moderately concave; teeth projecting dorsomedially, robust, supported by lateral margins of spondylium; muscle scars not observed.

Measurements

	Length	Width
CPC 8649	7.5	8
CPC 8650	6	6

Remarks: With only two silicified pedicle valves available for study, it is difficult to place this species in a genus. The structure of the spondylium cannot be observed in transverse section, and appears to be such that the species could belong to either *Coledium* Grant, 1965, or *Atribonium* Grant, 1965. Both of these genera are extremely variable externally (Grant, 1965), and in the type species are distinguished from one another externally by the possession in *Coledium* of a stronger anterior median deflection, and usually a stolidium or marginal frill. This species is tentatively referred to *Coledium* on the basis of its deep median sinus. It is well to note, however, that it is externally morphologically closest to the plicate species *Atribonium pauperum* (Belanski) described by Grant (1965, pp. 55-56, pl. 3, fig. 3) from the early late Devonian of Iowa; *A. pauperum* has two plicae on each ventral flank compared with three in this species. Species assigned to *Atribonium* by Grant range from the Middle Devonian to the early Mississippian, whereas those assigned to *Coledium* range from the Middle Devonian to the Permian.

Occurrence: Locality 108/0, Utting Calcarenite at Utting Gap.

Material: CPC 8649-8650 from locality 108/0. Two silicified pedicle valves only.

Age: Visean.

Order ATRYPIDA Rzhonsnitskaya, 1960
Superfamily ATRYPACEA Gill, 1871 (emend. Boucot et al., 1964)
Family ATRYPIDAE Gill, 1871

Two species of atrypids, one each of *Desquamatia (Synatrypa)* Copper and *Spinatrypa* Stainbrook, are described from Frasnian rocks of the Bonaparte Gulf Basin. Description of their internal morphology, determined mainly from acetate peels of serial sections, follows the terminology proposed by Copper (1967b, especially fig. 1). Copper's terminology for some external features (1967a), notably the attitude of the beak, is not used because procline, apsacline, orthocline, anacline, and hypercline are terms which have traditionally described the inclination of interareas (Williams & Rowell, 1965, fig. 61); they should not in my opinion be applied to other features.

Genus DESQUAMATIA Alekseeva, 1960
Subgenus SYNATRYPA Copper, 1966a

Type species: Desquamatia subzonata Biernat, 1964, from the Skaly Beds of Poland, by original designation of Copper, 1966a.

Diagnosis (from Copper, 1966a, pp. 10-12): 'Small *Desquamatia* rarely exceeding 25 mm width, finely ribbed and with numerous, even, close-spaced growth lamellae usually not more than 2 mm apart. In some aspects the general appearance is that of a miniaturized or dwarfed *Desquamatia* (although associated faunas are not small). Shell shape tends to be roundish or oval; extended hinge margins or planate valves are absent or uncommon. A small area and exposed delthyrium with deltidial plates and apical foramen are present. Frills are very short.'

Remarks: An emended diagnosis of the genus *Desquamatia* Alekseeva was given by Copper (1965a). The diagnosis emphasized the taxonomic importance of internal characters such as the hollow deltidial plates, the prominent lateral cavities or dental nuclei, and the hook-like jugal processes. Copper, however, did not make clear whether the details additional to those given by Alekseeva (1960) were observed on the type species or came from other species. With the exception of the hollow deltidial plates these characters are visible in serial sections of *D. khavae* Alekseeva, the type species of *Desquamatia* (Alekseeva, 1960, fig. 1; 1962, fig. 24); Alekseeva's sections of *Desquamatia magna* Grabau give the impression that the deltidial plates are hollow.

In a following paper, Copper (1966a) recognized two subgenera within *Desquamatia: Synatrypa* (type species *Desquamatia subzonata* Biernat) for small species having closely spaced frilled growth lamellae, fine tubular ribs, and a small ventral interarea with deltidial plates and an apical foramen; and *Variatrypa* (type species *Desquamatia ajugata* Copper) for large species with widely spaced frills, coarser tubular ribs, a wide ventral interarea, and extended hinge margins. *Atrypa desquamata kimberleyensis* Coleman, described below, is referred to *Desquamatia (Synatrypa)* on the basis of close similarities in ornament and the morphology of the posterior margin and ventral interarea. *Variatrypa* contains two morphological groups: the group of *ajugata* which has a single frill only, and the group of *zonata,* typified by *Desquamatia (Variatrypa) zonata* (Schnur), which has many widely spaced frills, and frequently extended hinge margins.

D. (Synatrypa) kimberleyensis (Coleman) has a number of similarities with species of *Desquamatia (Seratrypa)* Copper, 1967d (type species *D. (Seratrypa) pectinata* (Schroter)), which has its main development in the Frasnian (Copper, 1967d, p. 132).

Species of *Seratrypa* are large, have flattened ribs, widely spaced concentric lamellae, an incurved umbo, a small ventral interarea with reduced deltidial plates, and a poorly exposed foramen; internally the hinge plates and crura are thin.

Johnson & Boucot (1968), in a discussion on the external morphology of *Anatrypa* Nalivkin, considered that *Desquamatia* and *Anatrypa* were closely related, with *Desquamatia* possibly being a subgenus of *Anatrypa.* They concluded that the subgenus *Variatrypa* belonged to *Anatrypa* rather than *Desquamatia* because of a similarity in ornament. One of their main arguments was that the ribs on *Variatrypa ajugata* were interrupted by growth lamellae. Copper's illustration (1965a, pl. 27, fig. 1) of the holotype, however, clearly shows a well-defined frill. *Anatrypa* is characterized also by a reversed convexity (i.e. the pedicle valve is more convex than the brachial valve), and this combined with the frill in *Variatrypa* is, in my opinion, sufficient to separate them generically. The reversed convexity also distinguishes *Anatrypa* from *Desquamatia.* Johnson & Boucot (1968) pointed out that *Anatrypa micans* (von Buch) and *Desquamatia khavae* Alekseeva, both type species, have the same ventral beak structure and a similar ornament, but an inspection of *D. khavae* (Alekseeva, 1962, pl. 3, fig. 6) leads me to believe that the concentric lamellae on their specimen (Johnson & Boucot, 1968, pl. 160, figs 11-15) were largely removed by abrasion; growth lamellae are certainly visible on the ventral anterior surface (fig. 13). Because of the reversed convexity, the difference in concentric ornament, and the lack of internal information on *Anatrypa micans* (von Buch), I propose to retain *Desquamatia* as a separate genus.

Copper (1965a, p. 311) noted that *Desquamatia* had a world-wide distribution, but could not be recognized in published work in Australia. He considered that Australian atrypid species referred to *Atrypa desquamata* Sowerby belonged to *Spinatrypa* Stainbrook or *Spinatrypina* Rzhonsnitskaya, presumably on the basis of spines on a silicified specimen referred to *Atrypa desquamata kimberleyensis* Coleman by Veevers (1959a, p. 120, pl. 15, fig. 13). I agree with Copper that this spinose specimen is best placed in *Spinatrypa,* but as shown in this work, *kimberleyensis* belongs to *Desquamatia (Synatrypa).* My interpretation of *kimberleyensis* follows that originally given by Coleman (1951) rather than the broader and possibly anomalous sense of Veevers (1959a); before Veevers' lumping of several of Coleman's taxa into one species can be substantiated, internal as well as external structures must be examined in a manner similar to that used by Copper and Struve in their description of European atrypids. If recent European work is taken into account, it is likely that there are more than three atrypid species, one each of *Atrypa, Desquamatia,* and *Spinatrypa,* in the Givetian to Frasnian sequence in the Canning Basin of Western Australia.

In the Bonaparte Gulf Basin, *Desquamatia (Synatrypa)* is found together with *Spinatrypa* in a coquinite with a shaly limestone matrix in or closely associated with back-reef beds of a reef complex in the Westwood Member of the Cockatoo Formation. The association of *Desquamatia (Synatrypa)* with *Spinatrypa* is significant in view of Copper's (1965a, pp. 315-16) work in the Middle Devonian of Germany, where *Desquamatia* is rarely found with *Spinatrypa.* Copper (1966b) suggested seven ecological niches or biotopes for atrypid genera ranging from a nearshore littoral zone, through a sheltered or enclosed shallow shelf sea, and a reef or island barrier, to the open ocean. Both *Desquamatia* and *Spinatrypa* inhabited shaly sediment in a quiet, sometimes stagnant environment away from the reef, and because they are rarely found together, Copper (1965a, p. 316) suggested that they were incompatible. In an idealized cross section of atrypid biotopes of the Middle Devonian of Germany, Copper (1966b, fig. 2) placed *Desquamatia* in a biotope of 'frilled atrypids', and *Spinatrypa* in a biotope of 'isolated stable atrypids'.

He suggested (1967a) that atrypids with frills and spines inhabited quiet environments, whereas those secured by a pedicle lived in higher energy environments nearer the reef. Because both species of *Desquamatia (Synatrypa)* and *Spinatrypa* in the Westwood Member probably had functional pedicles (*Spinatrypa prideri larga* subsp. nov. has large pedicle muscles, a well-defined interarea, foramen, and deltidial plates, and in some ways resembles *Spinatrypina*) as well as stabilizing frills and spines, I suggest that they were able to inhabit a relatively high energy environment. It is unlikely that they were washed in from the deeper environments envisaged by Copper (1965a; 1966b). Further work, particularly in the Devonian reef complex in the Canning Basin, is required to fully understand the palaeoecology of the Western Australian atrypids.

L

DESQUAMATIA (SYNATRYPA) KIMBERLEYENSIS (Coleman)

(Pl. 38, figs 1-22)

1933 *Atrypa desquamata* Sowerby, Hosking, p. 72, pl. 7, figs 3, 4.
1951 *Atrypa desquamata kimberleyensis* Coleman, pp. 683-84, pl. 101, figs 7-19.
Not *Atrypa desquamata kimberleyensis* Coleman, Veevers, 1959a, pl. 15, fig. 13.
? *Atrypa desquamata kimberleyensis* Coleman, Veevers, 1959a, figs 73a, 73b.

Description: External. Shell round to subquadrate, equally biconvex in small specimens, moderately dorsally biconvex in adult; shell widest near midlength; commissure rectimarginate in small individuals, broadly and often strongly uniplicate in

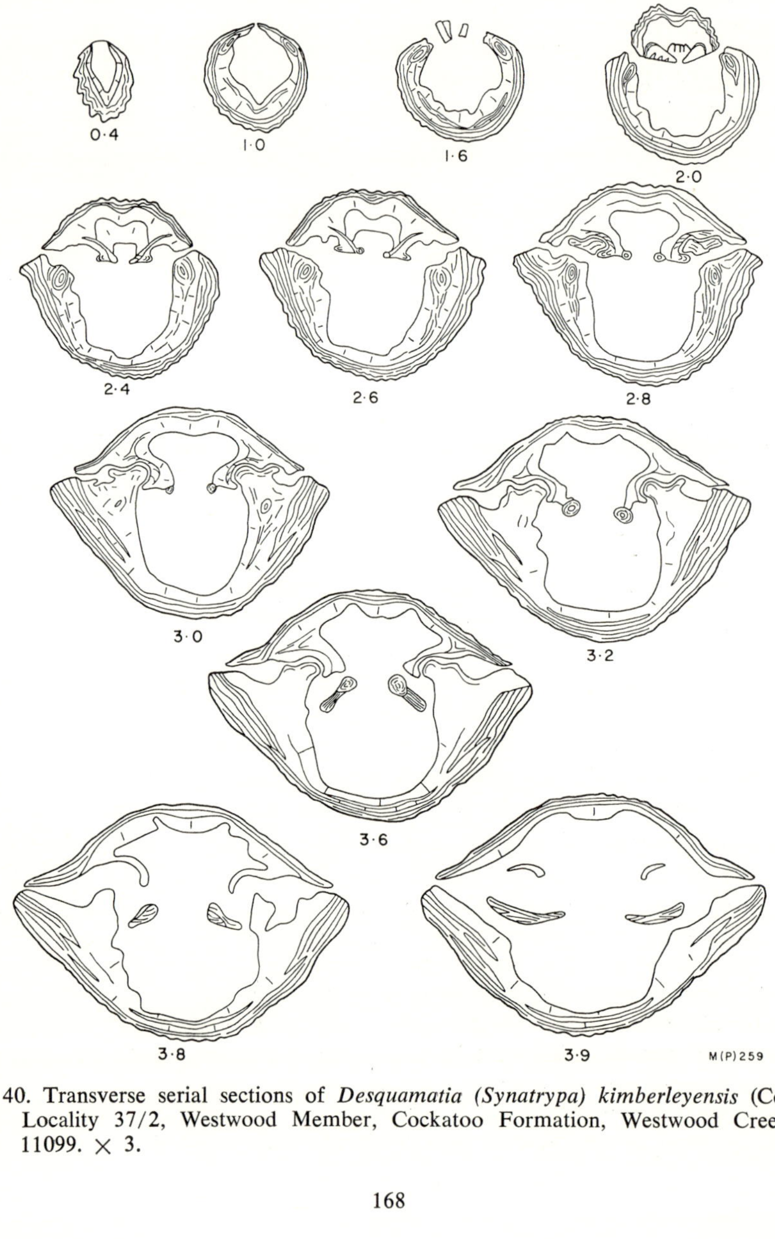

Figure 40. Transverse serial sections of *Desquamatia (Synatrypa) kimberleyensis* (Coleman). Locality 37/2, Westwood Member, Cockatoo Formation, Westwood Creek. CPC 11099. × 3.

large specimens; hinge approximately two-thirds width of shell; costae rounded, imbricating, separated by intercostal furrows of equal width, increasing by bifurcation and intercalation, and frequently passing on to a frill at anterior of each concentric lamella; density of costae fifteen to nineteen (usually seventeen to eighteen) per 10 mm at 10 mm from umbo; concentric lamellae widely spaced at posterior of shell, crowded at margins, and forming high erect to sub-erect ribbed frill.

Pedicle valve most convex at umbo, highest near midpoint of valve; umbo erect to sub-erect, bearing a round mesothyrid foramen, forming an apical angle of 110-130° (most specimens near 110°); cardinal margins usually inclined, but nearly straight in some individuals; lateral slopes extending evenly to margins; median sinus broad, shallow, confined to anterior half of mature individuals, and in some large specimens with a median dorsal deflection forming a strong lingual extension; interarea gently convex; delthyrium closed below foramen by two triangular disjunct deltidial plates.

Brachial valve moderately to strongly convex, highest medially, flattened on lateral and posterolateral margins; greatest convexity at umbo or at front of fold; umbo broad, not differentiated from lateral slopes; fold low, wide, confined to anterior half of large individuals.

Internal (Figs 40-41). Pedicle valve. Floor of pedicle cavity thickened with fibrous calcite, with a narrow median longitudinal groove; pedicle collar absent; deltidial plates solid apically, with long narrow cavities between thin inner and

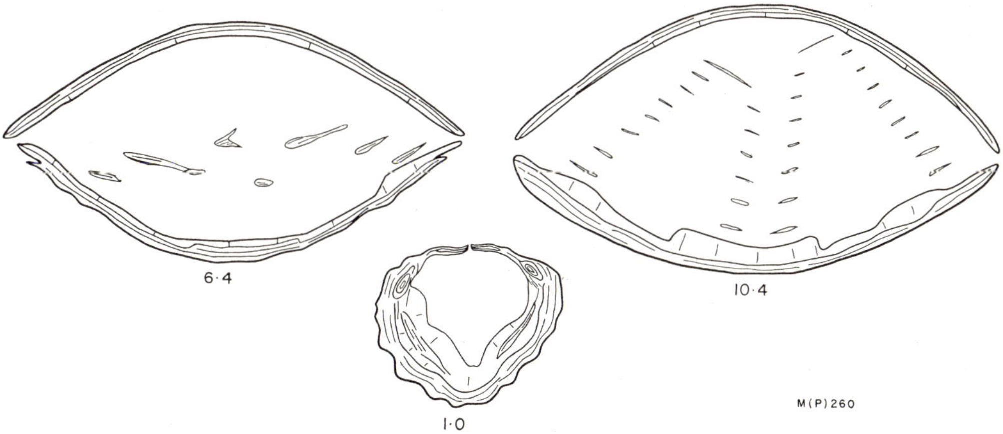

Figure 41. Transverse serial sections of *Desquamatia (Synatrypa) kimberleyensis* (Coleman) showing hollow deltidial plates at 1.0 mm, a hook-like jugal process at 6.4 mm, and the spires at 10.4 mm from the tip of the beak. Locality 37/2, Westwood Member, Cockatoo Formation, Westwood Creek, CPC 11100. 1.0 mm, × 5; 6.4 mm and 10.4 mm, × 3.

outer walls distally; lateral cavities absent in large specimens, but dental nuclei prominent; dental plates massive, with thick dental pads, and bearing large rounded teeth; teeth inwardly directed, in transverse section having a rounded main lobe, and a large lateral lobe; inner surface of valve, especially dental pads and muscle field, thickened with coarsely crystalline columnar calcite; muscle field having a high median platform bounded on either side by depressions.

Brachial valve. Cardinal process situated in notothyrial pit, consisting of about six irregular platelets; notothyrial pit small; hinge plates large, consisting of thick fibrous socket plates with more coarsely crystalline hinge pads on inner margins; sockets crenulated posteriorly, broadly rounded anteriorly, having large rounded inner socket ridges and smaller rounded middle socket ridges; crural bases arising from socket plates, round in transverse section posteriorly, anteriorly forming high ventral projections, then curving laterally and giving off elongate brushlike crura; jugal processes separate, arising from crura, long and thin, nearly horizontal in transverse section, terminating in small hooked jugal plates; spires dorsally directed, with up to nine volutions; posterior of valve thickened with columnar calcite; median part of valve bearing a low rounded median ridge branching anteriorly into two weaker ridges.

Measurements (dimensions of frills not included)

	Length (pedicle valve)	Length (brachial valve)	Width	Height
CPC 10969	23	21.5	25.5	13.5
CPC 10970	22	20.5	23.5	14.5
CPC 10971	17.5	16	17	10
CPC 10972	11.5	10.5	12	5
CPC 10973	19.5	18.5	20	12.5
CPC 10974	19.5	18.5	20.5	11

Remarks: Atrypa desquamata Sowerby, 1840, to which this material was originally referred as a subspecies by Coleman (1951), has been placed by Copper (1965b, pp. 364-69) in *Mimatrypa* rather than *Desquamatia. Atrypa desquamata kimberleyensis* Coleman, 1951, is therefore removed from *desquamata,* and assigned to *Desquamatia (Synatrypa).* Coleman's work on the atrypids of the Canning Basin was revised by Veevers (1959b), who included *Atrypa multimoda* Coleman and *Atrypa parva* Coleman in *kimberleyensis.* He regarded the swollen shape, incipient ears, and poorly defined interarea in *Atrypa multimoda* as ontogenetic characters only. In view of recent work in atrypid taxonomy, by for example Copper and Struve, these differences could well be significant, and until they and internal details of *multimoda* and *parva* can be studied in detail, I prefer to use Coleman's original interpretation of *kimberleyensis.*

Specimens of *D. (Synatrypa) kimberleyensis* from the Bonaparte Gulf Basin are distinguished from *Atrypa multimoda* Coleman (1951, pp. 682-83, pl. 100, figs 11-18; pl. 101, figs 1-6) by a less transverse shell, a more prominent and narrow ventral umbo, inclined cardinal margins, a smaller ventral interarea, and a more convex pedicle valve; *kimberleyensis* lacks ears on the posterolateral extremities.

D. (Synatrypa) subzonata (Biernat) from the Skaly Beds of Poland (Biernat, 1964, pp. 319-22, pl. 7, fig. 10; pl. 8, figs 1-7; pl. 9, figs 1-8), the type species of *Synatrypa,* resembles *kimberleyensis* in having fine tubular ribs crossed by closely spaced frilled growth lamellae, a small ventral interarea, and an exposed delthyrium having deltidial plates and an apical foramen. *D. (Synatrypa) subzonata* is distinguished from *kimberleyensis* by slightly finer ribbing, and in adult specimens by straighter cardinal margins and a slightly wider hinge.

In shape and external ornament, *kimberleyensis* is close to small specimens of *D. zonataeformis* Alekseeva (1962, pl. 3, figs 1-5) from the upper Givetian and lowermost Frasnian of the Kuznetsk Basin, USSR. *D. zonataeformis,* however, has lateral cavities rather than dental nuclei in the ventral valve, and adult specimens are larger, more elongate, and have a stronger median sinus in the pedicle valve when compared with *kimberleyensis.*

D. (?Desquamatia) schroeteri Copper (1967d, pp. 130-31, pl. 26, fig. 3; pl. 27, figs 1, 2) from the Frasnian of Bergisches Land, Germany, is morphologically related to *D. (Synatrypa) kimberleyensis,* having a closely comparable size, outline, and convexity, and fine tubular ribs. The ribs on *schroeteri* are slightly finer than those on *kimberleyensis,* but the species is mainly distinguished by the absence of frills on the growth lamellae. The internal morphology of *schroeteri* is undescribed.

D. prisca gladbachensis Jux (1965, pp. 158-59, pl. 20, figs 1-7), from the Givetian of Bergisch Gladback, Rheinisches Schiefergebirge, Germany, is externally close to *kimberleyensis,* having ribbing of comparable density, a similar shape, and ribbed frills. The ventral umbo of *prisca gladbachensis* may be more incurved than that in *kimberleyensis.* A closer comparison cannot be given because the interior of the German species remains undescribed.

D. ciliipes (Crickmay) (1957, p. 14, pl. 1, figs 1-8, 15; 1967, p. 5, pl. 1, fig. 3) from the Frasnian Grumbler Formation in the Northwest Territories of Canada resembles *kimberleyensis* in external outline, convexity of both valves, possession of an erect to sub-erect umbo, and in having a similar density of costae. Internally *D. ciliipes* has lateral cavities rather than dental nuclei. Crickmay (1967) considers *D. ciliipes* to be the youngest species of *Desquamatia* in Canada.

Occurrence: Localities 12/0A, 12/1, 12/3, 12/4, 12/135, 37/1, and 37/2 in the Westwood Member of the Cockatoo Formation at Westwood Creek.

Material: CPC 10969-10974, CPC 11099, CPC 11100. All from locality 37/2. Approximately 100 loose specimens.

Age: Frasnian.

Genus SPINATRYPA Stainbrook, 1951

Type species: Atrypa aspera var. *occidentalis* Hall, 1858, by original designation of Stainbrook, 1951.

Diagnosis: See Copper (1967c, pp. 494-95).

Remarks: Copper (1967c, p. 495) distinguished *Spinatrypa* Stainbrook from *Spinatrypina* Rzhonsnitskaya on the basis of 'its undulose, instead of tubular-imbricate ribs, its minute area, and small deltidial plates, rather than relatively large area and plates, and in its development of spines which are prominent and abundant when preserved. *Spinatrypa* also reaches a far greater size and shows a more prominent anterior fold.' Specimens from the Westwood Member of the Cockatoo Formation are large, have a prominent dorsal deflection of the commissure in the adult, fine but undulose ribs, and coarsely spinose concentric lamellae, all of which suggest *Spinatrypa.* The interarea and delthyrial plates, however, are well defined, and resemble those in *Spinatrypina.* Copper (pers. comm.) has observed similar features in species of *Spinatrypa* from the Frasnian of Boulonnais, France, and the Lime Creek Formation of Iowa, USA, and is in the process of designating a subgenus accommodating species which have some of the characteristics of *Spinatrypina.*

The association of *Spinatrypa* with *Desquamatia (Synatrypa)* in back-reef beds in the Westwood Member has already been mentioned on page 167. Species of both genera lived in an environment similar to that characterized by *Spinatrypina* in the Middle Devonian of Europe (Copper, 1966b, fig. 2), and exposure to a high energy environment, compared with the quiet environment in which *Spinatrypa* is usually found, may have led to development of a well-defined beak, interarea, and delthyrial plates in *Spinatrypa prideri larga* subsp. nov.

SPINATRYPA PRIDERI LARGA subsp. nov.
(Pl. 38, figs 23-45)

Diagnosis: Shell large, rounded, subquadrate to subrectangular, having straight cardinal margins; young individuals gently subequally biconvex and mature specimens moderately to strongly dorsally biconvex; commissure uniplicate in large individuals; ribs having a density of eight to eleven per 10 mm at 10 mm from the umbo, increasing by bifurcation and intercalation; concentric lamellae closely spaced, and crowded at the margins; ventral umbo long, and erect to sub-erect; delthyrium with disjunct deltidial plates; interarea long; sinus and fold developed on the posterior third of shell; brachial valve higher than pedicle valve, and gibbous in large specimens; ventral interior thickened apically; dorsal interior with a prominent posteromedian boss.

Description: External. Shell rounded, subquadrate to subrectangular, having nearly straight cardinal margins; young individuals subequally biconvex to nearly flat, mature specimens moderately to strongly dorsally biconvex; greatest width between one-third and one-half length of valve; hinge slightly shorter than greatest width; commissure rectimarginate in young individuals, slightly to moderately uniplicate in mature specimens; some large inflated specimens have a strongly uniplicate anterior commissure; shell strongly lamellose, regular undulations in each lamella forming broad rounded ribs; ribs having a density of eight to eleven per 10 mm at 10 mm from umbo; ribs increasing occasionally by bifurcation and intercalation; concentric lamellae closely spaced, having a density of three to five per 5 mm from 10 to 15 mm from umbo, and crowded at margins of large individuals; lamellae moderately frilled; spines approximately 1 mm in diameter at base at margin of a specimen 23.5 mm wide and 19 mm long, tapering distally, and prostrate along surface of valve.

Pedicle valve most strongly convex at umbo and at anterior margin, moderately convex to flat over greater part of valve; posterolateral extremities having rounded outlines; umbo sharply pointed, erect to sub-erect, extending 2 mm posteriorly from hinge of a specimen 24 mm wide and 21.5 mm long; apical angle approximately 100°; foramen mesothyrid, and ovoid; anterior part of delthyrium closed by triangular disjunct deltidial plates; interarea long for genus, and moderately concave; median sinus obsolete, when present forming a broad shallow depression on posterior third of valve.

Brachial valve higher and more convex than pedicle valve, and in large individuals frequently gibbous; greatest height at midlength, or in specimens with a strong dorsal median deflection near anterior margin of fold; fold extremely broad, poorly differentiated from lateral slopes; umbo low, broadly rounded, and usually most convex part of valve; posterolateral margins flat to slightly convex.

Internal (Fig. 42). Pedicle cavity broad, thickened ventrally with callus; adductor muscle scars diamond-shaped to narrowly elliptical, deeply impressed into callus anteriorly, bordered posterolaterally by flat platforms extending along lateral margins of pedicle cavity, anteriorly enclosed by posteromedian margins of diductor muscle scars; diductor muscle scars pointed posteriorly, having wider straight anterior margins, and longitudinal troughs carrying vascula myaria on lateral margins; median anterior parts of diductor muscle scars elevated on a callus of

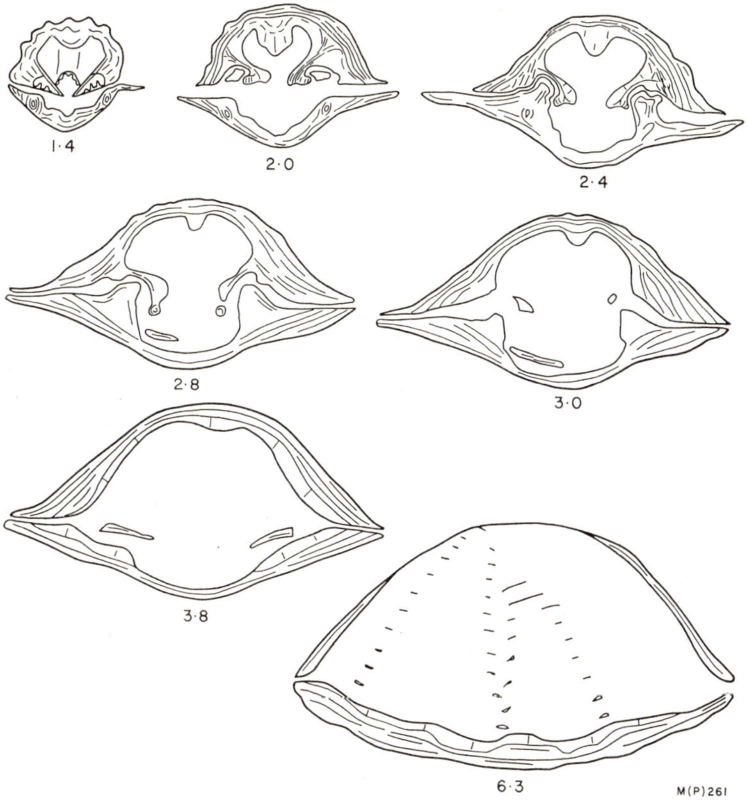

Figure 42. Transverse serial sections of *Spinatrypa prideri larga* subsp. nov. CPC 11101, locality 37/2, Westwood Member, Cockatoo Formation, Westwood Creek × 2.5.

varying height; callus convex or with a median longitudinal furrow constricted by inward curvature of vascula myaria near front of valve; accessory diductor muscle scars (or ventral pedicle muscles of Vandercammen & Lambiotte, 1962, fig. 3) narrow, and situated above and on lateral margins of diductor muscle scars in depressions in inner margins of callus; vascula myaria commencing from platform-like regions on posterolateral margins of diductor muscle scars, curving laterally near and in front of anterior margins of diductor scars, and curving medially near front of valve; in one specimen, vascula myaria give off four posterolaterally directed branches through a pitted area of vascula genitalia; pitted regions of vascula genitalia extending in narrow irregular strips from lateral margins of muscle field to near front of vascula myaria; in transverse section, teeth massive, having rounded main lobes, poorly defined lateral lobes, and small dental pads; dental nuclei prominent.

Brachial valve. Apex of valve thickened with coarsely crystalline calcite; cardinal process situated in base of small notothyrial cavity; hinge plates massive, curving medially, having narrow socket plates and wide hinge pads; inner socket ridges wide, especially posteriorly, and middle socket ridges low; crural bases arising abruptly from socket plates, round in transverse section at distal extremities of socket plates, and anteriorly giving off brush-like crural plates; jugal plates arising from crural plates, and nearly horizontal; posteromedian floor of valve bearing a high boss or septum; spires with approximately eleven volutions.

Measurements (dimensions of spines not included)

	Length (pedicle valve)	Length (brachial valve)	Width	Height
CPC 10976 Paratype	21.5	—	25	—
CPC 10978 Paratype	13	12.5	13	5
CPC 10979 Paratype	17	15.5	20.5	8.5
CPC 10980 Paratype	20.5	19	24	12.5
CPC 10981 Holotype	21.5	20	24.5	12

Remarks: Spinatrypa prideri larga subsp. nov. is characterized in adult growth stages by its dorsibiconvex profile, straight cardinal margins, gibbous brachial valve having a poorly defined anterior fold, broad anterior ventral sinus, and a strong anterior dorsal deflection producing a uniplicate commissure. Other features, particularly the ribbing and the morphology of the ventral umbo and the interarea, are closely comparable with those of *S. prideri* (Coleman) from the Canning Basin, Western Australia (Coleman, 1951, pp. 684-85, pl. 102, figs 1-10). With two exceptions, Coleman's type specimens of *prideri* are the same size and shape and have the same density of ribbing as small individuals in the Westwood Member of the Cockatoo Formation. These exceptions are Coleman's figure 6 in plate 102 which has finer ribbing, and figure 7 on the same plate which shows a large marginal flange; both features exclude these specimens from *prideri*. In addition, Veevers (1959a, p. 123) considered that a specimen included by Coleman in *Atrypa parva* Coleman (1951, pl. 102, figs 23, 24) was best placed in *prideri*. Coleman's type specimens are in most cases smaller than the average dimensions of the species (Coleman, 1951, p. 686). However, because they are representative of large collections from a number of localities, it is unlikely that they are juveniles; hence *S. prideri larga* is distinguished from *S. prideri prideri* essentially on the basis of its larger size, gibbous brachial valve, and stronger median dorsal deflection of the anterior commissure. Specimens of *S. prideri* from the Gneudna Formation in the Carnarvon Basin, Western Australia (Coleman, 1951, pl. 102, figs 11-17; Veevers, 1959b, pl. 2, figs 1-10), are less transverse and have more inclined cardinal margins than either *S. prideri prideri* or *S. prideri larga*. In other external features they are closest to specimens of *S. prideri prideri*. Transverse serial sections of *S. prideri prideri* from the Canning Basin figured by Veevers (1959a, fig. 74) do not show a well-defined posteromedian boss in the brachial valve, but the boss is present in a specimen from the Carnarvon Basin (Veevers, 1959b, fig. 11), and in *S. prideri larga* (Fig. 42).

S. prideri (Coleman), originally named *Atrypa aspera prideri* Coleman, is removed from the previously catch-all species *Spinatrypa aspera* (Schlotheim) and given specific status because of significant differences in morphology and age. *S. aspera* has become better known in recent years following work by Biernat (1964) on *S. aspera* from the Skaly Beds (early Givetian or late Eifelian) of

Poland, and Struve (1956) and Copper (1967c) on topotype material of *S. aspera aspera* from the late Eifelian Freilingen Beds of Germany. *S. aspera aspera* differs from *S. prideri prideri* in the possession of fewer and less frequently dividing ribs, a more subcircular outline, a more convex pedicle valve, a smaller and more incurved ventral umbo, and a shorter ventral interarea. Coleman (1951, p. 684) considered *prideri* to be less convex than *aspera. S. prideri prideri* is recorded from the Frasnian *torrida* Zone of Veevers (1959a) in the Canning Basin, compared with a late Eifelian age for *S. aspera aspera* (Copper, 1967c, fig. 1). It is possible, however, that one of the localities of the syntypes of *S. prideri prideri,* just south of Longs Wells, is younger than Frasnian; rocks from that general area are dated from conodonts as late Givetian by Glenister & Klapper (1966).

Coleman (1951, p. 684) considered that *Atrypa bodini* Mansuy (1912, p. 75, pl. 13, figs 10a-c; pl. 14, figs 1a-b) from the late Devonian of Ta-Hi-Ti, Vietnam, occupied a morphologically intermediate position between the European *S. aspera* (Schlotheim) and the type specimens of *S. prideri prideri.* Copper (1967d, p. 124), however, has referred *bodini* to *Spinatrypina* Rzhonsnitskaya, and compares it most closely with *Spinatrypina (Exatrypa) tubaecostata* (Paeckelmann). Veevers (1959a, p. 30) compared *S. (Exatrypa) tubaecostata* with *S. prideri prideri,* but as suggested by Copper (1967d, p. 125), the two species differ in morphology and belong to different genera.

S. prideri larga differs from *S. aspera aspera* (Schlotheim) (Copper, 1967c, pl. 77, figs 1-9) in its larger size, greater number of ribs, more transverse outline, straighter cardinal margins, more prominent and straighter ventral umbo, longer interarea, flatter pedicle valve, more gibbous brachial valve, and a stronger anterior fold and sinus.

S. planosulcata (Webster) figured by Fenton & Fenton (1924, pl. 27, figs 13-16) from the Cerro Gordo Member of the Lime Creek Formation of Iowa, USA, resembles *S. prideri larga* in having a comparable density of ribs, closely spaced concentric lamellae, relatively straight cardinal margins, a broad anteriorly developed fold and sinus, and a thickened pedicle valve having deeply impressed muscle scars. *S. planosulcata* is distinguished by its shorter beak.

Specimens referred to *Atrypa* cf. *albertensis* Warren from the Frasnian Macgea fauna of Canada (Warren & Stelck, 1956, pl. 18, figs 7-13) resemble *S. prideri larga* in having a dorsibiconvex profile, straight cardinal margins, a prominent ventral umbo, and a broadly uniplicate anterior commissure. The ribs on *Atrypa* cf. *albertensis* are tubular rather than undulose, which suggests that the species may belong to *Spinatrypina;* no spines are visible in Warren & Stelck's illustrations. The Canadian species is distinguished from *S. prideri larga* by its more quadrate outline and tubular ribbing. *Atrypa albertensis* Warren (1944, pp. 118-19, pl. 3, figs 13-15) from the Waterways Formation of Alberta is almost certainly a *Spinatrypa.* It is distinguished from *S. prideri larga* by coarser ribs, inclined cardinal margins, and a less transverse outline.

S. legayi (Rigaux) from the Frasnian of Ferques, France (Rigaux, 1908, p. 22, pl. 2, fig. 10) resembles *S. prideri larga* in having a large number of costae on the exterior of the shell, straight cardinal margins, and a strong dorsal median deflection of the anterior commissure in the adult. *S. legayi* is larger and has a deeper ventral sinus.

A species referred to *Atrypa spinosa* var. *chitralensis* by Reed (1922, pp. 54-55, pl. 9, figs 16-23; pl. 10, fig. 1) from Shugrum and Karagh in the Himalayas, has an ornament of similar density and morphology to that of *S. prideri larga.* It is distinguished on the basis of its rounded outline, shorter hinge, and inclined cardinal margins.

Occurrence: Localities 12/3, 12/4, 13/6, 37/1, and 37/2 (the type locality), Westwood Member of the Cockatoo Formation at Westwood Creek, and locality 64/1, Hargreaves Member of the Cockatoo Formation in the Pretlove Hills.

Material: CPC 10975-10981, CPC 11101. Holotype CPC 10981, paratypes CPC 10975-10980. All from locality 37/2. Approximately 200 loose specimens.

Age: Frasnian.

Order SPIRIFERIDA Waagen, 1883

Suborder RETZIIDINA Boucot, Johnson, & Staton, 1964

Superfamily RETZIACEA Waagen, 1883

Family RETZIIDAE Waagen, 1883

Genus HUSTEDIA Hall & Clarke, 1893

Type species: Terebratula mormoni Marcou, 1858, by original designation of Hall & Clarke, 1893.

Remarks: Recent diagnoses of Hustedia are given by Boucot et al. (1965, p. H652) and Carter (1967, p. 318). Boucot et al. consider the delthyrium to be closed by conjunct deltidial plates, whereas Carter described a smooth curved ventral interarea, and a delthyrium closed by a flat deltidial plate or ?symphytium. Dunbar & Condra (1932), in a description of the type species, demonstrate that the wide delthyrium is closed by a flat continuous plate or symphytium. Specimens from the Utting Calcarenite have a symphytium which is comparable to that of the type species.

HUSTEDIA PAULA sp. nov.

(Pl. 37, figs 12-26)

Diagnosis: Shell small for genus, and subovate; commissure rectimarginate; costae rounded, numbering sixteen on pedicle valve; median furrow on pedicle valve bearing a faint costella; symphytium long and prominent; pedicle collar indistinct; brachial valve with flattened posterolateral margins; dorsal diductor muscle field impressed into floor of valve.

Description: External. Shell small, rostrate, elongate-subovate, equally biconvex; greatest height at or slightly behind midlength; commissure rectimarginate; costae commencing at umbo, simple, rounded, wider towards margins, and separated by narrow rounded intercostal furrows; one fine costella in median longitudinal furrow of pedicle valve; sixteen costae on pedicle valve; concentric growth lines common at margins; punctae obscured by silicification.

Pedicle valve. Umbo well defined, flanked by concave umbonal shoulders, suberect, perforated apically by a round foramen in a mesothyrid position; delthyrium closed by a long flat symphytium, in some cases bearing a median depression, but never a join; valve evenly convex in lateral profile; flanks steep and in some cases asymmetrical at posterolateral margins.

Brachial valve most strongly convex at umbo; umbo short, bluntly rounded, flanked by concave umbonal shoulders extending on to nearly flat posterolateral extremities; flanks sloping evenly to lateral and anterior margins.

Internal. Pedicle valve. Pedicle collar indistinct; teeth and muscle scars not observed.

Brachial valve. Cardinal process bilobed ventrally, lobes having rounded anterior and concave posterior faces; lateral margins of process bearing prong-like crurae; lower part of cardinal plate bearing a posteriorly recurved median prong extending ventrally to level of main lobes of cardinal process; median septum wide, short, bearing a median keel, supporting cardinal plate at base of median prong; adductor muscle field rounded-rectangular, with deeply impressed inner scars, and less well defined outer scars; canals of vascula media arising from and branching just in front of distal end of median septum, extending along a median elevation to front of valve; vascula myaria consisting of many branching canals extending to anterior and lateral margins from adductor muscle field.

Measurements	Length (pedicle valve)	Length (brachial valve)	Width	Height
CPC 10964 Paratype	6.5	—	6	—
CPC 10965 Paratype	—	5	5	—
CPC 10967 Holotype	5.5	5	5	3
CPC 10968 Paratype	5	4.5	5	2.5

Remarks: Hustedia paula sp. nov. is characterized especially by its small size and long symphytium on the pedicle valve. Most described species of *Hustedia* are larger, but *H. pygmaea* Rowley from the lower Burlington Limestone of Missouri (Rowley, 1900, p. 266, pl. 5, figs 25-27) is comparable in size, shape, and density of costae. *H. pygmaea* is distinguished from *paula* by its more angular costae, the absence of a median costella in the pedicle valve, and a smaller symphytium; the interior of *pygmaea* has not been described.

H. radialis (Phillips) described by Davidson (1860, pp. 87-88, pl. 17, figs 19, 20) from the Visean of Great Britain resembles this species in shape, density of costae, and in having a straight anterior commissure. Some specimens of *radialis,* however, have an enlarged median costa in the pedicle valve, and a corresponding furrow in the brachial valve. *H. radialis* is generally larger than this species. Specimens from the Visean of Malaya referred to *H.* cf. *radialis* (Phillips) by Muir-Wood (1948, pp. 65-67, pl. 2, figs 1-3, 12) are larger, and have a narrow median sinus in the pedicle valve. Of Soviet species referred to *H.* aff. *radialis* those from the late Tournaisian Ruzakov Beds of northeastern Kazakhstan figured by Nalivkin (1937, pl. 34, figs 5, 6) are closest externally to *paula.* Specimens figured by Yanishevsky (1918, pl. 7, figs 11, 12) from the Visean of Fergana have fewer costae than *paula.* Visean material from Belgium referred to *H. radialis* by de Koninck (1887, pl. 22, figs 16-19) have fewer costae, and are larger and more elongate than this species.

H. paula may be close externally to North African Visean material referred to *H.* cf. *mormoni* (Marcou) by Termier & Termier (1950, pl. 122, figs 21-24). *H. mormoni,* the type species, from the Pennsylvanian of Nebraska (Dunbar & Condra, 1932, pp. 356-58, pl. 42, figs 9-11) has a similar shape and costate ornament, but is larger and has a shorter symphytium than *paula.*

The species is named from the Latin *paulus* — little, in reference to its size.

Occurrence: Locality 108/0, the type locality, Utting Calcarenite, Utting Gap.

Material: CPC 10964-10968. Holotype CPC 10967, paratypes CPC 10964, 10965, 10966, 10968. All from locality 108/0. Twelve silicified specimens.

Age: Visean.

Suborder ATHYRIDIDINA Boucot, Johnson, & Staton, 1964
Superfamily ATHYRIDACEA M'Coy, 1844
Family ATHYRIDIDAE M'Coy, 1844
Subfamily ATHYRIDINAE M'Coy, 1844
Genus CARDIOTHYRIS nov.

Type species: Cardiothyris bisulcata gen. et sp. nov. from locality 7/1 (Tournaisian) in the Ningbing Limestone, Ningbing Range.

Description: Shell round to oval, emarginate, biconvex, and bisulcate; shell material thin; ventral umbo incurved over delthyrium, bearing a small pedicle foramen; palintropes small; external ornament of prominent, simple, widely spaced capillae crossed by fine concentric growth lines; pedicle valve with dental plates extending along margins of pedicle cavity; pedicle cavity short; cardinal plate quadrate, perforated at apex, thickened, with a gently concave dorsal surface divided by a median furrow; brachial apparatus not observed; muscle scars poorly defined.

Remarks: Cardiothyris is characterized by distinctive radial capillae on the exterior of the shell, strong sulci on both valves which give an emarginate outline, and a thick cardinal plate in the brachial valve. The genus is morphologically close to the Lower and Middle Devonian genus *Leptathyris* Siehl, 1962, which is a small smooth athyridid having a thin deeply depressed medially crested cardinal plate. *Cardiothyris* is distinguished from *Leptathyris* by well-defined radial capillae on the exterior of the shell, strong sulci on both valves which give a deeply emarginate anterior, a wider hinge, and a cardinal plate in the brachial valve which is thick, gently concave, and has a median furrow.

Composita Brown, 1849, is distinguished from *Cardiothyris* by its smooth shell which is crossed by concentric growth lines only, terebratuliform shape, deeply impressed muscle fields in both valves, extended flanges on the cardinal plate, blunter ventral umbo, and in the possession of a fold on the brachial valve; many species of *Composita* have an accessory sinus on the midpart of the fold, but this is not comparable with the dorsal sinus in *Cardiothyris.*

The name of this genus is derived from the Greek *kardia* — the heart, and refers to the heart-shaped outline of the shell.

CARDIOTHYRIS BISULCATA gen. et sp. nov.
(Pl. 39, figs 1-21)

Diagnosis: Shell subequally biconvex, about as wide as long; sinus on each valve of equal depth, resulting in a rectimarginate commissure; pedicle foramen epithyrid; capillae with a density of three to five per 1 mm on the midpart of each lateral slope, but more widely spaced in the sulci; dental plates short, slender, and bearing large blunt teeth; inner socket ridges on the cardinal plate strongly developed; sockets broadly rounded to subquadrate in transverse section.

Description: External. Shell subequally biconvex, round to oval, strongly emarginate at front; shell nearly as wide as long, greatest width around midlength; hinge slightly curved, two-thirds the width of shell; commissure rectimarginate; ornament of simple widely spaced capillae having a density of from three to five per 1 mm on midpart of each lateral slope, but more widely spaced in sinus; capillae crossed by fine concentric growth lines.

Pedicle valve most strongly convex at umbo; umbo narrow, moderately incurved, obscuring delthyrium, and perforated by a small round to ovoid epithyrid foramen; umbonal shoulders short, extending into small concave palintropes; lateral slopes slightly flattened at posterolateral extremities, moderately convex and sloping evenly to margins over remainder of valve; sinus commencing immediately in front of umbo, extending to anterior of valve, shallower posteriorly, much deeper towards front, and evenly rounded in profile.

Brachial valve. Umbo shorter, blunter, more strongly incurved than on pedicle valve; umbonal shoulders and palintropes extremely small, and slightly concave; both sinus and lateral slopes have the same morphology as on pedicle valve.

Internal (Fig. 43). Pedicle valve. Dental plates short, extending 3 mm along margins of pedicle cavity on a valve 13.5 mm long, bearing large, blunt, almost subquadrate, inwardly projecting teeth; in transverse section, dental plates high, mainly slender, inner faces subparallel to slightly concave, thickened at floor of valve; pedicle cavity short, tapering posteriorly, having a flat floor; muscle scars weakly defined, situated on a slightly elevated subrectangular platform immediately in front of pedicle cavity.

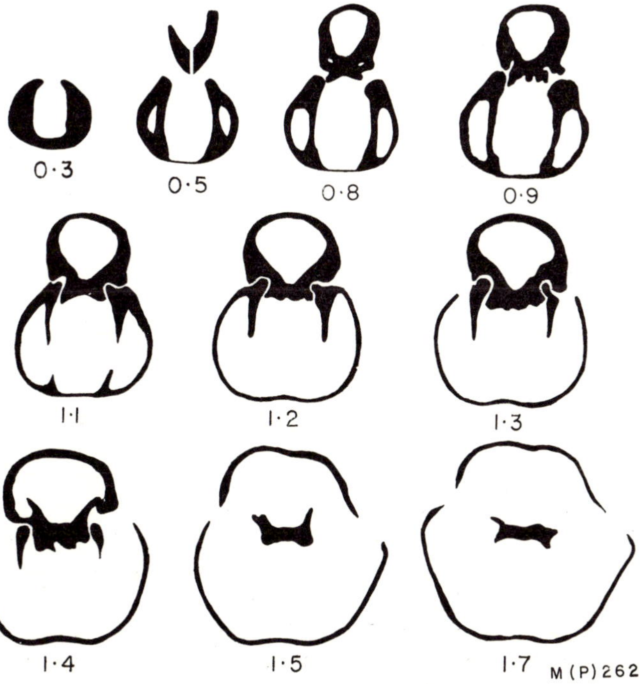

Figure 43. Transverse serial sections of *Cardiothyris bisulcata* gen. et sp. nov. CPC 11102, locality 7/1, Ningbing Limestone, Ningbing Range. × 4.

Brachial valve. Cardinal plate in same plane as commissure, robust, and thickened at junction with shoulders of valve; ventral surface of cardinal plate gently concave, median furrow separates two concave lateral portions; apical foramen small; in transverse section, sockets broadly rounded to subquadrate; muscle scars and details of brachial apparatus not observed.

Measurements

	Length	Width	Height
CPC 8245 Holotype	15.5	16	10
CPC 8242 Paratype	13.5	14	8
CPC 8243 Paratype	14	15	8
CPC 8244 Paratype	14 est.	14.5	8.5
CPC 8247 Paratype	12	12.5	7

Remarks: Emarginate athyridinoid species are relatively uncommon. A form from the Chester Group of Kentucky, *Composita laevis* Weller (1914, pp. 491-92, pl. 82, figs 14-20), has a shallow sinus on the pedicle valve, and a slight depression on the front of the brachial valve. It is distinguished from *Cardiothyris bisulcata* by much shallower sulci (particularly on the brachial valve), a more elliptical shape, and the apparent lack of radial ornament.

A form referred to *Athyris* cf. *protea* var. *globularis* Phillips by Patte (1926), from Na Tao, Tonkin, Vietnam, is more strongly biconvex, more elliptical in shape, and has smaller sulci on both valves; the sinus on the brachial valve commences at midlength, and is comparatively shallow. The internal morphology of both *Composita laevis* and *Athyris* cf. *protea* var. *globularis* has not been described, and hence detailed comparisons cannot be given.

The specific name refers to the bisulcate nature of the shell.

Occurrence: Locality 7/1 (Tournaisian part), Ningbing Limestone, Ningbing Range.

Material: CPC 8242-8247, CPC 11102. Holotype CPC 8245; paratypes CPC 8242-8244, 8246, 8247. All from locality 7/1. About 20 specimens obtained by calcining massive reef limestone.

Age: Tournaisian.

Genus Cleiothyridina Buckman, 1906

Type species: Atrypa pectinifera J. de C. Sowerby, 1841, from the Permian of England (*see* Carter, 1967, pp. 343-44).

Cleiothyridina gloveri Thomas

(Pl. 39, figs 22-33)

1970 *Cleiothyridina gloveri* Thomas, pp. 184-86, pl. 24, figs 1-10, 14.

Description: External. Shell small, round to semi-elliptical in shape, slightly wider than long, and dorsibiconvex; greatest width at midlength of shell; hinge curved anteroventrally, and almost three-quarters length of shell; commissure uniplicate, ornament of closely spaced, nearly regular imbricating concentric lamellae bearing dense sheets of erect to sub-erect flattened spines; spines arising from around the whole of each lamella, generally absent by abrasion from posterior part of shell, bearing radial lirae on their upper surfaces; concentric lamellae having a density of seven to ten per 3 mm at 5 mm from umbo.

Pedicle valve most convex near umbo; lateral slopes flat, flared ventrally and becoming gently concave in adult specimens; median sinus broad, shallow, commencing near midpoint, and deepest at anterior margin; umbo erect, perforated at tip by a round pedicle foramen; umbonal shoulders short and slightly concave; palintropes flat, well differentiated from body of valve; delthyrium open, very broad, and triangular.

Brachial valve globular, highly convex; umbo poorly defined, slightly incurved, filling delthyrium of pedicle valve; lateral slopes gently convex, sloping evenly to margins; fold poorly defined except at anterior margin; on some valves a weak secondary median sinus on fold often extends to anterior margin.

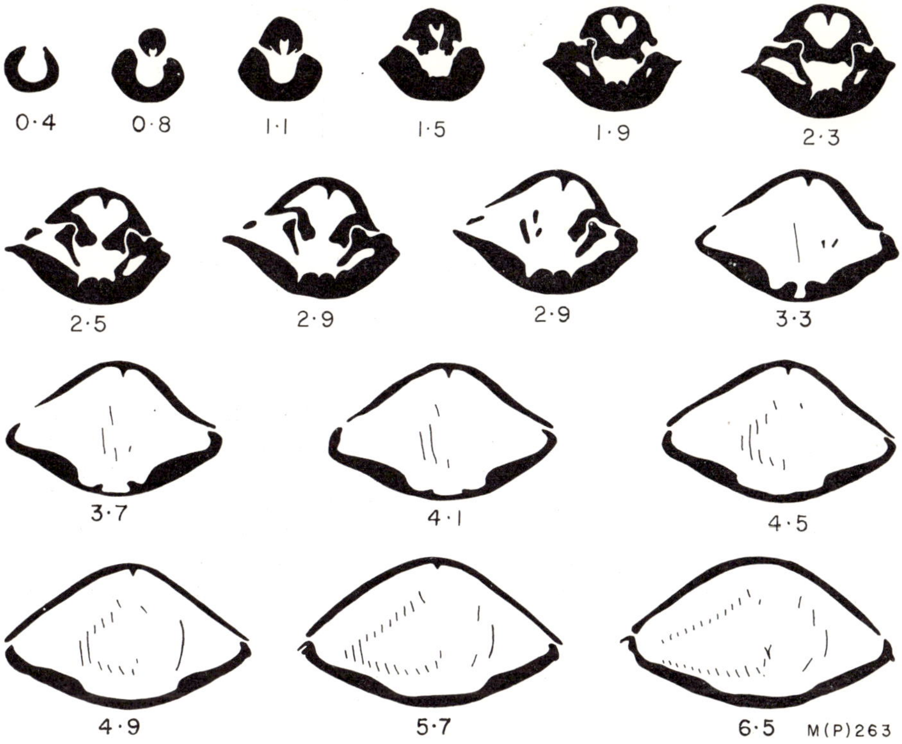

Figure 44. Transverse serial sections of *Cleiothyridina gloveri* Thomas. CPC 11103, locality 101/17, Burt Range Formation. × 2.

Internal (Fig. 44). Pedicle valve. Pedicle cavity narrow, expanding slightly anteriorly, with a rounded floor occasionally bearing a median groove; small ridges extending from anterolateral parts of pedicle cavity around posterolateral margins of diductor muscle scars; smooth convex triangular calluses at anterior end of pedicle cavity divided by a median furrow, separated from cavity walls by narrow grooves; teeth cone-like, projecting inwards, supported by diverging dental plates arising from lateral margins of pedicle cavity; dental plates supported dorsolaterally against shell wall; adductor muscle scars impressed into calluses at end of pedicle cavity, subrectangular, having slightly pointed anterior ends and rounded posterior margins, often separated by a low median ridge; diductor muscle scars broad, subovate, well impressed posterolaterally, with broadly rounded margins, separated posteriorly by adductor muscle scars and anteriorly by a low median ridge; two linear canals of vascula media arising from adductor muscle scars, and up to eight pinnate canals of vascula genitalia from diductor muscle scars (Fig. 45).

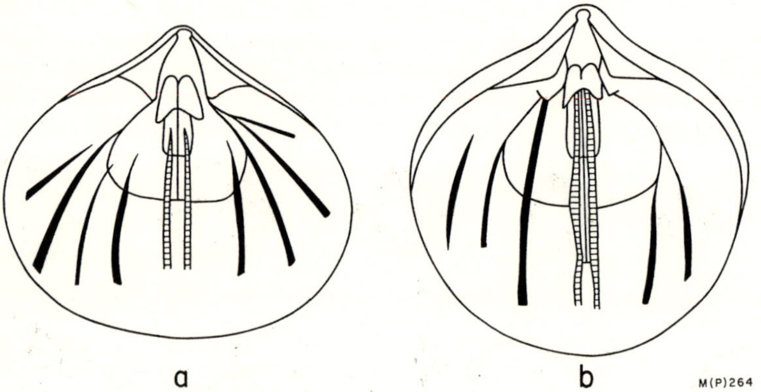

Figure 45. Pedicle valve interiors of *Cleiothyridina gloveri* Thomas showing the vascula media (lined), the vascula genitalia (black), the muscle field, and the calluses between the muscle field and the pedicle cavity. Locality 101/2, Burt Range Formation. a-CPC 8252, b-CPC 8251; both × 3.

Brachial valve. Sockets narrow, elongate, expanding anterolaterally, having rounded floors, and situated well below surface of cardinal plate; inner socket ridges high, especially at posterior margin of cardinal plate; cardinal plate triangular, perforated at apex by a small elongate foramen, and having a flat to slightly concave ventral face; median ridge triangular in cross section, commencing from beneath cardinal plate, extending two-thirds length of valve, and separating elongate sharply pointed adductor muscle scars; jugum simple, U-shaped; spires with up to twelve volutions.

Measurements	Length (pedicle valve)	Length (brachial valve)	Width	Height
CPC 8248	14	13	15	7.5
CPC 8249	15	14	18	8
CPC 8251	14	—	14	—
CPC 8252	—	—	14	—

Remarks: Thomas (1970) described *Cleiothyridina gloveri* Thomas from silicified and ferruginized limestone on the Carlton to Legune track at Sandy Creek crossing. He considered these rocks to be of probable Visean age. Veevers & Roberts (1968), however, identified the limestone at Sandy Creek as Burt Range Formation, and faunal studies in this work have dated the fauna as Tournaisian in age.

Specimens of *C. gloveri* from locality 101/2 in the Burt Range Formation are more finely silicified than the material from Sandy Creek. The pedicle valve interior has two callus-like structures which have not previously been described in either *C. gloveri* or in other species of *Cleiothyridina* (Fig. 45). The calluses are convex, triangular, and are located at the anterior end of the pedicle cavity between the adductor muscle scars and the deeply concave region of the cavity; they are separated by narrow furrows from the thickened dental plates. Pedicle adjustor muscles may have been attached to the calluses; however, because the calluses are smooth and convex, I suggest that they are unrelated to the musculature, and were deposited as a thickening by the mantle.

Because of its unusual reflexed pedicle valve, *C. gloveri* is comparable with relatively few species of *Cleiothyridina.* The shape of the pedicle valve resembles that of some specimens of *C. sublamellosa* (Hall) illustrated by Weller (1914, pl. 80, figs 51, 59) from the Chester Group of Illinois, USA. *C. gloveri* is distinguished by its more convex brachial valve.

The species referred to *Athyris parvirostra* Meek & Worthen by Nalivkin (1937, pl. 39, figs 10, 11), from the Visean Yagovkin Beds of Kazakhstan, has a similarly shaped pedicle valve, but differs in having a less convex brachial valve and a more acute apical angle. The former distinction applies also to specimens of *C. parvirostra* (Meek & Worthen) from the Keokuk Limestone of Illinois (Weller, 1914, pp. 481-82, pl. 80, figs 61-66).

Occurrence: Localities 100/17, 100/18, 100/20-100/22, 100/24-100/29, in an area between 6 and 7 miles northwest of Mount Septimus; localities 101/2, 101/5, 101/9, 101/15-101/18, in an area between 4.5 and 6 miles west of Mount Septimus; localities 103/2, 109/1, 109/2, 109/3, 128/1, and 128/6, middle part of Enga Ridge; localities 129/4, 421/5, 421/9, and 421/11, southern end of Enga Ridge; localities 114/9, 115/3, 115/7, and 117/1, Spirit Hill; locality 133/2, Alpha Hill; localities 122/2 and 122/3, Sandy Creek; Burt Range Formation.

Material: CPC 8248-8253, CPC 11103. All from locality 101/2. Several hundred silicified shells, loose specimens, and specimens preserved in limestone.

Age: Tournaisian.

CLEIOTHYRIDINA MINILYA Thomas

(Pl. 40, figs 1-15)

1970 *Cleiothyridina minilya* Thomas, pp. 180-83, pl. 23, figs 6-11, 13.

Description: External. Shell elongate to subrounded in juvenile growth stages, becoming gradually rounder and then transverse in adult stage; lateral margins rounded; greatest width at midlength; commissure rectimarginate in juvenile, strongly uniplicate to rarely rectimarginate in adult; concentric lamellae imbricating, giving off sheets of flat spines along posterior margins, becoming crowded at posterior, and having a density of about ten to fifteen per 5 mm at midlength of an adult shell.

Pedicle valve strongly to moderately convex, highest at midpoint; umbo prominent, incurved, projecting behind hinge, having a large round submesothyrid foramen at apex; cardinal margin submegathyrid in outline; umbonal shoulders gently concave, forming an apical angle of 100-115°; palintrope concave, well defined; lateral slopes moderately steep, becoming flatter towards posterolateral margins; median sinus commencing near midlength, rounded, varying from strongly developed to obsolete, and in some specimens forming a wide anterior lingual projection.

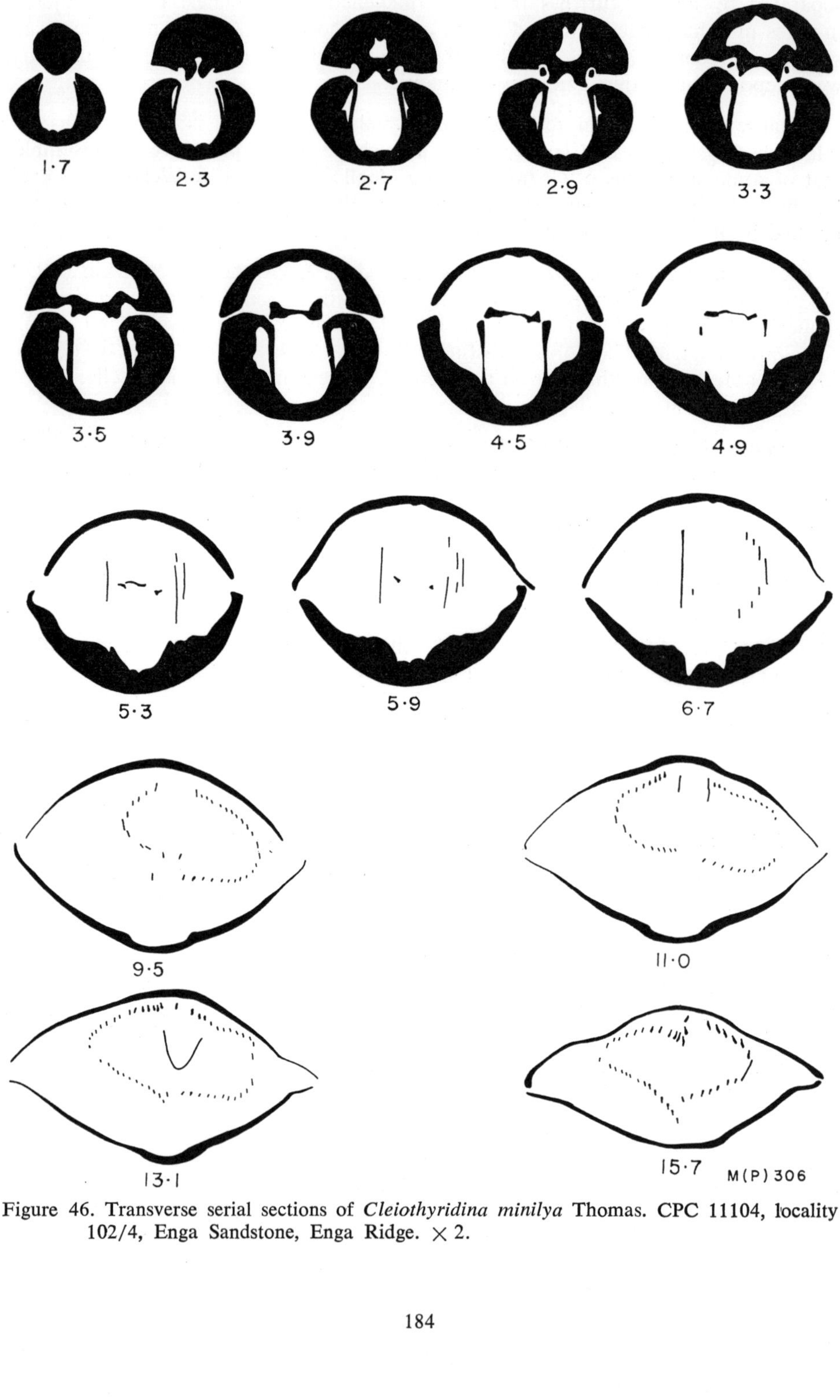

Figure 46. Transverse serial sections of *Cleiothyridina minilya* Thomas. CPC 11104, locality 102/4, Enga Sandstone, Enga Ridge. × 2.

Brachial valve usually less convex than pedicle valve, most convex at umbo, becoming flatter anterolaterally; umbo bluntly rounded, projecting a short distance behind hinge; lateral slopes gently convex; fold variably developed, obsolete in some specimens, or commencing at midpoint of valve, becoming broader and higher anteriorly; fold highest at or near front of valve in shells having a lingual extension of ventral sinus.

Internal (Fig. 46). Pedicle valve. Pedicle cavity long and narrow, with floor of cavity divided by a distinct median furrow; dental plates commencing at tip of umbo, flanking sides of pedicle cavity, separated from inner walls of shell, and in some valves projecting a short distance anteriorly into body cavity; dental plates thin near floor of valve, becoming stouter distally; teeth moderately large, pointed, and projecting dorsomedially; anterior of pedicle cavity bearing two rounded calluses divided by a median furrow, separated from walls of cavity by two deep troughs; front of troughs, at posterior margins of diductor muscle scars, swollen by two further callus-like structures; diductor muscle scars rounded to pointed and impressed posteriorly, becoming broader and less well impressed towards front; adductor muscle scars impressed and bluntly rounded posteriorly, tapering anteriorly, divided by a median furrow; vascula media consisting of four pairs of narrow linear canals arising from front of muscle field, extending to anterior margin; vascula genitalia narrow, having up to eight pairs of canals, and branching at margins; innermost canals of vascula genitalia arising from lateral margins of muscle field, and outermost (posterior) canals from posterolateral shoulders.

Brachial valve. Cardinal plate flat or keeled medially, triangular, having a small apical foramen, and bordered laterally by two narrowly triangular crural bases; inner socket ridges sometimes projecting ventrally below level of cardinal plate; inner socket ridges reaching floor of valve, greatly thickened dorsally; sockets narrow posteriorly, broadly rounded anteriorly; median ridge arising from beneath cardinal plate, broad and rounded posteriorly, becoming lower and narrower anteriorly, extending to about midlength of valve; adductor muscle scars elongate, pointed posteriorly, rounded anteriorly, impressed into floor, and divided by median ridge; jugum U-shaped in transverse section, spinose anteriorly; spires with up to thirteen volutions.

Measurements

	Length	Width	Height
CPC 8234	14	14	7
CPC 8238	29	41.4	25
CPC 8239	37	35.5	21

Remarks: Juvenile specimens of *Cleiothyridina minilya* Thomas from the Utting Calcarenite (localities 108/0, 108/3, and 108/4) have a shape and internal features which differ from those of mature individuals. These features, observed in silicified specimens, are briefly outlined in the following description.

Juvenile specimens of *Cleiothyridina minilya.* Shell longer than wide, attenuated posteriorly, but broadly rounded anteriorly, and usually biconvex; commissure rectimarginate; ornament of variably spaced concentric lamellae (no spines observed). Pedicle valve having a wide apical foramen; pedicle cavity poorly differentiated from body of valve; dental plates unthickened, and bearing large pointed inwardly directed teeth; muscle field not impressed. Brachial valve less convex than pedicle valve, having a small pointed umbo; cardinal plate short, flat, having a large triangular apical foramen, and supported on two high inner socket ridges; sockets deep, and well defined; adductor muscle scars subovate, slightly impressed into floor of valve, and separated by a median ridge.

The most important developments in the ontogeny of the species are the change from an elongate to a transverse outline, and the growth of a fold and sinus; many internal features which are not seen in juvenile shells, such as calluses and median ridges, develop as the shell becomes progressively thicker during ontogeny.

In the Bonaparte Gulf Basin, *C. minilya* is recorded from the Burt Range Formation, the Enga Sandstone, the Septimus Limestone, and the Utting Calcarenite. Specimens from the Enga Sandstone and the Septimus Limestone are closely comparable in size, external ornament, and shape with the type material from the Moogooree Limestone in the Carnarvon Basin (Thomas, 1970, pl. 23, figs 6, 7, 10). Specimens from the Utting Calcarenite are smaller than those from other formations; apart from this difference, they are closely comparable morphologically. A sectioned specimen from locality 102/4 in the Enga Sandstone (Fig. 46) has longer dental plates, and a more deeply impressed ventral muscle field than those in the specimen sectioned by Thomas (1970, fig. 70), but this variation is considered to be within that of the species.

Thomas (1970) considered that *C. minilya* was closest morphologically to *C. obmaxima* McChesney from the Fern Glen Formation of Illinois and the Keokuk Limestone of Iowa (Weller, 1914, pl. 79, figs 1-11). He separated the two species on the smaller size, weaker sinus, and less inflected anterior commissure on *C. minilya*. Specimens of *C. minilya* collected for this study from the Enga Sandstone and the Utting Calcarenite vary in the strength of the median septum, but some (*see* Pl. 4, figs 7-10) have a dorsal median deflection as strong as in *C. obmaxima*.

Specimens referred to *C. obmaxima* McChesney by Yang (1964, pl. 3, figs 23-34) from the Tournaisian of northeastern Gueizhou, China, are externally close to the more transverse individuals of *C. minilya*. The Chinese material is more transverse in young growth stages, and is distinguished by a consistently stronger anterior deflection, a deeper median sinus, and a higher fold.

Thomas (1970) also compared *C. minilya* with *C. glabistria* (Phillips) from Bolland, Yorkshire, and *Athyris ingens* de Koninck and *A. ornata* de Koninck from the Tournaisian of Belgium.

Occurrence: Localities 101/14-101/16, 101/18, 101/20, 101/21, in an area 4.5 miles west of Mount Septimus; localities 103/2, 103/7, 103/9, 103/11, 109/1-109/3, 109/5-109/7, 128/1, 128/2, 128/6, middle part of Enga Ridge; localities 113/5 and 113/7, Burt Range Amphitheatre; locality 133/2, Alpha Hill; locality 124/1, Flapper Hills; Burt Range Formation. Locality 102/4, northern end of Enga Ridge; localities 103/17B and 109/8, middle part of Enga Ridge; Enga Sandstone. Localities 104/12, 104/13, 104/15, 104/17, 150/6, Mount Septimus; locality 112/2, Langfield Point; Septimus Limestone. Localities 107/1, 107/7, 108/0, 108/3, 108/4, 108/4A, 143/1, Utting Calcarenite, Utting Gap.

Material: CPC 8234-8240, CPC 11104. CPC 8234-8236 from locality 102/4; CPC 8237-8238 from locality 108/3; CPC 8239 from locality 108/4; CPC 8240 from locality 108/0. Several hundred specimens.

Age: Tournaisian and Visean.

?CLEIOTHYRIDINA sp.

(Pl. 40, figs 16-22)

1970 *Cleiothyridina?* sp. nov. Thomas, pp. 189-91, pl. 23, figs 12a, 12b.

Description: External. Shell large, transversely subovate, and subequally biconvex; greatest width at midlength; hinge gently curved, approximately four-fifths the width of shell; commissure strongly uniplicate; ornament of regularly spaced concentric lamellae having a density of nine per 5 mm on midpart of a large individual; concentric lamellae strongly imbricating, smooth, lacking spines or other ornament.

Pedicle valve. Umbo blunt, usually poorly differentiated from umbonal shoulders, extending well behind hinge, moderately incurved, having a large apical foramen; apical angle 100-110°; lateral slopes most strongly convex at front of valve; palintrope moderately high, concave, well separated from umbonal shoulders; sinus commencing behind umbo, usually at one-third length of valve, having an evenly rounded floor and lateral margins, becoming wider and deeper and with a prominent lingual extension anteriorly; sinal angle 22-25°.

Brachial valve. Umbo small, pointed, extending a short distance behind hinge; lateral slopes most convex at front on either side of median fold, in some specimens becoming flattened on posterolateral extremities; valve highest at midlength, or along anterior part of fold; fold rising well above lateral slopes, poorly defined posteriorly, becoming higher anteriorly, and truncated by dorsal deflection of ventral sinus.

Internal. Pedicle valve. Pedicle cavity large, longer than wide, having a nearly flat floor bearing two longitudinal furrows, bordered laterally by thick anteriorly divergent dental plates; diductor muscle field subovate, bearing linear mantle canals, posteriorly divided by small anteriorly pointed adductor muscle scars; interior of valve with closely spaced linear mantle canals radiating from muscle field.

Brachial valve interior not observed.

Measurements

	Length	Width	Height
CPC 8241	32	45	—
CPC 8651	22	30 est.	—
CPC 8652	43.5	51.5	29.5

Remarks: This species is doubtfully assigned to *Cleiothyridina* because of the apparent lack of spines on the anterior of the concentric lamellae. On the few specimens displaying external ornament, the lamellae are more closely spaced than in *Athyris* M'Coy, 1844, and lack wide fringes with an ornament of radial grooves characteristic of *Actinoconchus* M'Coy, 1844; they do not appear to have been spinose on the available external moulds, but delicate extensions of the lamellae could have been removed by abrasion.

Thomas (1970, p. 191) compared *?Cleiothyridina* sp. most closely with *C. minilya* Thomas, distinguishing *?Cleiothyridina* sp. on the basis of its larger size and more pronounced fold. Specimens of *C. minilya* Thomas from the Bonaparte Gulf Basin (*see* Pl. 40, figs 1-15) have a greater variation in the strength of the fold than the type material from the Carnarvon Basin, and hence I tentatively differentiate the species on size, and a possible difference in the lamellose external ornament. Better preserved specimens of *?Cleiothyridina* are needed before a closer comparison can be made.

C. obmaxima (McChesney), figured by Weller (1914, pl. 79, figs 1-11) from the Fern Glen Formation of Illinois and the Keokuk Limestone of Iowa, resembles *?Cleiothyridina* sp. in its large size and well-defined fold and sinus; *C. obmaxima* has a more narrowly pointed ventral umbo.

Occurrence: Locality 305/1, 1.5 miles northeast of Ningbing Homestead; locality 25, 2.5 miles north-northeast of Point Spring, Weaber Range; locality 301, Milligans Hills; Burvill Beds. Locality 446/1, Waggon Creek breccia in Waggon Creek valley.

Material: CPC 8241, CPC 8651, CPC 8652. All from locality 305/1. Twenty-five specimens.

Age: Visean.

Suborder SPIRIFERIDINA Waagen, 1883

Superfamily SPIRIFERACEA King, 1846

Family CYRTOSPIRIFERIDAE Termier & Termier, 1949

Genus CYRTOSPIRIFER Fredericks, 1919

Type species: Spirifer verneuili Murchison, 1840, by original designation of Fredericks, 1919.

Remarks: The genus *Cyrtospirifer* is used in the same broad sense as proposed by Pitrat (1965), who included a large number of genera, including *Sinospirifer* Grabau (1931a), *Hunanospirifer* and *Centrospirifer* Tien (1938), *Liraspirifer* Stainbrook (1950), *Regelia* Crickmay (1952), and many of the genera proposed by Gatinaud (1949), in synonymy with *Cyrtospirifer* Fredericks. Most of these genera were previously separated from *Cyrtospirifer* on the basis of minor features such as micro-ornament, or the configuration of the sinal costae. In my opinion, these features need to be revised in considerable detail before they provide a satisfactory basis for separate generic status.

CYRTOSPIRIFER NINGBINGENSIS sp. nov.

(Pl. 41, figs 1-24; Pl. 42, figs 1-16; Pl. 43, figs 1-16)

Diagnosis: Shell of average size for genus; hinge extends into small pointed auricles; pedicle valve usually has a high flat or slightly convex catacline interarea, and a relatively straight umbo extending a short distance behind the hinge; sinus deep, with evenly rounded lateral margins; fold low; ventral adminicula parallel to subparallel, usually extending more than half the length of the valve.

Description: External. Shell subequally biconvex, wider than long, and pentagonal; hinge extending into pointed auricles; commissure uniplicate; costae simple, rounded, commencing at umbo and along median posterior part of umbonal shoulders, separated by intercostal grooves of same width; costae and intercostal grooves ornamented with fine radial lirae, and crossed by closely spaced concentric growth lines; twenty-four costae on each lateral slope of a shell 26 mm long and 41 mm wide.

Pedicle valve pentagonal, evenly convex, highest immediately above hinge; umbo pointed, almost straight, extending a short distance behind hinge; umbonal shoulders straight; lateral slopes running steeply to margins, flattening abruptly at auricles; sinus deep, broadly U-shaped in section, with evenly rounded lateral margins, forming a short lingual extension at front of valve, and having a sinal angle of 30-40°; sinus ornamented with bifurcating costae (Fig. 47); ventral

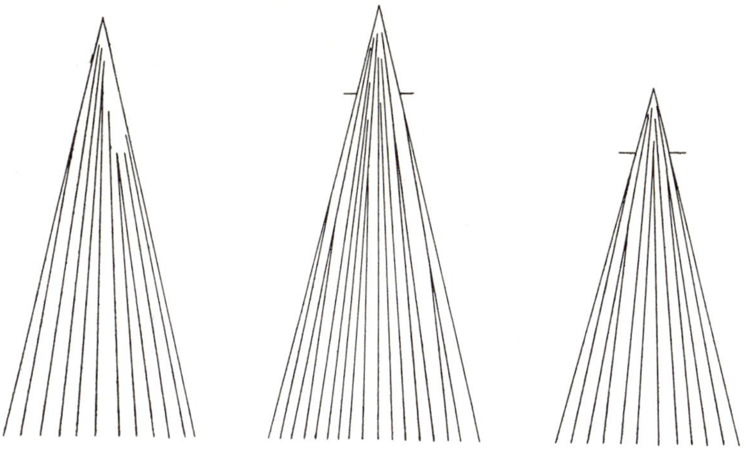

Figure 47. Pattern of sinal costae in *Cyrtospirifer ningbingensis* sp. nov. from the Ningbing Limestone. a-CPC 8141 from locality 140/1, Ningbing Range; b-CPC 11105, and c-CPC 8128, both from the locality 443/4, Jeremiah Hills. Both b and c are drawn from rubber casts in which the tip of the umbo is not preserved, and the posterior part of the sinus in a is poorly preserved. All × 2.

interarea almost catacline, high, flat towards hinge, but generally concave immediately beneath umbo; interarea 18.5 mm high on a valve 50 mm wide and 31.5 mm long, ornamented with denticle grooves having a density of five to six per 1 mm, and horizontal growth lines; delthyrium open, having narrow denticular cavities between lateral margins and posterior surfaces of dental plates; delthyrial angle approximately 40°.

Brachial valve subrectangular, moderately convex, highest towards middle or anterior part of valve; umbo broadly rounded, incurved, extending a short distance behind hinge; fold low, rounded, outlined by prominent furrows along its lateral margins, truncated anteriorly by lingual extension of ventral sinus; fold ornamented by narrow bifurcating costae; lateral slopes moderately convex, becoming flatter towards posterolateral extremities.

Internal (Fig. 48). Pedicle valve. Adminicula slender, straight, parallel to subparallel, joined at their posterodorsal extremities by a thickened horizontal delthyrial plate, and usually extending along floor of sinus; in some individuals adminicula more divergent, resting on floor of lateral slopes; subparallel adminicula extending more than half the length of valve (three-quarters length in some specimens), but divergent adminicula frequently shorter; adminicula straight to gently concave in transverse section; dental plates supported by adminicula, dorsally divergent in transverse section, thickened posteriorly, and subtending very large recurved club-shaped teeth; muscle field triangular, and longitudinally striated; adductor muscle scars narrow, elongate, flanked by broader triangular diductor muscle scars.

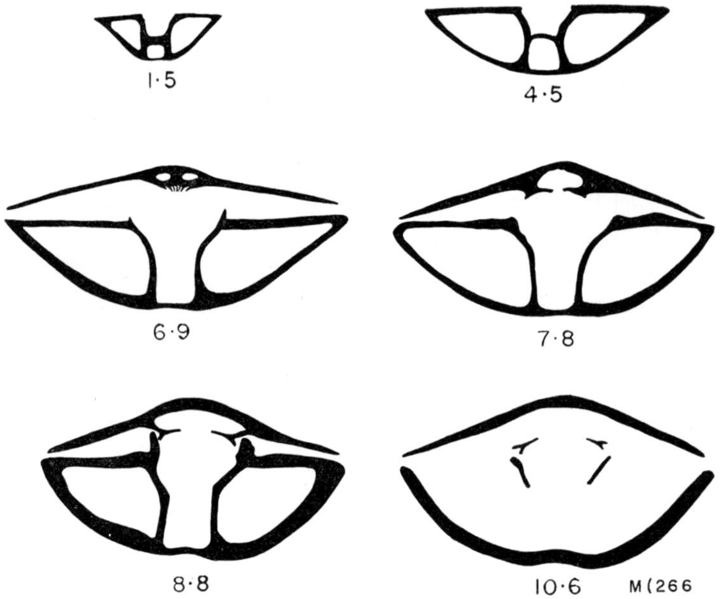

Figure 48. Idealized transverse serial sections of *Cyrtospirifer ningbingensis* sp. nov. CPC 11106, locality 140/1, Ningbing Limestone, Ningbing Range. × 1.

Brachial valve. Cardinal process resting on an apical callus, supported laterally by proximal parts of inner socket ridges, and having about ten small platelets; sockets deep, U-shaped in cross section, having high inner socket ridges; crura nearly horizontal, extending anteriorly from inner margins of inner socket ridges; median ridge originating from apical callus, extending more than three-quarters length of valve, and separating adductor muscle scars; adductor muscle scars varying from a single pair of very deeply impressed and narrowly triangular scars, to two pairs of broader, less well impressed scars; posterior adductor muscle scars subquadrate, with rounded posterior margins, and anterolateral margins draped over posterolateral margins of anterior scars; anterior adductor muscle scars subrectangular, rounded posteriorly, with tapering subrounded margins at front.

Remarks: The specimens of *Cyrtospirifer ningbingensis* described above are from the type locality (140/1), and are typical of specimens from all localities in the Ningbing Limestone except 141/1 and 443/4. Variation in the specimens from these localities is discussed below, together with additional details which can be added to the description of the species.

The pedicle valve of specimens from locality 141/1 in the Ningbing Range has an extremely variable morphology which ranges between two end members: *(a)* one in which the interarea is relatively low, concave, and trapezoidal in outline, the umbonal shoulders are concave, and the umbo is incurved and extends well behind the hinge (Fig. 49, b; Fig. 50, b); and *(b)* the typical form (described above), in which the interarea is high and triangular in outline, the umbonal shoulders are straight, and the umbo is small and relatively straight (Fig. 49, d). Most of the specimens from locality 141/1 belong to the first type *(a)*, but there is continuous gradation between the two end members.

Measurements	Length	Width	Height (of shell)	Height (of ventral interarea)
CPC 8141 Holotype	27	45 est.	—	14
CPC 8139 Paratype	23 est.	31.5	22	11
CPC 8140 Paratype	18	42	—	14
CPC 8142 Paratype	31.5	50	—	21.5
CPC 8145 Paratype (brachial valve)	21	36	—	—
		All from locality 140/1		
CPC 8125	36	1	19.5	4.5
CPC 8126	33	33	20	6
CPC 8127	28 est.	29.5	—	6.5
CPC 8128	23	29	—	—
CPC 8131 } brachial	27	38 est.	—	—
CPC 8132 } valves	27	38 est.	—	—
		All from locality 443/4		
CPC 8133	30	44	30	7
CPC 8134	24 est.	40 est	—	8
CPC 8135	31.5	47	—	8.5
CPC 8136	23	35	23	7.5
CPC 8137	33	54	—	12.5
CPC 8138	42	55	—	10
CPC 8147 }	26.5	41	—	—
CPC 8148 } brachial	36	52 est.	—	—
CPC 8149 } valves	27.5	50 est.	—	—
		All from locality 141/1		

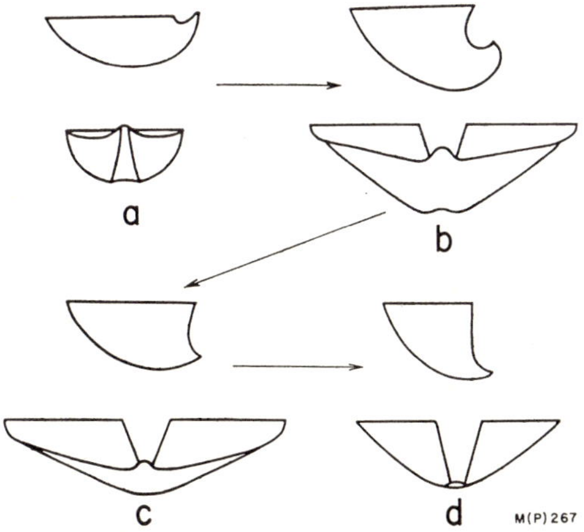

Figure 49. Lateral and posterior profiles of pedicle valves of *Cyrtospirifer ningbingensis* sp. nov. showing the range of variation in specimens from locality 141/1. Most of the specimens resemble b. Specimens at locality 443/4 are like a; all other localities in the Ningbing Limestone contain d-type valves.

Some brachial valves have an obsolete fold, and appear to have an almost completely smooth venter. Other features not observed on the type specimens which add to the description of the species include a micro-ornament of fine radial lirae having a density of eight to ten per 1 mm near the margins of the shell; and a delthyrium closed by a concave delthyrial plate extending almost its entire height (observed in one specimen only).

The collection from locality 443/4 in the Jeremiah Hills contains individuals which are generaly smaller than those from other localities. They are subovate to semicircular, have rounded posterolateral margins, and their greatest width is at the midlength (Fig. 50, a). In the pedicle valve the interarea is low, concave,

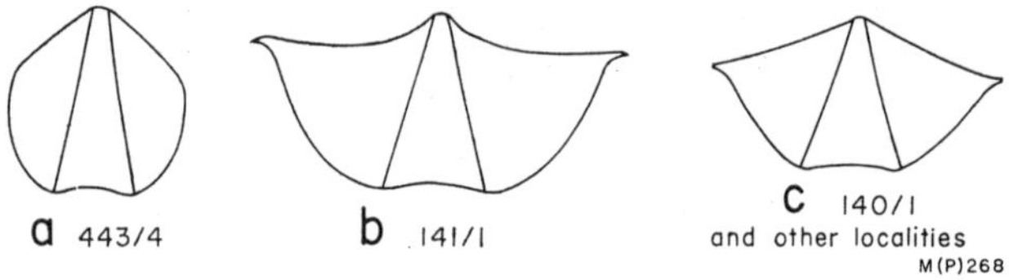

Figure 50. Ventral outlines of *Cyrtospirifer ningbingensis* sp. nov. a is from locality 443/4, b is from locality 141/1, and c is from locality 140/1; Ningbing Limestone.

and apsacline, the umbonal shoulders concave, and the umbo is incurved over the brachial valve and extends well behind the hinge. All the specimens have a similar shape, and, except for their apsacline interarea, they resemble one individual from locality 141/1.

Apart from the differences in shape, specimens from localities 141/1 and 443/4 are closely comparable morphologically with those from elsewhere in the Ningbing Limestone. The range of variation in the internal structures, such as the ventral adminicula, appears to be consistent with that in other collections, and the external ornament and micro-ornament, where observable, is the same on all specimens. Variation in the number of costae on the fold and sinus is to some extent dependent on the age of the specimens, and is unlikely to be of specific importance.

Because of continuous variation between 'end members' in the collection from locality 141/1, I feel it is realistic to refer all of the material from that locality to a single species; no formal designation is given to the apparently 'abnormal form' because it would require an arbitrary division in a continuously variable sequence of specimens from a single locality. It could be argued that, on the grounds of their different shape and the apparent absence of typical specimens, the material from locality 443/4 should be given subspecific status. However, because these specimens have the same external ornament and internal structures as other specimens, and because they have not been found elsewhere, I feel that it is best to regard them as a variant of *C. ningbingensis*. Their subovate shape may be the result of a slightly different environment, for example quieter water in which there was no need for the stability provided by an alate hinge typical of the other forms. It is noteworthy also that the age of the rocks at locality 443/4 is the same as that at the type locality (Famennian to IIβ-to IIIα, E. C. Druce, pers. comm.).

The two cases of variation discussed above are unlikely to have been the result of sorting by currents because the typical form is exclusively present in other localities containing *C. ningbingensis,* and because specimens similar to those at locality 443/4 have not been found elsewhere (with the exception of one specimen from locality 141/1). If sorting had controlled the distribution of the 'anomalous specimens' there would be a reasonable percentage of them in localities throughout the Ningbing Limestone.

The reason for the wide range of variation in specimens from locality 141/1 is difficult to explain. It could be argued that those shells living with their pedicle valves on the substrate developed a concave interarea and an incurved umbo (the anomalous type), whereas those living on their posterior margins developed a flat interarea for maximum stability. This is an unlikely hypothesis when it is considered that the typical form occurs exclusively at all other localities except 141/1 and 443/4.

The typical form of *C. ningbingensis* is present in reef, inter-reef, and back-reef environments within the Ningbing Limestone. Localities 141/1 and 443/4, containing the apparently abnormal forms, are both in back-reef limestone, and hence it is unlikely that their relative positions within the reef complex influenced the variation within the species.

Variation in the pedicle valve, particularly in the height and curvature of the interarea, length and curvature of the umbo, and length of the hinge, have been recorded in other species of *Cyrtospirifer* by, for example, Reed (1922), Nalivkin (1941), and Vandercammen (1959). Reed tackled the problem of nomenclature by dividing an intergrading assemblage into two subspecies (*C. pamiricus pamiricus* (Reed) and *C. pamiricus paralis* (Reed)), whereas both Nalivkin and Vandercammen placed a variable assemblage in one species (*C. schelonicus* Nalivkin, 1941, pp. 212-13, pl. 6, figs 1-5; *C. monticolaformis* Vandercammen, 1959, pp. 77-86, pl. 3, figs 1-7). Nalivkin recognized three informal morphological forms within *C. schelonicus,* and Vandercammen two informal forms in *C. monticolaformis.* Vandercammen (1959, pp. 15-16) suggested that the dimorphism shown in the *verneuili* group of spiriferoids was sexually controlled because each form was usually equally represented in the one bed of rock. The variation in *C. ningbingensis* is restricted to two localities, one of which contains an intergrading form, and the other a single form, and hence it does not follow this type of variation.

The typical form of *C. ningbingensis* resembles *C. pamiricus paralis* (Reed) (1922, pp. 110-11, pl. 16, figs 7-13) in overall shape, but is distinguished by a higher ventral interarea. Except for the lack of a wide auriculate hinge, *C. pamiricus pamiricus* (Reed) (1922, pp. 109-10, pl. 16, figs 1-6) resembles the abnormal form which is dominant in the collection from locality 141/1. *C. pamiricus* comes from the Famennian of the Pamirs, USSR.

Some specimens of *C. schelonicus* Nalivkin from the Shelon, Svinord, and Ilmen beds of the Russian Platform resemble the typical form of *C. ningbingensis (*Nalivkin, 1941, pl. 6, figs 2, 5). These specimens are distinguished from *C. ningbingensis* by their coarser external ornament. Another specimen of *C. schelonicus* figured by Krylova (1951, pl. 4, fig. 3) from the Devonian of the Stalingrad area, USSR, has an outline resembling that of the typical form of *C. ningbingensis.* This specimen differs from *C. ningbingensis* in having a coarser external ornament, a stronger fold and sinus, and a slightly lower ventral interarea.

The anomalous form from locality 141/1 has an outline comparable with that of the more transverse specimens of *C. calcaratus magnus* Martinova (1961, pl. 14, figs 1-5, especially fig. 2) from the Lower Famennian Meister Beds of Kazakhstan. Large specimens from locality 141/1 (e.g. Pl. 42, figs 11-13) are morphologically close to *C. verneuili* var. *gosseleti* (Grabau) from the Upper Devonian of Belgium (Grabau, 1931a, pl. 28, figs 2, 3).

The material from locality 443/4 resembles a specimen of *C. whitneyi* (Hall) figured by Kindle (1909, pl. 8, fig. 5) from the Famennian Ouray Limestone in New Mexico, which has a short hinge, rounded posterolateral extremities, and a low ventral interarea. Other specimens of *C. whitneyi* figured by Kindle (1909, pl. 8, figs 2-4) have wider hinges, ventral interareas of varying heights, and resemble both of the other forms of *C. ningbingensis*. Specimens of *C. whitneyi* from the Hackberry Formation, Iowa, in the U.S. National Museum, are smaller than both Kindle's specimens and *C. ningbingensis;* the height of the ventral interarea is variable, but never as high as in the typical form of *C. ningbingensis,* the umbo is usually well incurved, and the fold and sinus are more strongly developed.

Graphical methods of showing the variation in *C. ningbingensis* could not be used because most specimens were broken, or embedded in hard limestone. Most of the figured specimens were prepared by calcining blocks of limestone.

The specific name is taken from the Ningbing Range.

Occurrence: Localities 8/6, 21/16, 21/22, 140/1 (the type locality), 140/2, 140/3, 140/5, 141/1, 141/5, 149/2, 455/2, 455/3, Ningbing Range; localities 443/4, 443/13, 443/22, 447/10, Jeremiah Hills; Ningbing Limestone (Famennian part). Localities 105/735, 105/830, 105/895, Buttons Beds, Buttons Crossing on the Ord River.

Material: CPC 8125-8149, CPC 11105, CPC 11106. Holotype CPC 8141, paratypes CPC 8139, 8140, 8142, 8143, 8144, and 8145. CPC 8125-8132, 11105 from locality 443/4; CPC 8133-8138, 8146-8148 from locality 141/1; CPC 8139-8145, 11106 from locality 140/1. Several hundred specimens preserved in massive limestone.

Age: Famennian.

CYRTOSPIRIFER DEPRESSUS sp. nov.

(Pl. 44, figs 1-20)

Diagnosis: Shell small to average size for genus; wider than long, auriculate in adult specimens, but with rounded posterolateral margins in juvenile growth stages; costae evenly rounded posteriorly, becoming flatter anteriorly, and wider than the intercostal furrows; pedicle valve has two forms, one with an orthocline to very slightly apsacline interarea, and straight umbonal shoulders, and the other with a catacline to very steeply apsacline interarea, and concave umbonal shoulders; sinus shallow, narrow posteriorly, having a wide flat floor anteriorly; fold low and flat-topped; cardinal process supported by a short septum-like callus.

Description: External. Shell of small to medium size, subequally biconvex in forms with a low ventral interarea to unequally biconvex in forms with a high ventral interarea, wider than long, subrectangular to subovate; hinge extending into small flattened auricles in large specimens; small individuals with rounded posterolateral margins; commissure uniplicate; costae simple, originating from umbo and from along median parts of umbonal shoulders, evenly rounded at posterior of shell, becoming wider and flatter anteriorly, and separated by narrow intercostal furrows; twenty to thirty costae on each lateral slope on valves between 25 and 45 mm wide; micro-ornament of fine radial lirae having a density of ten per 1 mm at anterior margin of a valve 30 mm wide and 18 mm long, and concentric growth lines.

Pedicle valve has two forms: one with an orthocline to very slightly apsacline interarea, and straight umbonal shoulders, and the other with a catacline to very steeply apsacline interarea, and concave umbonal shoulders (Fig. 51); pedicle valve highest and most convex near umbo, the form with catacline interarea

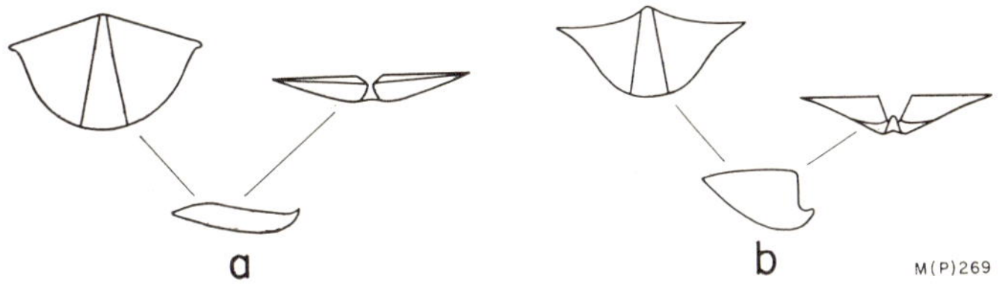

Figure 51. Ventral, posterior, and lateral profiles of the two forms of the pedicle valve of *Cyrtospirifer depressus* sp. nov. Locality 145/5, Buttons Beds. a has an orthocline to very slightly apsacline interarea, and straight umbonal shoulders; b has a catacline to steeply apsacline interarea, and concave umbonal shoulders.

being higher than the other; anterior and lateral margins evenly rounded; umbo small, narrow, and slightly incurved; lateral slopes moderately convex, becoming flattened at posterolateral extremities; interarea 4.5 mm high on a valve 14 mm long and 28 mm wide with a catacline interarea, and 7 mm long on a valve more than 21 mm long with an orthocline interarea; interarea ornamented with coarse denticle grooves having a density of about four per 1 mm; apex of delthyrium closed by a small flat recessed delthyrial plate having an arched dorsal margin; delthyrial angle ranging from 35° to 60°; sinus shallow, narrow and V- to U-shaped in cross section near umbo, gradually becoming wider and flat-bottomed anteriorly; sinus bordered by two prominent costae, and ornamented by up to nine mainly intercalating costae (Fig. 52); sinal angle 20-25°.

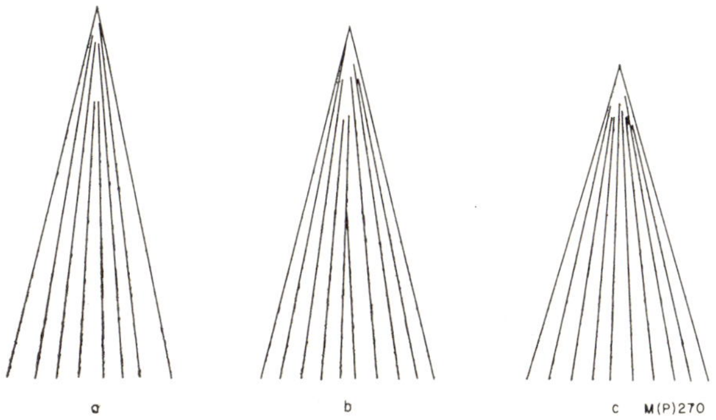

Figure 52. Pattern of sinal costae on *Cyrtospirifer depressus* sp. nov. Locality 145/5, Buttons Beds. a-CPC 8153; b-CPC 8151; c-CPC 8155. All × 2.

Brachial valve. Umbo small and barely extending behind hinge; lateral slopes evenly convex except for flattened posterolateral margins; fold low, flat topped, narrow towards umbo, much broader anteriorly, and ornamented by up to nine intercalating costae (Fig. 53); lateral margins of fold outlined by strong intercostal furrows.

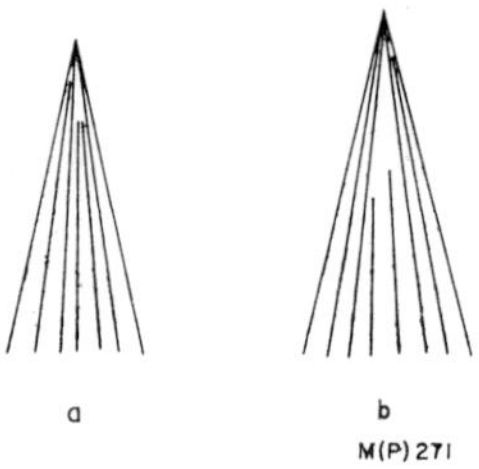

Figure 53. Pattern of costae on the fold of *Cyrtospirifer depressus* sp. nov. Locality 145/5, Buttons Beds. a-CPC 8150; b-CPC 8152. Both × 2.

Internal. Pedicle valve. Ventral adminicula slender, unthickened, subparallel to slightly divergent, usually diverging at 20-30°, and extending almost half the length of valve; inner floor of sinus steeply inclined, forming a narrow median 'ridge' towards posterior of valve; muscle scars and teeth not observed.

Brachial valve. Cardinal process resting on an apical callus usually in the form of a short septum-like structure; median ridge low, extremely narrow, arising from callus, extending almost to midlength of valve, separating adductor muscle scars; adductor muscle scars narrow, pointed to bluntly rounded posteriorly, expanding anteriorly, and having a straight anterior margin; cardinal process, sockets, and crura not observed.

Measurements		Length	Width	Height/Length of Ventral Interarea
CPC 8153 Holotype		20.5	35 est.	5
CPC 8154 Paratype		14	28	4.5
CPC 8155 Paratype		18	30 est.	5
CPC 8156 Paratype (brachial valve)		20 est.	40	—
CPC 8157 Paratype		21	—	7
CPC 8158 Paratype		14	28 est.	6.5
CPC 8159	pedicle valve	12.5	30	4.5
	brachial valve	20	32	—
CPC 8160	brachial valve	28	36 est.	—
	brachial valve	9	18 est.	—
CPC 8161 brachial valve		12	25	—
CPC 8162		15.5	28 est.	5

Remarks: The two distinct forms of *Cyrtospirifer depressus* sp. nov. are recognized in the same bed at locality 145/5. There are no specimens in the collection which intergrade between the two forms, and hence the dimorphism of *C. depressus* resembles that in some Belgian species of *Cyrtospirifer* described by Vandercammen (1959). Vandercammen suggested that the dimorphism in the Belgian species resulted from sexual differences. The description and diagnosis gives the impression that *C. depressus* is always dimorphic. However, because my material has been collected from one locality only, this may not be the case, and the variation could resemble that in *C. ningbingensis* (pp. 190-92), where a single 'typical' form is present in all except two of the localities in the Ningbing Limestone, and 'atypical' forms in these two localities are regarded as local variants of the species. The type of variation in *C. depressus* therefore cannot be determined conclusively until the species is collected from other localities.

The closest species is *C. kemerovensis* (Besnosova) (1959, pp. 51-53, pl. 2, fig. 1) from the Lower Tournaisian Abishevsky Beds of the Kuznetsk Basin, which resembles the form of *C. depressus* having an orthocline interarea. *C. kemerovensis* has a deeper median sinus, lacks a delthyrial plate, and is larger.

C. ziganensis Krestovnikov & Karpyshev (1948, pl. 1, fig. 7) from the Etroeungt of the Urals, USSR, has a shape and lateral profile similar to that of the form of *C. depressus* having an apsacline interarea. *C. ziganensis* is distinguished from *C. depressus* by the possession of a slightly finer costate ornament.

The specific name *depressus* (Latin for low or flat) refers to the low fold on the brachial valve.

Occurrence: Locality 145/5, Buttons Beds, 7.5 miles northwest of Mount Septimus.

Material: CPC 8150-8162. Holotype CPC 8153, paratypes CPC 8154-8158, and 8160. All from locality 145/5. Fifty specimens.

Age: Famennian.

Genus TENTICOSPIRIFER Tien, 1938

Type species: Spirifer tenticulum de Verneuil, 1845, from the late Devonian of the Lake Ilmen region, Leningrad, USSR, by original designation of Tien, 1938.

Diagnosis: See Pitrat (1965, p. H704).

Remarks: Chinese specimens of *Tenticospirifer,* for example *T. tenticulum* (Verneuil), were described by Tien (1938) as having a 'proper median septum' in the brachial valve. Besnosova (1959, p. 43) questioned the existence of the septum and illustrated (fig. 7) serial sections of topotype material in which a median septum was absent. In the specimen she illustrated in figure 7, section b is structurally close to those figured by Tien (1938, fig. 31) except that it contains a greater amount of apical shell material in the brachial valve; a smaller apical thickening in the Chinese specimens may have caused Tien to mistake the support for the cardinal process as a small median septum.

Sidyachenko (1962) pointed out that the internal structure of *Tenticospirifer* was identical with that of *Cyrtospirifer* Fredericks, and placed *Tenticospirifer* in synonymy with that genus because, in her opinion, the pyramidal shape of *T. tenticulum* was too variable to be of generic significance. Pitrat (1965, p. H704) noted the similarity between *Tenticospirifer* and *Cyrtospirifer,* and retained *Tenticospirifer* as a separate genus on the basis of its hemipyramidal pedicle valve with a large high interarea, without referring to the variability discussed by Sidyachenko. I agree with the retention of the genus for small to medium-sized individuals possessing a hemipyramidal pedicle valve, but point out that (1) in some species there is considerable variation in the shape of the pedicle valve, particularly in the height and convexity of the interarea, and the degree of incurvature of the umbo; and (2) there may be a gradation between large species of the genus and small species of *Cyrtospirifer* possessing high ventral interareas.

TENTICOSPIRIFER COLUMNARIS sp. nov.

(Pl. 45, figs 1-26)

Diagnosis: Shell small for genus, having a low fold and a narrow sinus bearing relatively few costae; sixteen to twenty-one costae on each lateral slope; pedicle valve high, and rounded pentagonal in outline; ventral umbo varying from nearly straight to moderately incurved; interarea high, flat to concave, and catacline to apsacline; delthyrial plate small, and confined to the apex of the delthyrium; brachial valve low, gently convex, and transverse; cardinal process resting on a swollen ridge between the inner socket ridges.

Description: External. Shell small, unequally biconvex, and rounded pentagonal; greatest width towards midlength or at hinge; lateral margins rounded; postero-lateral margins rounded to angular; commissure uniplicate; costae simple, arising from tip of umbo, and numbering from sixteen to twenty-one on each lateral slope; costae and intercostal furrows rounded, of equal width, ornamented by simple regularly spaced radial lirae having a density of twelve per 1 mm at 3 mm from umbo (measured on a brachial valve 8 mm long and 11 mm wide); radial ornament crossed by fine concentric growth lines.

Pedicle valve higher and more convex than brachial valve; umbo pointed, occasionally twisted laterally, and variably incurved, on most valves moderately incurved and extending well behind hinge (6.5 mm on a valve 21 mm long and 21 mm wide), and on others with extremely high interareas, nearly straight, and projecting only a short distance behind hinge (3 mm on a valve 16.5 mm long and 23 mm wide); umbonal shoulders flat to slightly concave; lateral slopes steep, extending uniformly from umbo to lateral margins; interarea usually apsacline and concave, greatest concavity immediately below beak, but on some large specimens straight except for a slight concavity immediately below beak, and almost catacline; interarea asymmetrical on valves with twisted beaks; interarea ornamented with denticle grooves having a density of eight to twelve per 1 mm, and horizontal growth lines; delthyrium open, with a delthyrial angle averaging 40°, and ranging from 25° to 50°; lateral margins of delthyrium separated from dental plates by deep denticular cavities; sinus shallow, evenly rounded, commencing at tip of umbo; sinal costae varying from six to twelve, increasing by bifurcation (Fig. 54).

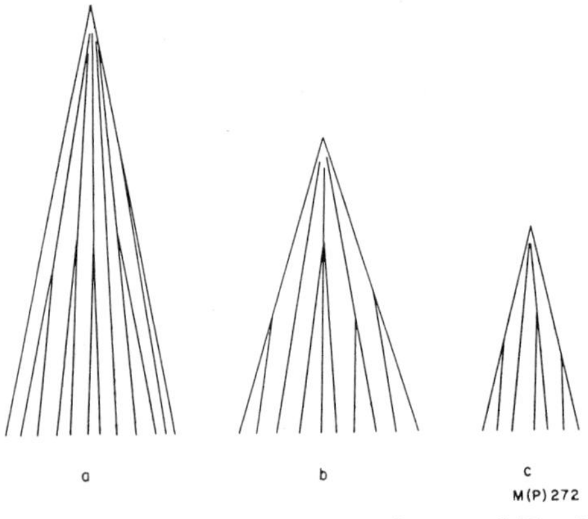

Figure 54. Pattern of costae in the sinus of three specimens of ***Tenticospirifer columnaris*** sp. nov. a-CPC 8179, locality 12/150; b-CPC 11107, locality 250/8; c-CPC 11108, locality 250/8. All × 2.

Brachial valve low, gently convex, wider than long, and subrectangular; umbo projecting a short distance behind hinge (1 mm on a valve 8.5 mm long and 12.5 mm wide) and having gently concave umbonal shoulders; lateral slopes evenly convex; interarea anacline to orthocline and 0.5 mm high on a valve 8.5 mm long and 12.5 mm wide; notothyrium broad, and bounded by outer socket ridges; fold low, frequently poorly differentiated from lateral slopes, and less commonly outlined by lateral furrows; fold of large specimens bearing up to twelve costae, outer costae simple, and inner two pairs of costae bifurcating posteriorly (Fig. 55).

Figure 55. Pattern of costae on the fold of *Tenticospirifer columnaris* sp. nov. CPC 8175, locality 12/4. × 3.

Internal (Figs 56, 57). Pedicle valve. Ventral adminicula usually straight, diverging at 25-35°, extending one-third length of valve; in transverse section adminicula subparallel or converging dorsally towards dental plates, slender near floor of valve, but becoming thicker dorsally; dental plates curving laterally from adminicula, swollen dorsally, supporting large blunt teeth; subdelthyrial plate at apex of delthyrium; muscle field divided by a low median ridge extending one-quarter length of valve; muscle scars not observed.

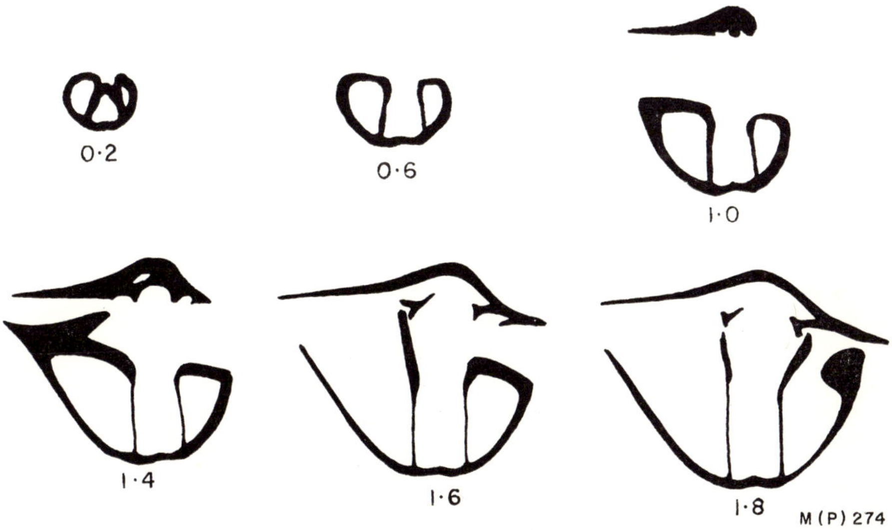

Figure 56. Transverse serial sections of *Tenticospirifer columnaris* sp. nov. CPC 11109, locality 12/4, Westwood Member, Westwood Creek. × 2.

N

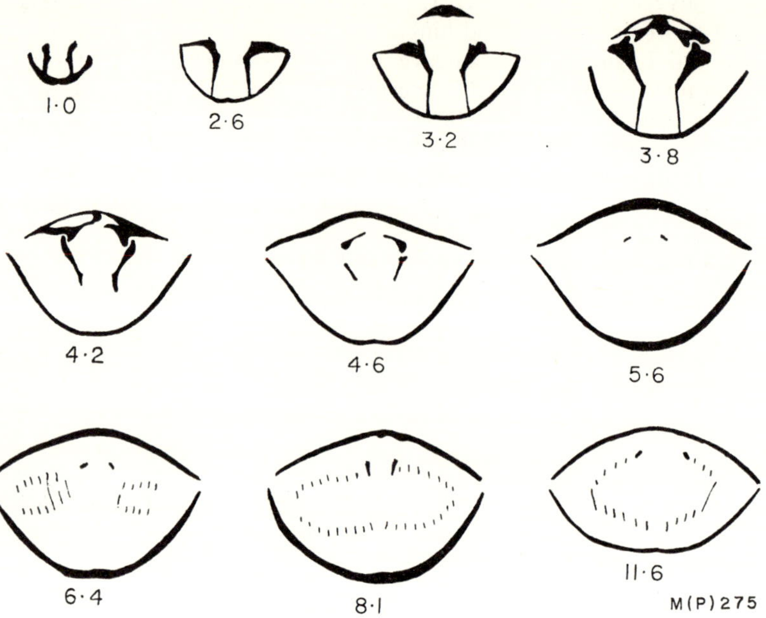

Figure 57. Transverse serial sections of *Tenticospirifer columnaris* sp. nov. CPC 11110, locality 12/150, Westwood Member, Westwood Creek. × 2.

Brachial valve. Cardinal process resting on a slender base between inner socket ridges, swollen ventrally, and having a myophore of more than six slender platelets; sockets strongly divergent, rounded to subangular, V-shaped in transverse section, having stout inner and outer socket ridges; crurae originating from anterior margins of inner socket ridges; spires projecting laterally, typically spiriferoid, and lacking a jugum.

Measurements	Length	Width	Height	Height/Length of Ventral Interarea
CPC 8137 Paratype	14.5	20	14.5	10
CPC 8174	15	22.5	17	13
CPC 8175 Holotype	8.5	11	6.5	2.5
CPC 8176	20.5	21	12.5	7
CPC 8177	11.5	23.5	—	12.5
CPC 8178	18.5	21	13	5.5
CPC 8180	13	18.5	—	6

Remarks: The variation within *Tenticospirifer columnaris* sp. nov. has already been briefly outlined in the description. Specimens from localities 12/4 and 12/150 in the Westwood Member of the Cockatoo Formation differ considerably in the height and concavity of the ventral interarea, the curvature of the umbo, the extension of the umbo behind the hinge, and the position of the widest part of the shell. Small specimens, between 4 and 6 mm wide, from locality 12/4 have a prominent incurved ventral umbo extending well behind the hinge, and a concave ventral interarea. They are widest a short distance behind midlength, and have rounded posterolateral extremities. In larger specimens, around 15 mm wide, the curvature of the umbo and the concavity of the interarea are either similar to those of the smaller individuals, or the umbo is straighter, and the interarea less concave. The shell is widest in front of the hinge, and the posterolateral margins are rounded. The largest specimens, 20 to 23 mm wide, have a high ventral interarea which is flat or slightly curved immediately beneath the umbo. The umbo is almost straight, and extends only a short distance behind the hinge. The shell is widest at the hinge, and the posterolateral margins are pointed.

Specimens from locality 12/150, 75 feet stratigraphically above locality 12/4, are lower and have more concave ventral interareas. Those between 10 and 15 mm wide have a concave ventral interarea, and a gently incurved umbo which extends well behind the hinge. The widest part of the shell is in front of the hinge. Larger individuals, 20 mm wide, retain these characteristics, and hence differ from the large specimens from locality 12/4.

A large collection from locality 250/8 in the Hargreaves Member of the Cockatoo Formation contains specimens resembling those from locality 12/4, suggesting that the form having a high nearly flat ventral interarea in the adult is typical of *columnaris,* and that the adult specimens from locality 12/150 retained the morphology of younger individuals. Collections from other localities are too small to determine whether or not the two variants of *columnaris* should be separated at the subspecific level. In view of the lack of variation in other features, and a similar variation in the pedicle valve of other species of *Tenticospirifer,* I prefer to group all of the specimens in one species. In this instance the variation in the shape of the pedicle valve may have been caused by changing environmental conditions within the Frasnian reef complex at Westwood Creek.

Variation similar to that in *T. columnaris* has been described by Nalivkin (1941, p. 212, pl. 6, figs 6, 7) in *Cyrtospirifer stolbovi* Nalivkin. This species appears to belong to *Tenticospirifer,* and differs from *columnaris* only in the possession of a wider hinge. *T. stolbovi* is known from the Frasnian Svinard Beds of the Russian Platform. Vandercammen (1959, pp. 69-76, pl. 2, figs 25-30) gave a more detailed description of the variation in specimens of *'Cyrtospirifer' stolbovi* Nalivkin from the Frasnian of Belgium, and recognized two forms within the species: (1) with a pyramidal pedicle valve having a flat or slightly incurved ventral interarea, and a slightly incurved umbo; and (2) with a lower and more convex ventral interarea and a more incurved umbo. These correspond closely to the larger specimens from localities 12/4 and 12/150 respectively. The Belgian material has a finer ornament, and pustules on the crests of the costae; this micro-ornament has not been described in the type specimens of *stolbovi.*

Paeckelmann (1942) used a similar dimorphism as the basis of two subspecies in the Upper Middle Devonian of Germany: *T. aperturata verneuiliformis* Paeckelmann (1942, fig. 24) which is morphologically close to specimens of *columnaris* from locality 12/150, and *T. aperturata cuspidata* (Archaic & Verneuil) which has a straight to slightly concave ventral interarea (Paeckelmann, 1942, fig. 28) and is similar to specimens from locality 12/4.

The form of *columnaris* from locality 12/4 is closely comparable with *T. cyrtinaformis* (Hall & Whitfield) from the Lime Creek Formation of Iowa (Fenton & Fenton, 1924, pp. 162-64, pl. 29, figs 1-7). *T. cyrtinaformis* has a comparable straight or slightly incurved ventral interarea, a similar external ornament, and a weak fold and sinus. The two species are distinguished by internal features of the brachial valve; the sockets and inner socket ridges in *columnaris* diverge at a larger angle than those in *cyrtinaformis,* and the cardinal process of *columnaris*

N2

consists of a high swollen ridge with a myophore of slender platelets, whereas in *cyrtinaformis* it consists of low platelets between the inner socket ridges. *T. rara* (Webster), also from the Lime Creek Formation (Fenton & Fenton, 1924, pl. 30, figs 23-26; pl. 31, figs 12-15), may be large individuals of *cyrtinaformis* possessing a well-defined fold and sinus. *T. rara* resembles specimens of *columnaris* from locality 250/8 which typically have a stronger fold and sinus than those from localities 12/4 and 12/150.

The type species, *T. tenticulum* (Verneuil) from the late Devonian of the USSR (Pitrat, 1965, fig. 570, 3a-c), is larger than this species. Specimens from the late Devonian of Yunnan referred to *tenticulum* by Mansuy (1912, pl. 13, fig. 5) are similar in size and shape to the typical form of *columnaris*. Grabau (1931a, p. 377) assigned these specimens to *Spirifer* cf. *tenticulum* var. *quadrangularis* Grabau.

Several species from Hunan, China, described by Tien (1938) can be compared with *columnaris*. *T. supervilis* Tien (1938, pp. 127-29, pl. 17, figs 1-3; pl. 19, figs 1-8) is larger, has a more convex brachial valve, a wider and more pronounced fold and sinus, slightly broader costae, a larger subdelthyrial plate, and a higher support beneath the cardinal process. *T. vilis* (Grabau) figured by Tien (1938, pl. 16, figs 8, 9) is generally more transverse, and has a wider sinus and a lower ventral interarea. *T. hsikuangshanensis* Tien (1938, pp. 131-36, pl. 18, figs 3-5) is slightly more transverse, has a much stronger fold and sinus, a lower ventral interarea, and a slightly more convex brachial valve. The beak of *hsikuangshanensis* is straight when young and curved when mature: the reverse of this species.

T. lictor (Nalivkin) from the Semiluki Beds of the Russian Platform is morphologically close to *columnaris*. A specimen illustrated by Nalivkin (1930, pl. 7, fig. 5) is morphologically midway between the forms of *columnaris* from localities 12/4 and 12/150, and a specimen illustrated by Lyashenko (1959, pl. 59, fig. 6) resembles the typical form of *columnaris*.

The specific name *columnaris* is Latin for 'of pillars', and refers to the many high pillars of limestone near the type locality.

Occurrence: Localities 12/3, 12/4 (the type locality), 12/150, 12/9, 13/1, 13/6, 37/1, 37/2, Westwood Member, Westwood Creek; locality 250/8, Hargreaves Member, Hargreaves Hills; locality 64/1, Hargreaves Member, Pretlove Hills; Cockatoo Formation.

Material: CPC 8173-8181, CPC 11107-11110. Holotype CPC 8175, paratype CPC 8173. CPC 8173-8175, 8177, 11109 from locality 12/4; CPC 8176, 8178-8179, 11110, from locality 12/150; CPC 8180, 11107-11108 from locality 250/8; CPC 8181 from locality 37/2. Approximately 80 loose specimens, and many specimens in blocks of quartz sandstone.

Age: Frasnian.

Family SPIRIFERIDAE King, 1846

Genus PROSPIRA Maxwell, 1954

Type species: Prospira typa Maxwell, 1954, pp. 35-36, pl. 4, figs 6-8 (*see also* Maxwell, 1961, pp. 89-91, pl. 20, figs 10-14), from the Tournaisian Neil's Creek Clastics of Queensland, by original designation of Maxwell, 1954.

Description: Shell moderately transverse in young growth stages, quadrate at maturity, having mucronate posterolateral extremities; pedicle valve more strongly convex than brachial valve; ventral interarea moderately high, with blunt lateral extremities, bearing indistinct denticle grooves; ventral sinus narrow, particularly posteriorly, and shallow and smooth for the first 10 mm from umbo; in front of smooth area, sinus has a flattened floor, and extremely weak costae consisting of a median costa or pair of costae, and up to two pairs of costae branching from bordering costae; fold low, with a shallow median furrow, bordered by wide intercostal furrows; costae rounded, separated by narrow intercostal furrows, having a density of five per 5 mm at front of mature specimens; micro-ornament of strong radial lirae, and concentric growth lines; pedicle valve interior extensively thickened at apex, callus material almost completely obscuring short ventral adminicula; muscle field deeply impressed, consisting of narrow longitudinally striated adductor muscle scars, and striated elongate-ovoid diductor muscle scars; brachial valve interior with small circular to elongate adductor muscle scars; crura widely divergent, extending from inner margins of inner socket ridges.

Remarks: The description of *Prospira* is compiled from descriptions and figures of the type species given by Maxwell (1954; 1961), and from a description of topotype specimens by R. McKellar (pers. comm.), then of the Geological Survey of Queensland.

Prospira is distinguished from *Unispirifer* Campbell, 1957, by having fewer costae on the lateral slopes, an indistinct or very simple pattern of costae in the sinus, a more quadrate shape in the adult, mucronate rather than alate postero-lateral extremities, shorter ventral adminicula, and smaller size. Maxwell's suggestion (1961, p. 89) that species of *Unispirifer* are early members of the *Unispirifer-Prospira* complex of species cannot be substantiated because species of both genera are found in sediments of Tournaisian and Visean age. The relationship between *Prospira* and *Fusella* M'Coy, 1844, has been adequately discussed by Maxwell (1961).

PROSPIRA INCERTA Thomas

(Pl. 45, figs 27-38)

1970 *Prospira? incerta* Thomas, pp. 94-96, pl. 9, figs 1-5.

Description: External. Shell transverse, triangular, approximately half as long as wide, and moderately and equally biconvex; shell widest at hinge, which extends into narrow pointed extremities; small growth stages extremely wide, with sharp mucronate lateral extremities; larger growth stages less transverse because of an increasing anterior growth vector (Fig. 58), and having blunter lateral extremities and truncated interareas; commissure weakly uniplicate; twelve to fifteen costae on each lateral slope; costae simple, rounded, becoming obsolete laterally, and separated by narrow rounded furrows half the width of one costa; micro-ornament of very fine radial lirae and concentric growth lines; shell becoming lamellose anteriorly.

Pedicle valve most strongly convex at umbo; umbo small, poorly differentiated from lateral slopes, moderately incurved over ventral interarea; lateral slopes convex medially, becoming flatter towards lateral margins, ending abruptly at sharp cardinal ridge posteriorly; interarea apsacline, gently concave, especially near umbo, or flat, high along its entire width, truncated at lateral extremities, and 3.5 mm high at umbo on a valve 28.5 mm wide and 18.5 mm long; interarea ornamented with fine horizontal growth lines and vertical denticle grooves, grooves sometimes fusing on younger parts of interarea (i.e. the part nearest brachial valve); denticle grooves having a density of approximately three per 1 mm on younger parts of interarea; delthyrium open, delthyrial angle 50°, bordered by narrow denticular cavities and ridges formed by posterior surfaces of dental plates; sinus narrow, bordered by two strong costae, having a smooth flat to rounded floor, and steep sides, and extending into a narrow lingual projection anteriorly; sinal angle 12-15°.

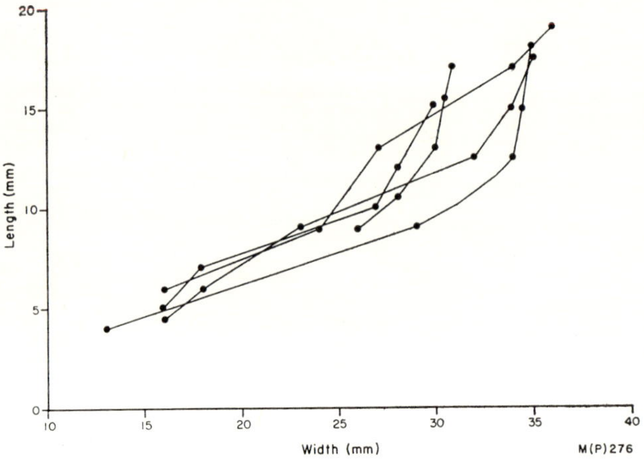

Figure 58. Length/width ontogenies of five pedicle valves of *Prospira incerta* Thomas from locality 102/4, Enga Sandstone, showing an increase in length with respect to width in the later stages of growth.

Brachial valve most strongly convex at umbo, having almost same convexity as pedicle valve; umbo small, projecting a short distance behind hinge, and slightly incurved over dorsal interarea; lateral slopes moderately convex, extending into flat lateral extremities; valve highest at midpoint; fold low posteriorly, slightly higher towards front, projecting a short distance anteriorly, and bearing a narrow median furrow; interarea orthocline, concave, marked with denticle grooves and horizontal growth lines, and 1 mm high at umbo on a valve 30 mm wide and 15 mm long.

Internal (Figs 59, 60). Pedicle valve. Apical callus thickening the posterior of valve, obscuring posterior parts of dental plates and adminicula; most valves aseptate, but one individual with a small low median septum at posterior of muscle field (Fig. 60); septum very short, extending less than 2 mm in front of apical callus in a valve 26 mm wide; muscle field narrow, deeply impressed posteriorly, becoming broader and shallower anteriorly, and flanked posteriorly by short adminicula; dental plates broad, and in transverse section diverging at 45-50°; teeth broken and not observed.

Measurements	Length	Width	Height	Height (of ventral interarea at umbo)
CPC 8184	18	30	15	—
CPC 8185	18.5	28.5	12.5	3.5

Brachial valve having a smaller apical thickening than pedicle valve; cardinal process small, consisting of from six to ten vertical lamellar plates; inner socket ridges high, divergent, and enclosing narrow rounded sockets; crura forming vertical plates on ventral sides of inner socket ridges, having straight or slightly concave inner faces; spires with up to fourteen volutions; muscle field narrow and impressed posteriorly, becoming wider and sometimes elevated on a platform anteriorly.

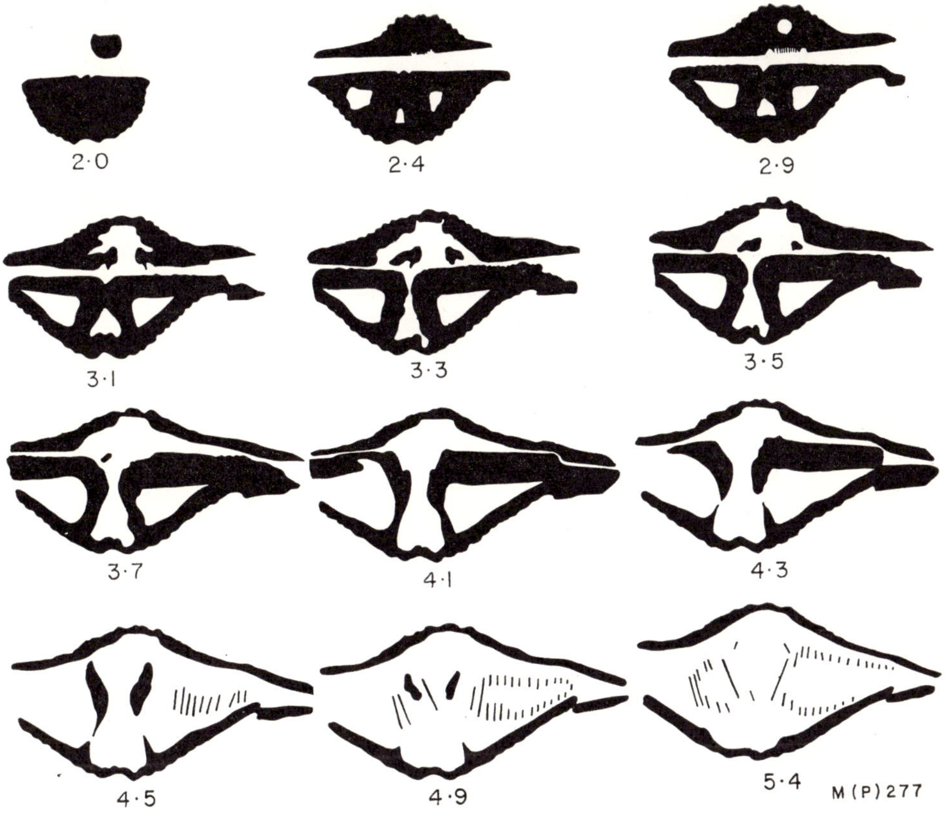

Figure 59. Transverse serial sections of *Prospira incerta* Thomas showing the typically aseptate pedicle valve. CPC 11111, locality 102/4, Enga Sandstone, northern end of Enga Ridge × 2.

Remarks: Externally *Prospira incerta* Thomas is morphologically close to *P. prima* Maxwell (1954, pp. 36-37, pl. 4, figs 4, 5; 1961, pl. 20, figs 8-10) from the Fairy Bower Beds, Rockhampton, and the lower part of the late Tournaisian Bancroft Formation in the Canindah Creek area, Queensland (McKellar, 1967). *P. prima* has a comparable shape, external ornament, a median sinus which usually lacks costation, and a fold bearing a median furrow. Maxwell (1954, p. 36) noted that no median septum or ridge was preserved in his material, but Dear (1963, unpubl. Ph.D. thesis) has assigned a form having a short ventral median septum to *prima.* The small septum in one specimen of *incerta* (Fig. 60) is probably an aberrant

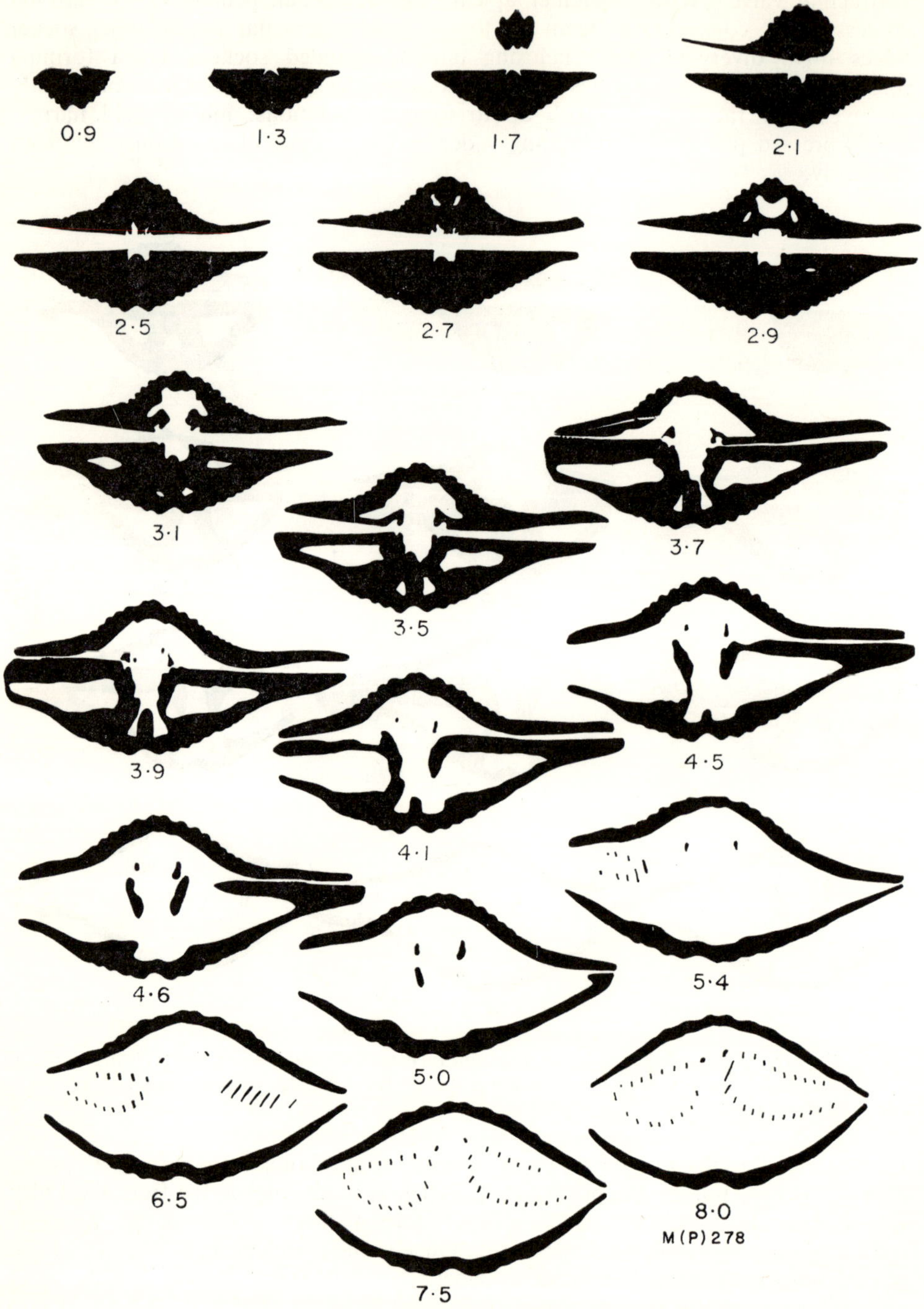

Figure 60. Transverse serial sections of a specimen of *Prospira incerta* Thomas showing the thickened apex on the pedicle valve, and the blunt, apparently aberrant, ventral median septum. CPC 11112, locality 102/4, Enga Sandstone, northern end of Enga Ridge. × 2.

structure. *P. burnettensis* Maxwell (1961, pp. 91-92, pl. 20, figs 1-7) from the Tellebang Formation, Old Canindah, Queensland, has slightly coarser costae than those of *incerta,* and a sinus which contains from three to seven sinal costae.

Spirifer piersonensis Moore (1928, pp. 274-75, pl. 12, figs 20-22) from the Pierson Limestone of Missouri is similar in shape, and has a comparable ventral interarea which is of nearly uniform height along its entire width, a smooth median sinus, and a fold bearing a median furrow. *S. piersonensis* is smaller than *P. incerta,* and has fewer costae on the lateral slopes. Branson (1938, p. 58, pl. 6, figs 8-10) refers *piersonensis* to *Delthyris novamexicana* (Miller), the type species of *Amesopleura* Carter, 1967. Carter (1967, p. 366), however, indicates that the holotype of *piersonensis* is not sufficiently well preserved for the species to be definitely assigned to *Amesopleura.* This author believes *Amesopleura* to be a junior synonym of *Voiseyella* Roberts, 1964.

Spirifer biplicoides Weller (1914, pp. 323-24, pl. 39, figs 27-30) from the Kinderhook at Burlington, Iowa, is more transverse, shorter, and has wider posterolateral extremities than *P. incerta;* the median sinus bears up to three sinal costae.

Spirifer roemerianus de Koninck (1887, p. 125, pl. 29, figs 22, 23, 25-27) from the Tournaisian of Belgium is more alate, has a more prominent umbo on the pedicle valve, and a wider fold and sinus when compared with *P. incerta.* The fold of *S. roemerianus* lacks a median furrow, and in some individuals there is a median costa in the sinus.

The type specimens of *P. incerta* are coarsely silicified and do not show all the characters of the species. Additional details are available from well-preserved material in the Burt Range Formation and Enga Sandstone in the Burt Range area. Thomas's material comes from silicified and ferruginized limestone at Sandy Creek Crossing on the track from Carlton Hill to Legune, which he (1962a) suggested was late Carboniferous in age. The limestone was re-examined in 1963 and 1965, and mapped as Burt Range Formation by Veevers & Roberts (1968). With the exception of *P. incerta* the fauna from the limestone suggests a correlation with the middle part of the Burt Range Formation (*aquila* Zone) in the Burt Range area. *P. incerta* is recorded in the upper part of the Burt Range Formation and in the lower part of the Enga Sandstone in the Burt Range area (Fig. 5), but occurrences at Sandy Creek and with other older faunas in isolated outcrops between Sandy Creek and Legune suggest that the species ranges from the *aquila* Zone to the *amnicum* Zone (Fig. 4).

Occurrence: Thomas's (1970) locality BW 5 at Sandy Creek Crossing on the track from Carlton Hill to Legune (the type locality); locality 122/2, Sandy Creek; localities 123/14 and 123/16, 11 miles southwest of Legune Homestead; locality 133/2, Alpha Hill; locality 109/5, middle part of Enga Ridge; Burt Range Formation. Locality 102/4, Enga Sandstone, northern end of Enga Ridge. Locality 210/6, unnamed Tournaisian breccia, Waggon Creek valley.

Material: CPC 8182-8185, CPC 11111, CPC 11112. All from locality 102/4. Forty additional free specimens, and a similar number preserved as casts and moulds in fine-grained quartz sandstone.

Age: Tournaisian.

Genus UNISPIRIFER Campbell, 1957

Type species: Spirifera striatoconvoluta Benson & Dun, 1920, p. 350, pl. 20, figs 7, 8, by original designation of Campbell, 1967.

Remarks: Unispirifer Campbell has been placed in synonymy with *Fusella* M'Coy, 1844, by Besnosova (1959), and in tentative synonymy with *Fusella* by Ivanova (*in* Sarycheva, 1960) and Carter (1967). Maxwell (1961, pp. 89-90), however, put forward a convincing argument that the true nature of *Fusella* cannot be established because the type species, *Spirifer fusiformis* J. Sowerby, is known only from a partly decorticated shell, and that M'Coy's (1844) original description of the genus and Buckman's (1908) revision were inaccurate because neither resembled Phillips's (1836, pl. 9, figs 10, 11) original illustrations of *Spirifer fusiformis.* Besnosova (1959) in her study of spiriferoids from the Kuznetsk Basin included *Spirifer tornacensis* de Koninck and many species resembling *Unispirifer* or *Prospira* Maxwell, 1954, in *Fusella,* whereas Carter (1967) followed M'Coy's (1844) original description, which is closest to *Spirifer rhomboidea* Phillips, and referred to *Fusella* his species *llanoensis* (Carter, 1967, pl. 38, figs 1-9). *Unispirifer tornacensis* is a transverse spiriferoid possessing a large number of fine costae on the lateral slopes and on the fold and sinus, whereas *Fusella llanoensis* and *Spirifer rhomboidea* are subelliptical in outline, and are ornamented by a smaller number of wider costae.

This obvious confusion justifies Maxwell's (1961) contention that the characters of *Fusella* will remain unknown until well-preserved topotype material of *Spirifer fusiformis* is studied. I maintain that until this revision is carried out, the genus *Fusella* should not be used*.

A detailed discussion on the relationships between *Unispirifer* Campbell and *Prospira* Maxwell (1954) has been given by Maxwell (1961, pp. 89-91). He concluded that *Unispirifer* was closely related to *Prospira,* but persisted in using the two genera: *Prospira* for forms with a coarser costate ornament and a smooth or very weakly costate fold and sinus; and *Unispirifer* for forms with a greater number of fine costae on the lateral slopes and on the fold and sinus (but with fewer costae in the sinus than *Cyrtospirifer* Fredericks, 1924). I distinguish *Unispirifer* from *Prospira* by its more transverse shell which has alate rather than mucronate posterolateral extremities in adult specimens, its larger size, longer ventral adminicula, greater number of finer costae on the lateral slopes, and a much more complex pattern of costae on the sinus and fold.

UNISPIRIFER LAURELENSIS (Thomas)

(Pl. 44, figs 21-38)

1970 *Prospira laurelensis* Thomas, pp. 84-89, pl. 2, figs 8-13.

Description: External. Shell subtriangular, nearly equally biconvex, more than twice as wide as long, with acutely angular lateral extremities; large specimens alate, and small growth stages strongly mucronate; shell widest at hinge; commissure weakly uniplicate; costae simple, well rounded, twice as wide as intercostal

---

*Waterhouse (1970) has re-examined the type specimen of *Spirifer fusiformis* Phillips, the type species of *Fusella* M'Coy, and contends it is related to the Syringothyrididae rather than to *Unispirifer* Campbell. (Waterhouse, J. B., 1970, The lower Carboniferous brachiopod genus *Fusella* M'Coy 1844. *Life Sci. Occ. Pap., R. Ont. Mus.,* 15, 1-12.

furrows, becoming weaker laterally and almost obsolete on posterolateral margins; between twenty and thirty costae on each lateral slope; one very wide brachial valve having two costae parallel with hinge on posterolateral margins; microornament of radial lirae having a density of approximately twenty per 1 mm.

Pedicle valve moderately convex, greatest convexity near umbo; umbo small, blunt in ventral view, moderately incurved, projecting a short distance behind brachial valve; umbonal shoulders gently curved; lateral slopes evenly convex medially, becoming flat on posterolateral margins; highest part of valve immediately in front of umbo; interarea apsacline, mainly flat, but with a slight concavity beneath umbo, 4 mm high on a valve 26 mm wide and 14 mm long, and ornamented with vertical denticle grooves having a density of five per 1 mm; delthyrial angle approximately 30°; median sinus shallow, becoming broader anteriorly, produced into a short lingual extension; sinal angle 20-25°; sinus ornamented with a single median costa or a pair of median branching costae surrounded by one or two pairs of simple lateral costae arising from margins of sinus (Fig. 61).

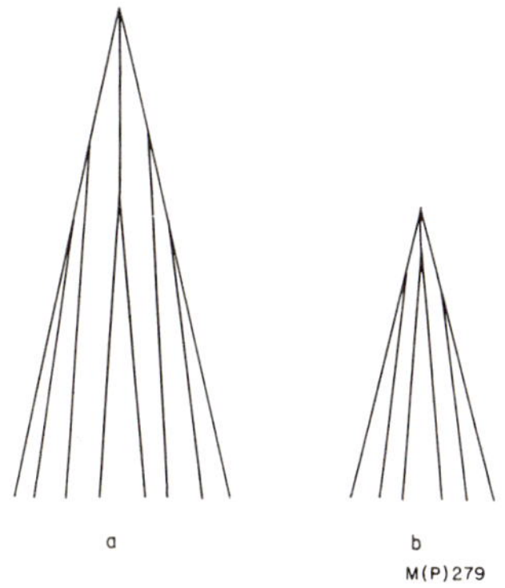

Figure 61. Pattern of costae in the sinus of *Unispirifer laurelensis* (Thomas). a-CPC 8169, b-CPC 8167; locality 101/18, Burt Range Formation, 4.5 miles west of Mount Septimus. × 2.

Brachial valve evenly convex in lateral profile, highest at midpoint; umbo small, barely incurved, projecting a short distance behind hinge; lateral slopes gently convex, swollen towards front of valve on either side of fold, becoming flatter

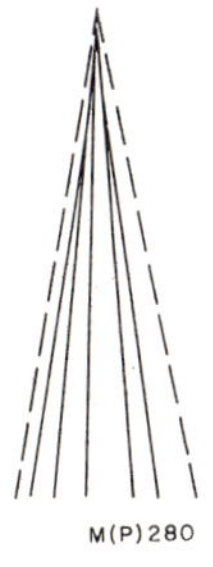

Figure 62. Pattern of costae on the fold of *Unispirifer laurelensis* (Thomas). The dashed lines indicate the furrows at the margins of the fold. CPC 8172, locality 128/2, Burt Range Formation, middle part of Enga Ridge. × 1.5.

towards hinge and lateral margins; fold narrow and well defined at umbo, broader and lower anteriorly, outlined by two moderately deep curving furrows, and ornamented with up to six costae consisting of a twin median pair giving rise by bifurcation to up to two lateral pairs (Fig. 62).

Internal (Fig. 63). Pedicle valve. Muscle field ovoid to quadrangular, tapering posteriorly, expanding anteriorly, and having rounded extremities; diductor muscle scars longitudinally striated, impressed, broad, and tapering posteriorly; adductor muscle scars poorly defined, situated on a low rounded ridge; adminicula short, straight, diverging at approximately 70°, enclosing posterior third of muscle field, 5 mm long on a valve 28 mm wide and 15 mm long; dental plates extending vertically for half their length from floor of valve, then curving laterally around margins of delthyrium; teeth short, blunt, and projecting inwards; shell thickened apically, especially at umbo and on posterolateral margins of muscle field.

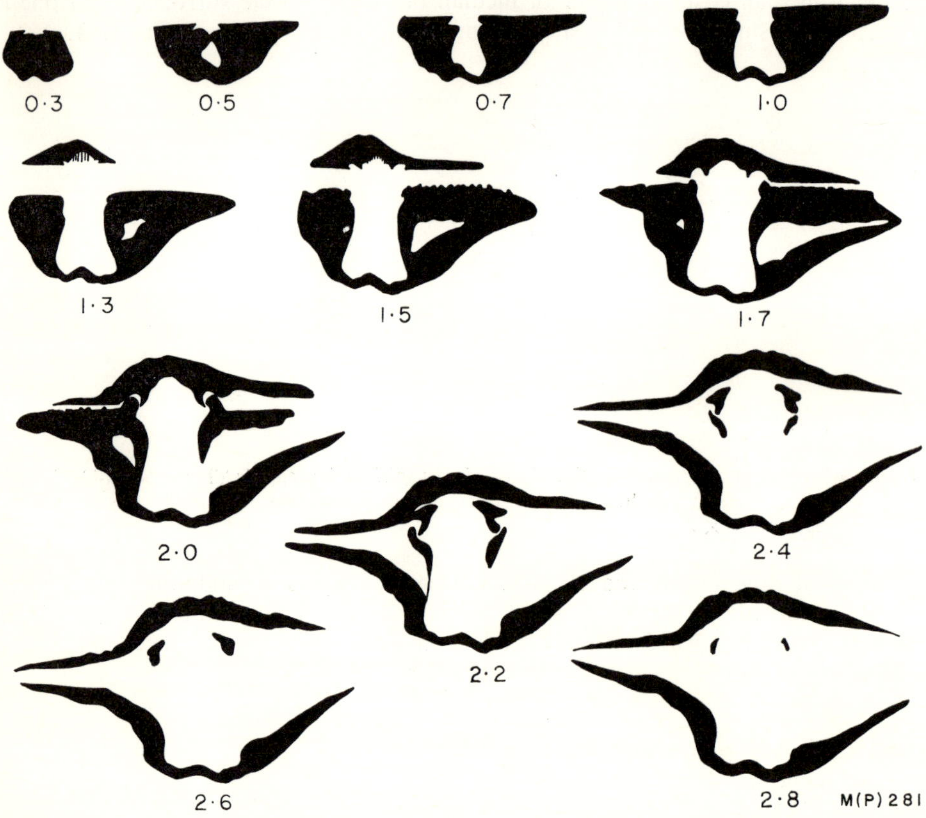

Figure 63. Transverse serial sections of *Unispirifer laurelensis* (Thomas). CPC 11459, locality 101/16, Burt Range Formation. × 3.

Brachial valve. Adductor muscle scars impressed, rectangular, tapering slightly posteriorly, blunt anteriorly, and separated by a broad rounded median ridge originating in front of cardinal process; inner socket ridges arising from dorsolateral margins of cardinal process, and giving rise dorsally to crura; in transverse section, crura diverging at approximately 45°, having a sinuous inner face with a median concavity, and laterally inflected ventral tips; crura 2 mm long on a valve 18.5 mm wide; sockets well rounded; cardinal process triangular, with a myophore of approximately ten small vertical lamellar plates up to 0.5 mm long; apex of valve thickened.

Measurements	Length	Width (at hinge)	Height
CPC 8163	15	27.5	10
CPC 8164	15	31	11
CPC 8165	14.5	—	13
CPC 8166	16 est.	38 est.	—
CPC 8168	20 est.	55	—
CPC 8169	20	41	—
CPC 8170	18.5	—	13
CPC 8171 (brachial valve)	16	42	—
CPC 8172 (brachial valve)	16	38	—

Remarks: The type specimens of *Unispirifer laurelensis* from the Laurel Formation, Canning Basin, Western Australia (Thomas, 1970, pl. 2, figs 8-13), are slightly smaller and frequently less auriculate than the material from the Burt Range Formation in the Bonaparte Gulf Basin. The small growth stages in both collections are mucronate, but larger specimens from the Laurel Formation tend to have less auriculate lateral extremities than large individuals from the Burt Range Formation, although some broken specimens (Thomas, 1970, pl. 2, fig. 11) appear to have had moderately wide lateral extremities.

Externally, *U. laurelensis* is extremely close morphologically to *U. platynotus* (Weller) (1914, pp. 317-19, pl. 39, figs 1-10), having a comparable outline, a blunt ventral umbo, a weak fold and sinus, and a similar costate ornament on the lateral slopes and on the fold and sinus. The interior of *U. platynotus* has not been described. *U. platynotus* is recorded from the upper three beds of the Kinderhook at Burlington, Iowa (Weller, 1914, pp. 318-19).

The form referred to *Fusella ussiensis* (Tolmachoff) by Besnosova (1959, pp. 63-65, pl. 2, fig. 6) and Sarycheva et al. (1963, p. 288, pl. 51, fig. 8), and originally described by Tolmachoff (1924, p. 560, pl. 11, figs 7-11), has a similar shape, and a comparable ornament on the lateral slopes when compared with *U. laurelensis;* it has a simple median costa in the sinus similar to that in Thomas's specimens of *U. laurelensis,* and in some specimens from the Burt Range Formation. Because of the uncertainty of the characters of *Fusella,* I believe Tolmachoff's species is best referred to *Unispirifer*. *U. ussiensis* is distinguished from *U. laurelensis* only by its more convex brachial valve and shorter ventral adminicula (Besnosova, 1959, fig. 18). *U. ussiensis* (Tolmachoff) is known from the Tournaisian Taidon and nt_1 Beds of the Kusnetsk Basin, Siberia.

The specimen referred to *Spirifer* aff. *clathratus* M'Coy by Vaughan (1905, pl. 26, fig. 5), from the Z_1 near Bristol, England, is similar in shape to the more auriculate individuals of *U. laurelensis,* and has a comparable ornament on the lateral slopes and on the sinus.

Unispirifer minnewankensis (Shimer) described by Brown (1952, pp. 100-1, pl. 4, fig. 6) from the Upper Tournaisian portion of the Rundle Formation, Mount Greenock, Alberta, Canada, has a narrower and more pronounced fold and sinus, and a slightly coarser external ornament. The internal characters of *U. minnewankensis* have not been described.

Occurrence: Localities 101/15, 101/16, 101/18, 4.5 miles west of Mount Septimus; localities 103/2, 109/1-109/3, 128/1, 128/2; middle part of Enga Ridge; and possibly at locality 122/3, Sandy Creek; Burt Range Formation.

Material: CPC 8163-8172, CPC 11459. CPC 8163-8166, 11459 from locality 101/16; CPC 8167-8169 from locality 101/18, CPC 8170 from locality 109/2; CPC 8171-8172 from locality 128/2. Approximately 40 specimens.

Age: Tournaisian.

Genus Spirifer Sowerby, 1816

Type species: By suspension of the Rules by the International Commission on Zoological Nomenclature, Opinion 100, *Conchyliolithus Anomites striatus* Martin, 1793, from Derbyshire, England.

Remarks: A complete description of the genus *Spirifer* is unavailable because the type specimen of *Spirifer striatus* is apparently lost. Browne (1953a, p. 58), however, reported a specimen at Manchester University which may be the type. Information on the morphology of *S. striatus* has been given by Semichatova (1941), Harrington & Leanza (1952), Browne (1953a), and Cvancara (1958). Both Pitrat (1965, p. H704) and Carter (1967, pp. 375-76) compiled information on *S. striatus* from many sources, and gave more comprehensive descriptions of the genus; Carter pointed out that his description was incomplete because the brachial valve interior remains undescribed. Until the uncertainty surrounding the type specimen is resolved and the species redescribed, either from the type or a neotype, I propose to follow Pitrat's and Carter's diagnoses of the genus.

Spirifer otwayi sp. nov.

(Pl. 46, figs 1-17)

Diagnosis: Shell subequally biconvex, subrectangular, and frequently transverse; costae having a density of eight per 10 mm at the anterior margin; sinus very wide anteriorly, having a strong median dorsal deflection; fold broad, truncated anteriorly by the tongue of the sinus; pedicle valve with a slightly incurved umbo, and a mainly flat, catacline interarea.

Description: External. Shell subequally biconvex, subrectangular, tapering only slightly towards lateral margins; greatest width at about one-third length of a mature valve; hinge straight, almost equal to greatest width; lateral margins broadly rounded in large specimens; posterolateral extremities slightly mucronate in earlier growth stages; commissure strongly uniplicate, mature shells having a marked median dorsal deflection; costae evenly rounded, broader towards front, separated by slightly narrower troughs, and increasing by bifurcation; eight costae per 10 mm on anterior margin of a shell 24 mm long and 50 mm wide; micro-ornament of radial lirae having a density of about ten per 1 mm, and fine closely spaced concentric growth lines.

Pedicle valve most strongly convex at umbo; umbo small, pointed, projecting 2.5 mm over interarea on a valve 24 mm long and 50 mm wide, and only slightly incurved; umbonal shoulders flat to gently concave; lateral slopes evenly convex medially, becoming flatter on posterolateral margins; sinus commencing at tip of umbo, broadening anteriorly into a dorsally directed tongue, having a sinal angle of 30-40°; floor of sinus narrow, flat to rounded, and trough-like; lateral margins of sinus straight, moderately inclined, becoming broader towards front; median costa in sinus arising from tip of umbo, and bifurcating once or twice anteriorly; flanks of sinus ornamented with three or four pairs of simple or bifurcating costae branching from marginal costae (Fig. 64); interarea catacline to very slightly apsacline, concave immediately beneath beak but flat for remainder of its width, and 5 mm high on a valve 24 mm long and 50 mm wide; interarea ornamented with wavy denticle grooves having a density of about three per 1 mm; delthyrium open, with depressed dental flanges along lateral margins, having a delthyrial angle of 60°.

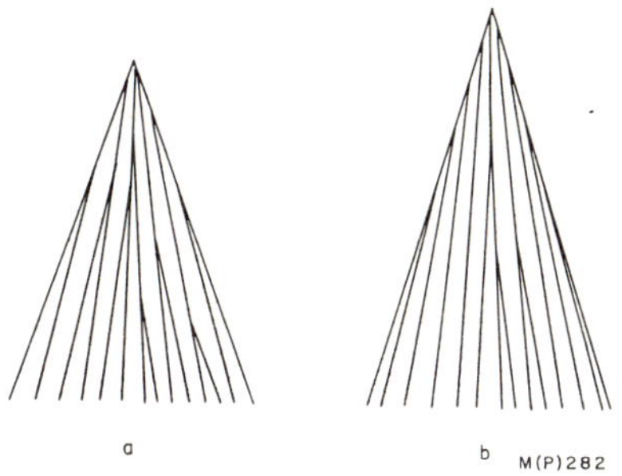

Figure 64. Pattern of costae in the sinus of *Spirifer otwayi* sp. nov. a-CPC 8186, b-CPC 8479; locality 7/1, Ningbing Limestone (Tournaisian part), Ningbing Range. × 1. Both specimens are damaged in the left anterior parts of the sinus.

Brachial valve umbo small, incurved, protruding 1.5 mm behind hinge on a valve 23 mm long and 50 mm wide, bluntly rounded, having small concave umbonal shoulders; lateral slopes mainly evenly convex, becoming flatter on posterolateral margins; fold commencing at umbo, high and wide anteriorly, and truncated by dorsally curving sinus on pedicle valve; about ten costae on fold, increasing by intercalation (Fig. 65).

Figure 65. Pattern of costae on the fold of *Spirifer otwayi* sp. nov. The curvature at the anterior represents the truncation of the fold by the dorsally deflected sinus, and the dashed lines indicate furrows at the margins of the fold. CPC 8186, locality 7/1, Ningbing Limestone (Tournaisian part), Ningbing Range. × 1.

Internal. Pedicle valve (Fig. 66). Dental plates straight and divergent in transverse section, moderately thickened, forming flanges on margins of delthyrium, and supporting strong pointed teeth; adminicula slightly thickened posteriorly, linked by a small callus or transverse plate at apex of delthyrium, and in transverse section straight or moderately concave outwards; diductor muscle scars narrow, rounded, subrectangular, forming two deep pits on either side of adductor muscle scars; adductor muscle scars narrow, elongate, situated in shallow depressions on margins of sinus.

Brachial valve (Fig. 67). Cardinal process strongly convex posteriorly, buttressed laterally by socket plates, and having a myophore of twelve to fifteen lamellar plates; sockets small, rounded, and supported by moderately thick socket plates; inner socket ridges giving rise to slender crura; spires not observed.

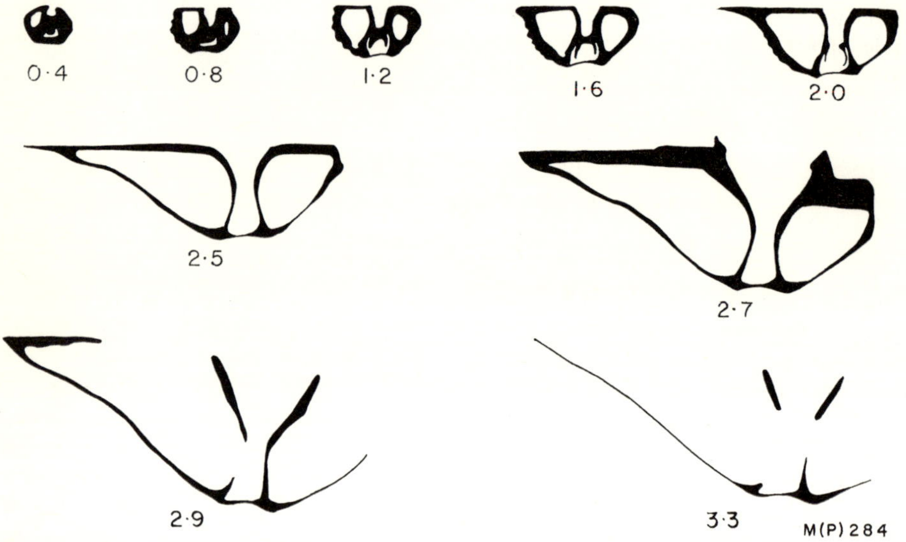

Figure 66. Transverse serial sections of the pedicle valve of *Spirifer otwayi* sp. nov. CPC 11113, locality 7/1, Ningbing Limestone (Tournaisian part), Ningbing Range. × 2.

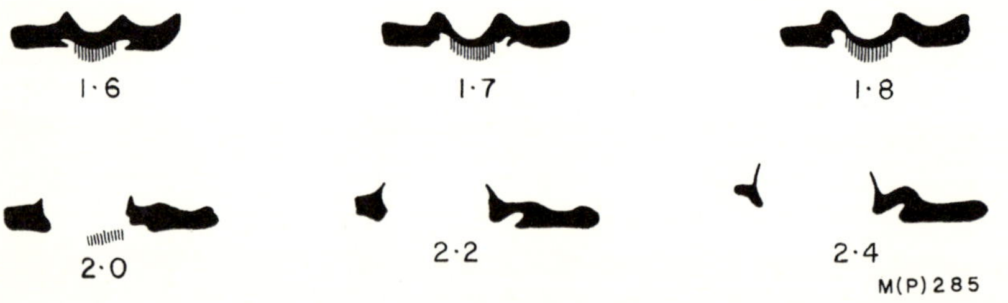

Figure 67. Transverse serial sections through the apical part of the brachial valve of *Spirifer otwayi* sp. nov. CPC 11114, locality 7/1, Ningbing Limestone (Tournaisian part), Ningbing Range. × 3.

Measurements

	Length	Width	Height	Length (of ventral interarea)
CPC 8186 Holotype	24	50	20	5
CPC 8187 Paratype	19.5	43	—	6.5
CPC 8479 Paratype	33	42	—	—
CPC 8188 Paratype	21	44	—	—
CPC 8189 Paratype	27	49	—	—

Remarks: The species closest to *Spirifer otwayi* is *S. chappelensis* Carter (1967, pp. 377-82, pl. 36, figs 1-4; pl. 37, figs 1-4) from the Chappel Limestone of Texas. *S. chappelensis* has a comparable range of variation in shape, a similar umbo on each valve, a wide fold and sinus which bear similar costal patterns, and a costate ornament in which the costae increase by bifurcation. From serial sections, both species have a thickening or transverse plate at the apex of the delthyrium; the other internal structures at the apex of the shell are also morphologically close.

Large specimens of *S. missouriensis* Swallow, figured by Weller (1914, pl. 39, figs 11-23) and Branson (1938, pl. 7, figs 11-14) from the Chouteau Limestone of Missouri, have an outline similar to that of *S. otwayi. S. missouriensis* is distinguished from *S. otwayi* by coarser, higher costae, and a narrower fold and sinus; it is probably best referred to *Anthracospirifer* Lane, 1963.

S. carinatus Rowley, described by Weller (1914, pp. 336-38, pl. 46, figs 13-17) from the lower part of the Burlington Formation, Missouri, differs from *S. otwayi* in having more inflated lateral slopes and a narrower sinus on the pedicle valve, and a more carinate fold on the brachial valve.

S. tersiensis Rotai from the Taidon and lower Netersinsky Beds (Tournaisian) of the Kusnetsk Basin (Sarycheva et al., 1963, p. 293, pl. 55, figs 1-3) has a comparable outline and external ornament, but is distinguished from *S. otwayi* by its narrower and less dorsally inflected sinus. The pattern of sinal costae illustrated by Besnosova (1959, p. 96, fig. 39) for *S. kasachstanensis* Simorin, a junior synonym of *S. tersiensis,* is close to that of this species, and differs only in having more simple median costae and fewer lateral costae; the fold (Besnosova, 1959, pl. 7, fig. 4) appears to bear bifurcating costae in contrast to the intercalating costae in *otwayi.* Internally the pedicle valve of the Russian species is more thickened apically, and in transverse section has straight rather than outwardly concave to straight ventral adminicula.

The species is named in honour of Mr R. Otway.

Occurrence: Locality 7/1, Ningbing Limestone (Tournaisian part), Ningbing Range.

Material: CPC 8186-8192, CPC 8479, CPC 11113, CPC 11114. Holotype CPC 8186, paratypes CPC 8187-8192, 8479. All from locality 7/1. Twenty specimens preserved in reef limestone.

Age: Tournaisian.

Spirifer sp.

(Pl. 47, figs 1-6)

Description: External. Shell spiriferoid, equally biconvex, twice as wide as long, with rounded lateral margins meeting hinge at an acute angle; commissure uniplicate; costae rounded, separated by intercostal furrows of equal width; up to twenty-three costae on each lateral slope, with many of the first six costae on either side of fold and sinus bifurcating (one trifurcating), and remaining lateral costae simple; fold and sinus with accessory costae; micro-ornament of radial lirae having a density of sixteen per 1 mm on umbo of one pedicle valve.

Pedicle valve umbo narrow, short, and extending just behind hinge; umbonal shoulders concave, sloping on to smooth flat or slightly convex posterolateral extremities; interarea apsacline, subequal in height for almost its entire length, concave beneath beak, flatter laterally, and bearing faint vertical striations (observed near delthyrium only); hinge denticulate; delthyrium open; lateral slopes convex medially, but becoming lower laterally; sinus broad anteriorly, bearing a dividing median costa bordered by approximately three pairs of accessory costae.

Brachial valve rounded-rectangular, high posteriorly and along crest of fold; umbo small, convex, barely extending behind hinge, and having gently concave umbonal shoulders; lateral slopes high medially, becoming lower laterally, and grading on to smooth flat posterolateral extremities; fold having a median costa flanked by two pairs of accessory costae, the outermost branching from primary marginal costae (origin of innermost accessory costae not observed).

Internal. Pedicle valve. Dental plates extending down inner margins of delthyrium and supported by short subparallel to divergent (50°) ventral adminicula; adminicula partly buried by callus in apically thickened specimens; muscle field situated towards front of adminicula and in front of an elongate convex callus emerging from between adminicula; adductor muscle scars narrow, elongate, striated, and situated on sides of or divided by a low median ridge; diductor muscle scars subovate to subrectangular, longitudinally striated, or partly dendritic posteriorly and longitudinally striated anteriorly, and impressed into posterior callus; floor of valve surrounding muscle field pitted, bearing irregular radially arranged mantle canals (vascula genitalia) originating from muscle field or from posterior parts of valve.

Brachial valve interior not observed.

Remarks: This material appears to be closest to the more transverse specimens of *Spirifer striatus* (Martin) from the Visean of Great Britain and Ireland figured by Davidson (1858, e.g. pl. 3, fig. 4, a specimen from near Richmond, Yorkshire). The few specimens of *Spirifer* sp. available for study suggest that it has fewer costae on the lateral slopes and probably on the fold and sinus when compared with Davidson's specimen. Typical specimens of *S. striatus* are less transverse, and often are widest in front of the hinge.

Spirifer sp. bears a close resemblance to *S. haydenianus* Girty (1927, pp. 416-17, pl. 24, figs 18-21) from the Visean Brazer Limestone of Idaho, USA. *S. haydenianus* has a comparable transverse outline, and morphologically similar ventral umbo, ventral interarea, fold, and sinus. The external ornament of *Spirifer* sp. is known from one poorly preserved exterior, and internal moulds; these suggest that *Spirifer* sp. is more coarsely costate than *S. haydenianus*.

Occurrence: Locality 306, Burvill Beds, Burvill Point.

Material: CPC 11988-11990. All from locality 306. Four ferruginized specimens.

Age: Visean.

Genus PODTSHEREMIA Kalashnikov, 1966

Type species: Podtsheremia prima Kalashnikov, 1966, from the Tulsk horizon (Visean), Kirpich-Kyrta, Podcherem River, Northern Urals, USSR, by original designation of Kalashnikov, 1966.

Diagnosis (from Kalashnikov, 1966, pp. 50-51, translated by Mrs G. A. Cooper): 'Shells of medium size. Hinge line shorter than the greatest width. Ears rounded. Sinus triplicate — with a median. Ribs flatly convex, numerous. On the flanks they branch and form fascicles of 2 or 3 ribs. Microsculpture is in the form of fine longitudinal stria (difficult to distinguish) and fine concentric lines of growth.

Delicate teeth and delthyrial plates appear a few millimetres from the tip of the beak. A delthyrial plate is preserved for about half of the delthyrium. The shell is thin, apical filling is absent. In the dorsal valve are a wide cardinal process, high crural plates and a short septal ridge'.

Remarks: The genus *Podtsheremia* was designated by Kalashnikov (1966) to include spiriferoids of moderate size having fasciculate costae on the lateral slopes, costae on the fold and sinus, a short hinge, rounded posterolateral extremities, and a thin shell. Kalashnikov separated *Podtsheremia* from *Cyrtospirifer* Fredericks, 1924, and *Tenticospirifer* Tien, 1938, on the basis of the pattern of sinal costae and the possession of fasciculate costae on the lateral slopes, a short hinge, and a low ventral interarea; from *Fusella* M'Coy, 1844 (and presumably *Unispirifer* Campbell, 1957, also) and *Spirifer* Sowerby, 1816, he distinguished *Podtsheremia* by its rounded posterolateral margins, short hinge, fasciculate costae, absence of apical thickening, and presence of a delthyrial plate. I consider the last two features to be of doubtful generic significance; the apical thickening in spiriferoids varies with the age of the individual and the type of environment it inhabited; and the delthyrial plate is, in most cases, probably a thickening which bordered the muscles at the apex of the delthyrium.

Anthracospirifer Lane, 1963, a characteristic spiriferoid from the upper Mississippian and Pennsylvanian of USA, is morphologically the closest genus to *Podtsheremia. Anthracospirifer* is distinguished by its higher, coarser costae on the lateral slopes, fold, and sinus, more transverse outline, and wider hinge. The costae on the inner margins of the lateral slopes bifurcate, but are not grouped into a fasciculate pattern as in the type species of *Podtsheremia,* and there are fewer costae on the fold and sinus.

Material from the Utting Calcarenite and the Burvill Beds of the Bonaparte Gulf Basin is, on the basis of similarity in ornament, tentatively referred to *Podtsheremia.* When compared with the type species, *P. prima* Kalashnikov, *P? humilicostata* sp. nov. and *P? thomasi* sp. nov. have comparable low rounded costae, but both are more transverse in outline, have wider hinges, pointed to somewhat rounded posterolateral extremities, and in adult specimens are thickened at the apices of the pedicle valves. Although many of these latter characters suggest an affinity with *Anthracospirifer,* I consider the morphology of the costae to be generically more important.

PODTSHEREMIA? HUMILICOSTATA sp. nov.

(Pl. 48, figs 1-27)

Diagnosis: Shell trigonal to transversely trigonal in outline, widest at the hinge, and with pointed posterolateral extremities; costae angular at umbo, but low, wide, and rounded over remainder of shell; eight to sixteen costae on each lateral slope, the first four on either side of the fold and sinus frequently bifurcating, but not fasciculate; sinus with a low median costa commencing in front of the umbo, and two pairs of simple accessory costae branching from costae bordering the sinus; fold with two median costae, and two pairs of simple accessory costae similar to those on the sinus; in 'mature' specimens the apical interior of the pedicle valve thickened with callus, and the muscle field situated at the front of the adminicula; ventral adminicula diverging at 30-40°; sinal angle approximately 25°.

Description: External. Shell trigonal to transversely trigonal, auriculate, widest at hinge, and moderately biconvex; commissure narrowly uniplicate; costae angular posteriorly, and wider, rounded, and separated by rounded intercostal furrows anteriorly; eight to sixteen costae on each lateral slope, the first four often bifurcating but never fasciculate, and the remainder simple; micro-ornament cross-hatched, with fine radial lirae having a density of sixteen to twenty per 1 mm crossed by regular concentric growth lines.

Pedicle valve highest at or behind one-third length of valve; umbo narrow, pointed, and incurved over hinge; umbonal shoulders short, steep, passing laterally on to convex to flat posterolateral extremities; posterolateral margins rounded in juveniles (6 mm wide and 4 mm long), flat and narrowly auriculate in larger specimens; interarea apsacline to catacline, moderately high, convex especially near umbo, bearing well-marked denticle grooves; delthyrial angle 60°, delthyrium open in thin-shelled specimens, closed apically by callus (sometimes forming an apical hood-like cover) in thickened specimens; sinus narrow, U-shaped in cross section, outlined by two moderately high bifurcating primary costae, becoming deeper anteriorly, having a sinal angle of about 25°; floor of sinus with a low simple median costa commencing in front of umbo (Fig. 68); sides of sinus bearing two (occasionally one on a small specimen) stronger simple accessory costae, each branching from marginal costae bordering sinus; primary costae bordering sinus giving rise by bifurcation to one and sometimes two costae on lateral slopes, remaining lateral costae arising from umbo; second lateral costa bifurcating, occasionally trifurcating or simple; third and fourth lateral costae either simple or bifurcating, and remaining lateral costae always simple.

Brachial valve transverse, rounded-rectangular to subovate, moderately to gently convex, and highest near midpoint or towards front of fold; umbo low, straight, projecting a short distance behind hinge, having slightly concave umbonal shoulders extending on to flat posterolateral extremities; interarea low; fold narrow at umbo, becoming broader and higher anteriorly; two simple costae separated by a narrow furrow extending from umbo to anterior margin, accompanied on each side by two simple accessory costae branching from primary costae bordering fold (Fig. 68); lateral costae arising from umbo; first four costae simple or bifurcating, first lateral costa frequently simple, and second and third costae bifurcating; second lateral costa trifurcating in one specimen; remaining lateral costae simple.

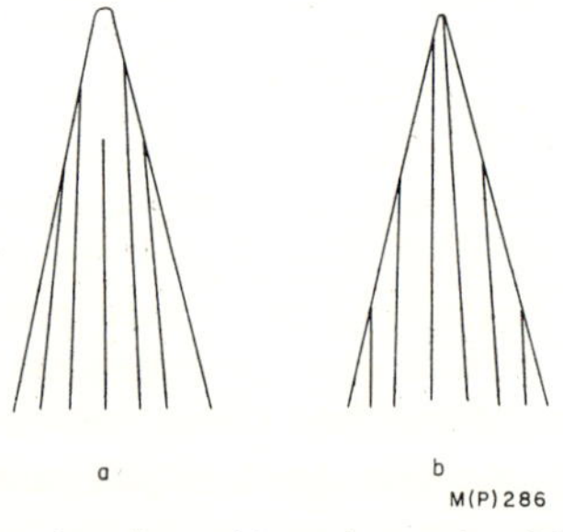

Figure 68. Pattern of costae in the sinus (a) and on the fold (b) of *Podtsheremia? humilicostata* sp. nov. a-CPC 10993, b-CPC 10995; locality 108/0, Utting Calcarenite, Utting Gap. × 1.5.

Internal. Pedicle valve. Dental plates extending down inner margins of delthyrium, supported by outwardly concave adminicula, and bearing long pointed teeth; inner margins of teeth grooved; hinge coarsely denticulate; ventral adminicula joined by a small concave apical plate; adminicula extending up to one-third length of valve in large specimens, shorter in small individuals, and diverging at 30-40°; in unthickened specimens, muscle field located at umbo, divided by a narrow myophragm; adductor muscle scars narrow, elongate, bordering myophragm; diductor muscle scars pointed posteriorly, wider anteriorly, and faintly striated; some valves (presumably older specimens) thickened posteriorly by a callus extending anteriorly from umbo between adminicula; in the latter specimens, muscle field situated towards midpart of valve near ends of adminicula, and posteriorly deeply impressed into callus; adductor muscle scars in narrow linear pits divided by a myophragm, and bordered by elongate ovoid striated or partially dendritic diductor muscle scars; umbonal shoulders thickened also, bearing a large number of deep circular pits (?vascula genitalia).

Brachial valve. Cardinal process triangular, consisting of twelve to fifteen lamellar plates, and buttressed by crura and inner socket ridges; inner socket ridges stout, having elevated anterior extremities; crura arising from dorsal parts of inner socket ridges, having gently concave inner faces; sockets wide distally, with a rounded floor; adductor muscle scars in two pairs; inner scars confined to fold, pointed posteriorly, having wide irregular anterior margins, and divided by a narrow myophragm; outer scars trigonal, impressed and narrowly rounded posteriorly, and with wide striate margins anteriorly; mantle canals arising from beneath sockets, forming a radiating linear pattern on margins of muscle field.

Measurements	Length (pedicle valve)	Length (brachial valve)	Width	Height	Height (of ventral interarea)
CPC 10991	15	13.5	23.5	11.5	4.5
CPC 10993 Holotype	16.5	—	38 est.	—	4

Remarks: Differences with *Podtsheremia prima* Kalashnikov, the type species, have already been discussed in the remarks on the genus. The only species of *Podtsheremia* resembling *P? humilicostata* is *P. acutisimilis* Semichatova (1943, pl. 4, figs 1-6) which has a wide hinge and a trigonal outline. *P? humilicostata* is larger and has slightly finer costae on the lateral slopes than *acutisimilis.*

The external ornament of *humilicostata* appears to be close to that of the *Spirifer duplicostatus* Phillips group of species from the Visean of Great Britain, which Kalashnikov (1966) refers to *Podtsheremia. S. duplicostatus* Phillips from the Visean of Ireland, figured by Davidson (1858, pl. 3, figs 7-10; pl. 4, figs 3-11) has rounded costae similar to those of this species; the division of the lateral costae near the fold and sinus is comparable also. *S. duplicostatus* is distinguished from *humilicostata* by its subovate outline, shorter hinge, and different arrangement of sinal costae. Specimens of *S. duplicostatus* from the D_2 of Derbyshire, England, in the U.S. National Museum collection, are more strongly biconvex, less transverse, have wider flatter costae separated by narrow furrows, and have bifurcating accessory costae in the sinus; one specimen has a bifurcating median costa. On the fold there is a single median costa, often weakly defined, bounded by two or three pairs of dividing costae. *P? humilicostata* does, however, resemble smaller transverse specimens identified by L. G. de Koninck as *Spirifer duplicostatus* which are now in the collection of the Museum of Comparative Zoology, Harvard University (Number 1767); de Koninck's label states that they are young individuals from Vise, Belgium. These specimens are smaller and more transverse than typical examples of *duplicostatus,* and have a costal pattern on the fold and sinus similar to that of *P? humilicostata;* they most closely resemble specimens of *humilicostata* from locality 306 in the Burvill Beds. The Belgian specimens are associated with subovate *duplicostatus*-like forms, and hence they may only be variants of the species.

North American species of *Anthracospirifer* are commonly ornamented with coarser, higher, and simpler costae than *humilicostata. Anthracospirifer birdspringensis* Lane, the type species, has similar costal patterns on the fold and sinus, and morphologically similar internal structures, but is distinguished from *P? humilicostata* by higher, more angular costae, a much higher median costa in the sinus, and a posteriorly unthickened shell.

A. shoshonensis (Branson & Greger) (1918, pl. 18, figs 26, 27), from the early Meramecian (lower or middle Salem) of Wyoming, has comparable dividing costae on the inner margins of the lateral slopes, and a similar pattern of accessory costae in the fold and sinus. *A. shoshonensis* has a shallower median sinus with a stronger median costa, and is larger; the large specimens are less transverse in outline, although they may be transverse at growth stages comparable in size with *P? humilicostata,* and have a greater number of costae on the lateral slopes. One of a group of cotypes from the University of Missouri collection has a split median costa in the sinus. Specimens of *A. shoshonensis* figured by Easton (1962, pl. 9, figs 21, 22) probably are variants of *A. curvilateralis* (Easton) (1962, pl. 9, figs 14-19) (Mackenzie Gordon Jr, pers. comm.). Easton's specimens come from the Heath (late Mississippian) and Cameron Creek (early Pennsylvanian) formations of Montana, from the same localities as *A. curvilateralis. A. curvilateralis* is distinguished from *P? humilicostata* by its larger size, less transverse outline, more strongly incurved ventral umbo, and stronger costae, especially the median accessory costa in the sinus; the costae adjacent to the fold and sinus divide less frequently than those in *P? humilicostata.*

A. missouriensis (Swallow), described by Weller (1914, pl. 39, figs 11-23) from the Chouteau Formation, Missouri, has a large number of narrower, sharper costae on the lateral slopes. The internal morphology of *A. missouriensis* has not been described.

The name of this species is derived from the Latin *humilis,* low, and *costa,* rib; it refers to the low costae, especially to the median costa in the sinus.

Occurrence: Localities 107/1, 108/0 (the type locality), 108/3, and 143/1, Utting Calcarenite, Utting Gap. Locality 306, Burvill Beds, Burvill Point.

Material: CPC 10991-11005. Holotype CPC 10993, paratypes CPC 10995, 10996, 10999, 11002. CPC 10991-10992 from locality 306; CPC 10993-11002 from locality 108/0; CPC 11003-11005 from locality 107/1. More than 70 specimens.

Age: Visean.

Podtsheremia? thomasi sp. nov.

(Pl. 48, figs 28-30; Pl. 49, figs 1-24)

1970 Spiriferidae gen. et sp. nov. cf. *'Spirifer' duplicostatus* Phillips, Thomas, pp. 101-105, pl. 8, figs 9-11; pl. 29, fig. 8.

Diagnosis: Large for genus, trigonal, widest at the hinge; costae coarse, numbering fourteen to twenty on each lateral slope; the first five costae on each side of the fold and sinus commonly increasing by bifurcation or trifurcation, some having incipient fasciculation; median costa in the sinus strong, flanked by up to three pairs of accessory costae branching from the primary marginal costae; twin median costae on the fold bifurcating anteriorly in larger specimens, flanked by one or two pairs of costae branching from primary marginal costae; innermost costae on the flanks arising from the umbo; ventral umbo blunt, and poorly differentiated from the flanks; ventral muscle field deeply impressed between the adminicula.

Description: External. Shell large for genus, trigonal, subequally biconvex, greatest width at hinge; posterolateral margins pointed to slightly auriculate; commissure strongly uniplicate; costae coarse, broadly rounded, and separated by narrow intercostal furrows; fourteen to twenty costae on each lateral slope, the first five costae on either side of fold and sinus frequently increasing by bifurcation or trifurcation, and the remainder being simple; in some individuals, dividing costae have incipient fasciculation; micro-ornament of fine longitudinal lirae crossed by regular concentric growth lines.

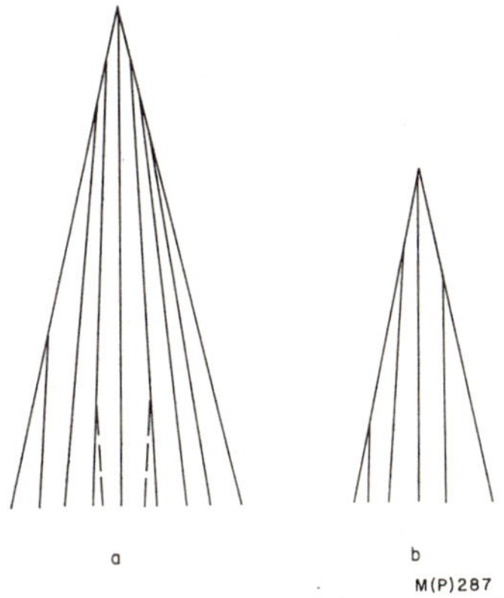

Figure 69. Pattern of costae in the sinus of *Podtsheremia? thomasi* sp. nov. a-CPC 11009, b-CPC 11012c; locality 301, Burvill Beds, Milligans Hills. × 1.5.

Pedicle valve highest at about one-third its length, most convex at umbo; umbo wide, poorly differentiated from flanks, extending behind and incurved over hinge; umbonal shoulders moderately concave; flanks broadly arched medially, flatter towards lateral margins; interarea moderately high, apsacline, concave near umbo, flatter laterally, bearing denticle grooves; delthyrium open, filled apically with callus, delthyrial angle 70°; hinge denticulate; sinus commencing at umbo, U-shaped in cross section, deepest at anterior margin, sinal angle 20°; sinal costae (Fig. 69) consisting of a strong simple median costa arising from umbo, and one to three pairs of accessory costae branching from primary costae bordering sinus; in larger specimens, innermost pair of accessory costae have an incipient anterior bifurcation; several earlier costae on lateral slopes, usually numbers 2, 3, and 4, bifurcating or trifurcating; remaining costae simple.

Brachial valve less convex than pedicle valve, highest towards posterior or along crest of fold; lateral slopes gently convex, becoming flatter posterolaterally; umbo small, extending a short distance behind hinge; interarea short and anacline; fold narrow posteriorly, having concave lateral margins, becoming wider and sometimes higher anteriorly; twin median costae on fold arising from umbo and either simple or, in large specimens, bifurcating or trifurcating anteriorly (Fig. 70); median costae bordered by a single pair of costae along margins of fold in small specimens, and by one or two additional pairs of costae branching from primary costae in large specimens; inner costae on lateral slopes, usually numbers 1, 2, 3, and 4, bifurcating or trifurcating, and remaining costae simple.

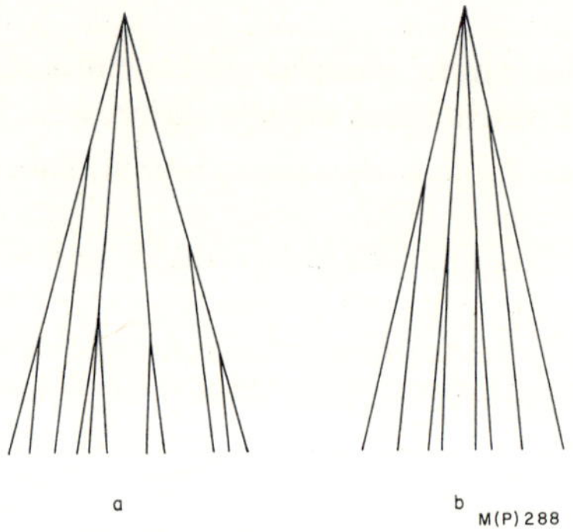

Figure 70. Costae on the fold of *Podtsheremia? thomasi* sp. nov. a-CPC 11016, b-CPC 11013; locality 301, Burvill Beds, Milligans Hills. × 1.5.

Internal. Pedicle valve. Dental plates extending down inner margins of delthyrium, and subtending thickened, pointed teeth; adminicula supporting dental plates, short, diverging at 30-40°; adminicula partly buried by callus in apically thickened valves; muscle field divided by a low median ridge; adductor muscle scars narrow, elongate, and on low platforms; in apically unthickened specimens, diductor muscle scars narrowly rectangular and striated; in thickened individuals, muscle field deeply impressed, situated between adminicula, and diductor muscle scars broader, subovate to trigonal, and striated or dendritic; floor of valve on either side of cardinalia pitted and bearing narrow branching ridges (?vascula genitalia) arising from margins of muscle field and from umbo.

Brachial valve. Cardinal process triangular, consisting of about twenty to twenty-five thin lamellar plates resting on an apical callus; inner socket ridges diverging at 90°, narrow posteriorly, with high distal extremities; sockets slightly sigmoidal, deep, rounded, and very broad; crura extending dorsally from and forming a thickening along inner margins of inner socket ridges; crura usually short, but in some specimens reaching floor of valve and forming true crural plates; adductor muscle scars narrow, elongate, with rounded extremities, and divided by a long low myophragm.

Measurements	Length (pedicle valve)	Length (brachial valve)	Width	Height	Height (of ventral interarea)
CPC 11009 Holotype	39.5	—	41	—	—
CPC 11014 Paratype	—	23	36 est.	—	—
CPC 11016 Paratype	—	25	48.5	—	—
CPC 11017	16	13	27 est.	12	3.5
CPC 11020	28	22	45 est.	18	7

Remarks: Podtsheremia? thomasi sp. nov. occurs with a morphologically closely related species, *P? humilicostata* sp. nov., in the Utting Calcarenite and the Burvill Beds. When compared with *humilicostata, thomasi* is larger, less transverse, and has coarser costae, especially on the fold and sinus. The median costa in the sinus is high, begins at the umbo, and is flanked by up to three pairs of accessory costae branching from the primary costae, whereas in *humilicostata* the median costa is low, begins well in front of the umbo, and is flanked by up to two pairs of accessory costae. Differences in the configuration of the costae on the fold can be seen by comparing Figures 68 and 70. The innermost costae on the flanks of the pedicle valve arise from the umbo, in contrast to those of *humilicostata* which branch from the primary costae bordering the sinus. Additional differences in the pedicle valve of *thomasi* include the possession of a more swollen umbo, a flatter interarea, a narrower median sinus, and a muscle field which in apically thickened specimens is between instead of in front of the ventral adminicula.

Except for a lesser degree of fasciculation, the costae on the lateral slopes of *P? thomasi* resemble those of the type species, *P. prima* Kalashnikov, 1966; the pattern of costae on the fold and sinus of *prima* cannot be accurately determined from Kalashnikov's photographs. *P? thomasi* is distinguished by its more transverse outline, wider hinge, and the lack of a long lingual extension at the front of the sinus.

The closest overseas affinities of *P? thomasi* are with specimens of the *Spirifer duplicostatus* Phillips group of species from the Visean of Belgium. A collection identified by L. G. de Koninck as *S. duplicostatus* in the Museum of Comparative Zoology, Harvard University (Numbers 1756 and 2649), contains specimens which resemble *P? thomasi* in shape, and in the morphology of the costate ornament on the fold, sinus, and lateral slopes. The Belgian specimens are usually transverse, subrectangular, and widest at or just in front of the hinge, but they are associated with subovate individuals similar to the British type (Davidson, 1858, pl. 3, fig. 8); their costae range from high and angular to relatively low and broad, which again is typical of British specimens of *S. duplicostatus.* Davidson's illustrations of *S. duplicostatus* (1858, pl. 3, figs 7-10; pl. 4, figs 3-11) show it to be subovate in outline, with a short hinge and a narrow ventral umbo. The style of ornament is comparable with that of *P? thomasi,* but the precise costal configuration cannot be determined from Davidson's figures; it is clear, however, that the costae adjacent to the fold and sinus divide, and that the latter structures have a comparable number of accessory costae. Thomas's (1970, p. 104) suggestion that *P? thomasi* was closest to British specimens of the *S. duplicostatus* group was made on the basis of mostly broken material. A comparison of *S. duplicostatus* with complete valve of *P? thomasi* from the Burvill Beds at Milligans Hills shows that the latter is distinguished by its wide hinge, transverse shape, and broader ventral umbo.

Anthracospirifer shoshonensis (Branson & Greger) (1918, pl. 18, figs 26, 27) from the early Meramecian (lower or middle Salem) of Wyoming, is morphologically the closest North American species to *P? thomasi. A. shoshonensis* has a similar shape and size, and an external ornament which has a comparable costal configuration in the sinus and probably on the flanks, although the flanks may have fewer dividing costae.

A. curvilateralis (Easton) (1962, pp. 68-70, pl. 9, figs 14-19), from the late Mississippian and early Pennsylvanian of Montana, is similar in outline, and in the shape of the ventral umbo. It has coarser costae, and those adjacent to the fold and sinus branch infrequently; the ventral umbo is more strongly incurved, the sinus wider, and the brachial valve is more convex and has a higher fold.

Occurrence: Locality 301, Milligans Hills (the type locality); locality 306, Burvill Point; Burvill Beds; locality 107/1, Utting Calcarenite, Utting Gap.

Material: CPC 11006-11020. Holotype CPC 11009; paratypes CPC 11010, 11011, 11013, 11014, 11016. CPC 11006-11008, 11017-11020 from locality 306; CPC 11009-11016 from locality 301. Thirty-five specimens.

Age: Visean.

Family MUCROSPIRIFERIDAE Pitrat, 1965

Genus TYLOTHYRIS North, 1920

Type species: Cyrtia laminosa M'Coy, 1844, from the Carboniferous of Hook Head, Ireland, by original designation of North, 1920.

Remarks: According to North (1920, p. 196) the posterior part of the median septum of *Tylothyris* is high, reaching nearly the level of the interarea, but is embedded in apical callus which infills most of the delthyrium. Anteriorly, the septum is lower and terminates near the midpart of the valve. A topotype specimen of *Tylothyris laminosa* (M'Coy) from the Carboniferous Limestone at Hook Head, Wexford (specimen no. 28424 in the collection of the Institute of Geological Sciences, Leeds), kindly loaned by Dr W. H. C. Ramsbottom, has an extensive callus infilling the apex of the pedicle valve; from beneath this callus, just behind the front of the adminicula, a low median ridge emerges. No details of the buried part of the septum are visible. Other specimens of *Tylothyris laminosa* have a similar apical callus, but there is some variation in the height of the anterior portion of the median septum. This ranges from greater than 2.5 mm high in a specimen 54 mm wide and 29 mm long from the Carboniferous Limestone in the north of Ireland (Institute of Geological Sciences specimen no. 73830), to ridge-like in a specimen of the same species from Tournai, Belgium, 37 mm wide and 28 mm long in the collection of the U.S. National Museum.

Specimens of *Tylothyris* from locality 102/4 in the Enga Sandstone are of approximately the one mature growth stage. Thin sections show a median septum embedded in the apical callus, and in pedicle valve interiors there is a low median ridge emerging from the front of the callus (Pl. 50, figs 17, 18; Fig. 72). The posterior part of the septum is lower than that described by North (1920, p. 196), and is much lower than the septum in *T. subconica subconica* North (1920, pl. 13, fig. 15). North's description of a high posterior region on the septum may well have been influenced by the latter specimen, because his sections of *T. laminosa* (1920, p. 166, figs 1m, 1n) do not show an exceptionally high septum.

TYLOTHYRIS TRANSVERSA sp. nov.

(Pl. 50, figs 1-19)

Diagnosis: Shell usually transverse, and twice as wide as long, but occasionally subquadrate; lateral slopes with eight to twelve simple rounded plicae, having sharply pointed or mucronate distal terminations; ventral interarea apsacline, concave beneath the umbo, and relatively low; delthyrial angle 50-55°; median sinus narrow, having a rounded floor, and a sinal angle of 20°; fold narrow; ventral median septum low, obscured posteriorly by an extensive apical callus, and anteriorly ridge-like.

Description: External. Shell usually transversely trigonal, twice as wide as long, but some individuals subquadrate; juvenile stages with rounded cardinal margins; lateral extremities mucronate in young individuals, becoming more pointed in adults; length/width ratio decreasing in adult stages of growth (Fig. 71); com-

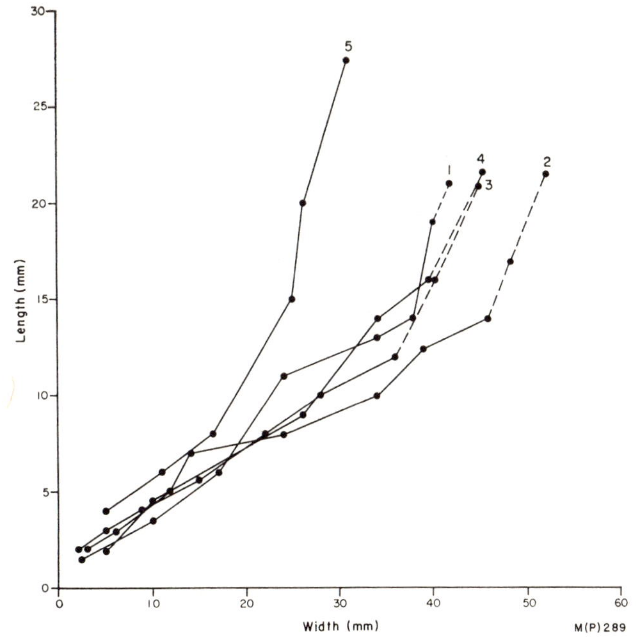

Figure 71. Length/width ontogenies of five pedicle valves of *Tylothyris transversa* sp. nov. from locality 102/4, Enga Sandstone, comparing the development of a subquadrate specimen (steepest curve on the left) with those of typically transverse individuals.

missure uniplicate; lateral slopes with eight to twelve simple plicae; plicae rounded, up to 2 mm high medially, becoming lower laterally, and obsolete on posterolateral margins; concentric lamellae regularly imbricating, having a density of ten to twelve per 5 mm on midregion of shell, and bearing fine concentric growth lines.

Pedicle valve moderately convex, greatest convexity at umbo; umbo narrow, sharply pointed, projecting well behind hinge, and recurving over ventral interarea; umbonal shoulders sloping steeply to flattened posterior margin; cardinal ridges almost straight; lateral slopes moderately steep medially, becoming flat distally; sinus deep, bordered by steep plicae, having a smooth curving floor, and extending anteriorly into a lingual projection; sinal angle 20°; ventral interarea apsacline, moderately concave beneath beak, becoming flatter laterally, 5 mm high at umbo on a shell 46 mm wide and 22 mm long; interarea ornamented with horizontal growth lines; delthyrium open except for a plug of apical callus, delthyrial angle 50-55°.

Brachial valve flatter than pedicle valve, highest at midpoint or towards front of fold; fold narrow, rounded, becoming higher anteriorly, often truncated by lingual projection from sinus; lateral slopes gently convex medially, becoming flat distally; umbo small and straight; interarea orthocline, narrow, slightly concave, and ornamented with parallel growth lines.

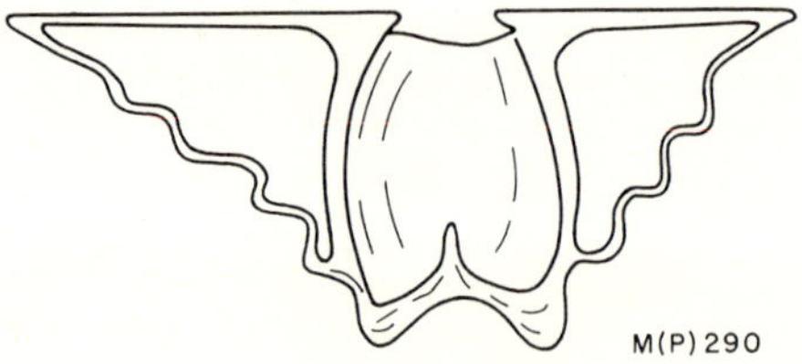

Figure 72. Transverse section through the apical part of the pedicle valve of *Tylothyris transversa* sp. nov. showing the low median septum embedded in secondary tissue. CPC 11115, locality 102/4, Enga Sandstone, northern end of Enga Ridge. × 5.3.

Internal (Figs 72, 73). Pedicle valve. Median septum low posteriorly, becoming ridge-like anteriorly between adductor muscle scars, and extending to midpoint of valve; posterior part of septum obscured by an extensive apical callus, leaving anterior portion only visible in mature specimens; dental plates enlarged by callus, supported by thickened ventral adminicula; teeth short and blunt; adductor muscle scars smooth, narrow, elongate, and deeply impressed into apical callus posteriorly, elevated on sinus anteriorly; diductor muscle scars dendritic, ovoid, situated on flanks of sinus, having obscure anterior extremities; floor of valve ornamented with fine spinules.

Brachial valve. Cardinal process on an apical callus, having a myophore with up to twenty-five extremely thin vertical plates approximately 1 mm long; sockets long, expanding anteriorly, broadly rounded, and bordered by stout inner socket ridges; crura short, divergent, slightly concave on inner surfaces, and arising from inner socket ridges.

Measurements	Length (pedicle valve)	Length (brachial valve)	Width	Height
CPC 8203 Holotype	20	16	43	18
CPC 8204 Paratype	19.5	18	43 est.	15
CPC 8205 Paratype	26.5	22	32.5	15
CPC 8206	—	14	35 est.	—

Remarks: Specimens illustrated on Plate 50, figures 1-16 show that there is considerable variation in the length/width ratio of *Tylothyris transversa*. In the collection from locality 102/4 in the Enga Sandstone most specimens are transverse (Pl. 50, figs 1-10), but there are more subquadrate individuals (Pl. 50, figs 11-14) which are consistently longer and less transverse throughout their ontogeny. This variation is also illustrated in the length/width ontogenies of pedicle valves (Fig. 71). One of the specimens (the subquadrate individual in Pl. 50, figs 11-14) has, overall, a much steeper curve in its ontogeny than the other typically transverse specimens. A steepening in the curves during the final stages of growth of the transverse specimens to nearly match that in the subquadrate form indicates an increase in length as opposed to width, and suggests that subquadrate specimens matured earlier.

Tylothyris transversa is closest to *T. laminosa* (M'Coy), as described by North (1920, pp. 197-98) and de Koninck (1887, pp. 103-5, pl. 22, figs 44-50). *T. transversa* is distinguished from *T. laminosa* by a lower ventral interarea, a much wider delthyrial angle (50-55° compared with 35° in *T. laminosa*), and a narrower fold and sinus. In addition, most specimens of *T. transversa* are shorter and more transverse. The occasional subquadrate individuals, which somewhat resemble *T. laminosa* in shape, are distinguished by their narrower fold and sinus, lower ventral interarea, and less acute apical angle.

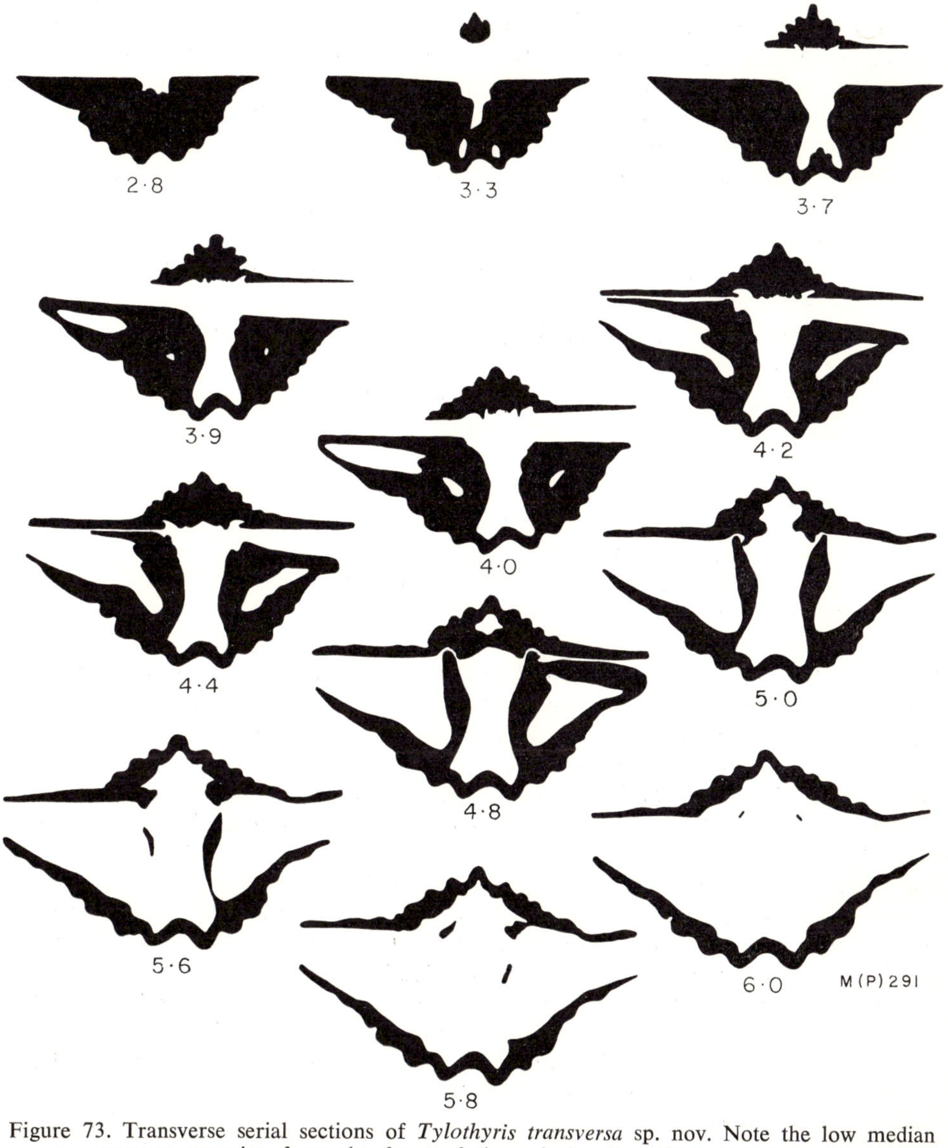

Figure 73. Transverse serial sections of *Tylothyris transversa* sp. nov. Note the low median septum emerging from the front of the apical callus. CPC 11116, locality 102/4, Enga Sandstone, northern end of Enga Ridge. × 2.

In England, *T. laminosa* (M'Coy) is recorded from the Z and C Zones, and apparently is rare above the C (North, 1920, p. 200); Brunton (pers. comm.) has a possible specimen of *laminosa* in the D Zone of Ireland. In Belgium, *laminosa* is recorded by Demanet (1958) from the Tn2c and Tn3b.

Tylothyris planimedia Cvancara (1958, pp. 876-79, pl. 113, figs 1-5) from the Buckets Gap Formation (late Visean to early Namurian) at Barrington, New South Wales, is distinguished from *T. transversa* by a more trigonal outline, more angular plicae, a flat floor on the median sinus, a less extensive ventral apical callus, and a higher ventral median septum.

A small lamellose spiriferoid from the lower part (Tournaisian) of the Pond Argillite of Queensland, referred to *Mucrospirifer kennedyensis* Maxwell (1954, pp. 18-19, pl. 2, figs 4-7) is externally somewhat similar to the younger stages of growth of subquadrate specimens of *T. transversa.* Maxwell's species, however, lacks a ventral median septum, and has more rounded plicae; on one topotype (no. 15138, University of Queensland collection, illustrated in reverse by Maxwell, 1954, pl. 2, fig. 5) one plica bifurcates, and there is a narrow median costa in the sinus.

The specific name refers to the transverse outline of all except several specimens from the type locality.

Occurrence: Locality 102/4, Enga Sandstone, northern end of Enga Ridge.

Material: CPC 8203-8209, CPC 11115, CPC 11116. Holotype CPC 8203; paratypes CPC 8204, 8205, and 8207. All from locality 102/4. One hundred and fifty free specimens.

Age: Tournaisian.

Genus VOISEYELLA Roberts, 1964

1967 *Amesopleura* Carter, pp. 363-64.

Type species: Strophopleura anterosa Campbell, 1957, from the Visean at Babbinboon, Werrie Syncline, New South Wales, by original designation of Roberts, 1964.

Diagnosis: See Roberts (1964, p. 187).

Remarks: Amesopleura Carter, 1967, based on *Spirifera novamexicana* Miller, is placed in synonymy with *Voiseyella* Roberts, 1964, on the basis of its transversely triangular outline, acutely angular posterolateral extremities, costae in which those bounding the sinus are noticeably larger than others on the shell, concentric growth lamellae, capillate micro-ornament, narrow fold and sinus, denticle grooves on the ventral interarea, short closely set ventral adminicula, and the presence of two ridges bordering the dorsal muscle field. Carter (1967, p. 364) in a footnote acknowledged the existence of *Voiseyella,* but claimed that because serial sections of *V. anterosa* were not given the internal structure of *Voiseyella* was obscure; most features were, however, described by Campbell (1957, pp. 79-80), and Carter was presumably referring to the ridge-like thickenings on either side of the dorsal muscle field. All individuals of *anterosa* so far collected from New South Wales are casts and moulds unsuitable for sectioning or, because of their small size, for making rubber casts showing apical structures. Carter noted specifically that illustrations of *anterosa* (Campbell, 1957; Roberts, 1964) gave no indication that the brachial valve had ridges of tissue on either side of the muscle field; these are visible on either side of the interior of the fold in internal moulds of the brachial valve of *anterosa* (Roberts, 1964, pl. 3, figs 1a, 3). The type specimens of

V. anterosa have been destroyed by fire, and the type locality at Babbinboon, New South Wales, will have to be re-collected before a more detailed comparison of the brachial valve interiors can be given.

Species assigned to *Voiseyella* now include *Spirifera novamexicana* Miller, 1881, from the Lake Valley Formation of New Mexico, *Amesopleura texana* Carter, 1967, from the Chappel Limestone of Texas, and probably *Spirifera mundula* Rowley, 1893, from the lower Burlington Limestone of Missouri, USA.

VOISEYELLA? sp.
(Pl. 51, figs 17-23)

Description: External. Pedicle valve transversely triangular, more than twice as wide as long, with sharply pointed posterolateral extremities and a rounded anterior margin; valve moderately convex medially, becoming flat on flanks; hinge straight, the widest part of valve, and coarsely denticulate; umbo pointed, nearly straight, short, and flanked by small concave umbonal shoulders; interarea high, gently concave beneath beak, but flat towards lateral margins, bearing prominent denticle grooves and parallel growth lines; delthyrium narrow, delthyrial angle 25°, plugged apically by a recessed subdelthyrial plate; sinus bounded by two prominent rounded costae, having a rounded floor, becoming shallower towards front of a large individual, and producing a small anterior tongue; sinal angle 20-25°; costae commencing at umbo or from adjacent posterior margin, simple, rounded, increasing in width anteriorly, relatively low in amplitude, becoming obsolete towards lateral margins; intercostal furrows rounded, and narrower than costae; up to ten costae on each flank; concentric ornament lamellose, lamellae varying from crowded and short to widely spaced and long; micro-ornament not observed.

Internal. Pedicle valve unthickened; dental plates supported on floor of valve by short straight divergent adminicula, and joined with one another in apex of valve by a robust subdelthyrial plate; muscle field with elongate narrowly triangular diductor muscle scars, obscure adductor muscle scars, and divided by a low myophragm originating at apex of valve.

Measurements	Length	Width	Height (of ventral interarea)
CPC 11028	—	18 est.	3.5
CPC 11029	9.5	28 est.	5.5
CPC 11030	4	11	—
CPC 11031	5.5	12 est.	3

Remarks: Because the brachial valve has not been collected, and the pedicle valve has an unusually high ventral interarea and a short umbo, this species is tentatively referred to *Voiseyella* Roberts, 1964. The type species, *V. anterosa* (Campbell) described by Campbell (1957, pp. 79-80, pl. 15, figs 19-23) and Roberts (1964, pp. 187-88, pl. 3, figs 1-7) from the Visean of New South Wales, is similar in size and outline to this species. It differs in having a more prominent and incurved umbo, fewer costae, fewer concentric lamellae on the posterior of the shell, a shorter ventral interarea, and subparallel adminicula in the pedicle valve.

Occurrence: Locality 108/0, Utting Calcarenite, Utting Gap.

Material: CPC 11028-CPC 11031. All from locality 108/0. Four silicified pedicle valves.

Age: Visean.

Family BRACHYTHYRIDIDAE Fredericks, 1919 (1924)

Subfamily CHORISTITIDINAE Waterhouse, 1968

Genus AUSTROCHORISTITES nov.

Type species: Austrochoristites levisulcatus gen. et sp. nov. from locality 100/30, Burt Range Formation, 6.5 miles northwest of Mount Septimus, Bonaparte Gulf Basin.

Description: Shell triangular in early growth stages, ovoid in adult, resembling *Choristites* Fischer in shape; shell thickened at apex, especially in the pedicle valve; commissure uniplicate to nearly rectimarginate; costae numerous on lateral slopes, bifurcating and well defined posteriorly, becoming lower and wider, and often indistinct anteriorly; micro-ornament of radial lirae; costae restricted to margins of fold and sinus, leaving a smooth median portion; fold and sinus well defined posteriorly, but frequently becoming indistinct anteriorly; hinge denticulate; ventral interarea truncated laterally, trapezoidal, and bearing denticle grooves; ventral adminicula subparallel, moderately long, and extensively thickened; brachial valve interior with short crural plates at apex; muscle field divided by a low median ridge.

Remarks: Characters of this material indicating a choristitid affinity are the elongate-ovoid shape of the adult, which developed from a triangular juvenile, the long subparallel ventral adminicula made up of a thin lamellar inner layer and a coarsely crystalline outer layer, the extensively thickened apex of the shell, the short crural plates in the brachial valve, and the low margins on the anterior of the sinus. The morphology of the ventral interarea, the brachial valve interior, and the bifurcating costae on the posterior of the shell, resembles that of *Ectochoristites* Campbell, 1957, from the Visean of eastern Australia. *Ectochoristites,* however, has a more finely costate ornament in which the costae maintain their strength anteriorly instead of tending to become obsolete as in *Austrochoristites,* more complex bifurcating costae on the fold and sinus, as opposed to the simple costae on the flanks of the fold and sinus in *Austrochoristites,* shorter ventral adminicula, and a less thickened pedicle valve.

Choristites Fischer, 1825, which is common in the Upper Carboniferous of Russia, Asia, and North Africa, and has also been reported from the Lower Carboniferous and Lower Permian by Pitrat (1965), has a closely comparable outline, denticulate hinge, and trapezoidal ventral interarea. *Choristites* is distinguished from *Austrochoristites* by its longer, more closely spaced subparallel ventral adminicula, which in many cases intersect the muscle field, well-defined costae on the fold and sinus as well as on the front of the shell, and the lack of crural plates in the brachial valve.

Eochoristites Chu, 1933, from the Lower Carboniferous of China, differs from *Austrochoristites* in its external ornament, non-denticulate hinge, triangular ventral interarea ornamented with horizontal growth lines only, and longer crural plates in the brachial valve; the ornament of *Eochoristites* is well defined over the entire shell, and when compared with that of *Austrochoristites* consists of a smaller number of simple robust costae on the lateral slopes, and much stronger costae on the fold and sinus.

The Russian Tournaisian genus *Palaeochoristites* Sokolskaya, 1941, is separated from *Austrochoristites* by its stronger and more simple costate ornament, triangular ventral interarea, transversely oval outline, and lack of a fold and sinus.

Austrochoristites levisulcatus sp. nov.

(Pl. 43, figs 17-29; Pl. 46, figs 18-24; Pl. 47, figs 7-12; Pl. 50, figs 20-23)

Diagnosis: As for the genus, with up to twenty-two costae on each lateral slope; micro-ornament of radial lirae having a density of five per 0.5 mm; sinus always shallower anteriorly, having a sinal angle of 17-20°, and one or usually two costae on each flank; ventral interarea apsacline to orthocline, bearing denticle grooves having a density of five to six per 1 mm.

Description: External. Shell subovate, subequally to unequally biconvex, with greatest width usually at one-third length; hinge slightly less than greatest width; commissure uniplicate to almost rectimarginate; costae rounded, well defined and separated by deep intercostal furrows at posterior, becoming broader, lower, and less distinct anteriorly; costae increasing by bifurcation near posterior of shell, usually simple anteriorly, and numbering as many as twenty-two on each lateral slope on a large shell (32 mm wide); micro-ornament of radial lirae having a density of five per 0.5 mm; distinct concentric lamellae occurring at well-spaced growth pauses.

Pedicle valve varying in shape from triangular in young growth stages to ovoid with rounded lateral margins in adults; valve most strongly convex at umbo, becoming less convex anteriorly, highest near midpoint; umbo prominent, pointed, incurved over interarea, extending well behind hinge, the most convex part of valve; umbonal shoulders broad, and flat to slightly concave; lateral slopes moderately convex in young individuals, to steep in adults; interarea apsacline to orthocline, concave beneath umbo to flat or slightly concave laterally, with truncated lateral margins; interarea 5 mm high on a valve 32 mm wide and 31 mm long, ornamented with deep denticle grooves having a density of five to six per 1 mm; delthyrium open, delthyrial angle 35-40°; sinal angle 17-20°; sinus narrow, well defined by high costae on juvenile parts of valve, becoming broader, shallower, and bordered by lower wider costae at front; sinal costae simple, located on sides of sinus only, leaving a smooth central trough (Fig. 74).

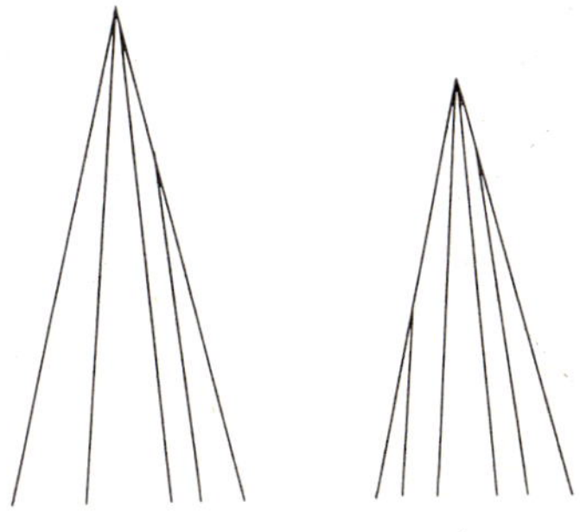

Figure 74. Pattern of costae in the sinus of *Austrochoristites levisulcatus* gen. et sp. nov. a-CPC 8201, b-CPC 8194; locality 100/30, Burt Range Formation, 6.5 miles northwest of Mount Septimus. × 1.5.

Brachial valve elliptical to subovate, wider than long or having subequal dimensions, usually lower and less convex than pedicle valve; umbo small, projecting a short distance over hinge; lateral slopes gently convex to almost flat; fold always well defined and outlined by deep furrows towards umbo, in some specimens becoming obsolete anteriorly; sides of fold ornamented with low simple costae, in most specimens having a median depression or flattening along crest.

Internal (Fig. 75). Pedicle valve greatly thickened by coarsely crystalline calcite at apex; dental plates short, buttressed by adminicula consisting of a very narrow fibrous layer surrounded by a coarsely crystalline sheath; adminicula subparallel, concave in transverse section, usually extending one-third length of valve, and abutting against one another at apex; teeth small and rounded; muscle field impressed into floor of valve between anterior extremities of adminicula, pointed posteriorly, and with a bluntly rounded anterior margin; diductor muscle scars broad, subrectangular, and tapering posteriorly; adductor muscle scars narrow and slightly elevated; mantle canals narrow, linear, extending anteriorly from front of muscle field.

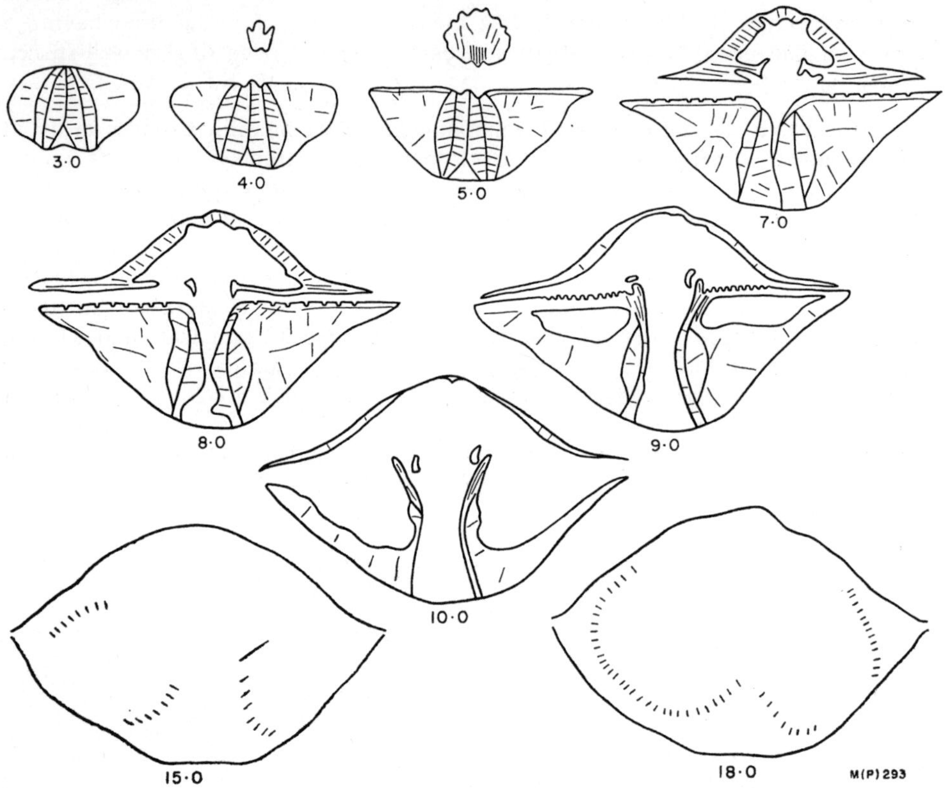

Figure 75. Transverse serial sections of *Austrochoristites levisulcatus* gen. et sp. nov. CPC 11117, locality 109/2, Burt Range Formation, middle part of Enga Ridge. × 2.

Brachial valve. Cardinal process triangular, having a myophore with approximately sixteen vertical lamellar plates; crura reaching floor at apex of valve, forming short crural plates, but anteriorly subtended by inner margins of inner socket ridges; inner socket ridges enclosing small sockets; adductor muscle scars in two pairs; inner adductor muscle scars impressed, narrowly triangular, tapering posteriorly, and broad and slightly curved anteriorly; outer adductor muscle scars less well impressed, narrow, slightly divergent, and subtending linear mantle canals extending along margins of fold; median ridge narrow, commencing a short distance in front of umbo, extending to front of inner adductor muscle scars.

Measurements

Pedicle valve	Length	Width	Height	Locality
CPC 8201 Holotype	23.5	29.5	11	100/30
CPC 10982	33	32	12.5	100/30
CPC 10983 Paratype	20	26	8.5	100/30
CPC 10984	29 est.	27 est.	11	100/30
CPC 8211 Paratype	30 est.	32 est.	10	100/30
CPC 8199	12	22.5	6	100/24
CPC 8200	28.5	28	11	100/24
CPC 8202	26	28	10	100/24
CPC 10986	36	33	15	101/10
Brachial valve				
CPC 8193 Paratype	25.5	31	8.5	100/30
CPC 8198	28	30.5	10	100/30
CPC 8202	20.5	28	5.5	100/24
CPC 10985	26	34	5.5	100/24
CPC 10987	30	33	5	101/10

Remarks: The specimens of *Austrochoristites levisulcatus* available for study are not sufficiently well preserved to plot length/width ontogenies. Young growth stages are triangular in shape, usually wider than long, and have rounded or slightly mucronate posterolateral extremities; during later stages there is a marked increase in anterior growth and a diminution of lateral growth, resulting in an ovoid adult shell.

Specimens from locality 100/24 in the Burt Range Formation, tentatively assigned to this species, have a low fold with smooth flanks and a prominent median furrow, in contrast to the usually well-defined fold with a faint median furrow or flattening, and two accessory costae on the flanks of the type specimens. Some individuals are subcircular in outline, but others are ovoid, and resemble specimens from the type locality. In the Burt Range Formation, locality 100/24 is at approximately the same stratigraphic level as locality 101/10, which has specimens of *Austrochoristites levisulcatus* identical with those from the type locality; this suggests that the variation is localized. Further collections must be made before the full range of variation of the species at locality 100/24 can be adequately documented.

Ectochoristites arenatus Thomas (1970, pp. 113-16, pl. 7, figs 1-11) bears a superficial resemblance to *Austrochoristites levisulcatus.* However, because Thomas's species is based on extremely poorly preserved material, I believe its use should be restricted to material from the type locality at Sandy Creek. In the pedicle valve, *E. arenatus* is similar in shape and in internal features, but has a finer costate ornament on the flanks, and more numerous costae in the sinus. The brachial valve is less convex than that of *A. levisulcatus,* and has a lower median fold, in some cases with a suggestion of a median furrow which resembles that of specimens from locality 100/24 in the Burt Range Formation. *Ectochoristites arenatus* var. *latus* Thomas (1970, pp. 116-19, pl. 7, figs 12-16), from the same locality at Sandy Creek, is usually wider than *A. levisulcatus,* and has a fold and sinus which are scarcely differentiated from the lateral slopes.

Choristitid species from elsewhere in Australia and overseas do not appear to be closely related morphologically to this species.

The name of the species is derived from the Latin *levis* — smooth, and *sulcus* — furrow, and refers to the smooth floor on the ventral sinus.

Occurrence: Locality 100/30 (the type locality), and possibly at localities 100/20-100/24, in an area between 6 and 7 miles northwest of Mount Septimus; localities 101/7B, 101/7C, 101/10, 101/15, 101/16, between 4.5 and 5 miles west of Mount Septimus; localities 115/3 and 117/1, Spirit Hill; localities 103/2, 109/1, 109/2, 128/2, 128/6, middle part of Enga Ridge; locality 124/1, Flapper Hill; Burt Range Formation. Locality 136/1, a possible equivalent of the Enga Sandstone, at Spirit Hill.

Material: CPC 8193-8202, CPC 8210-8211, CPC 10982-10987, CPC 11117. Holotype CPC 8201; paratypes CPC 8193, 8195, 8211, 8479, 10983. CPC 8193-8195, 8197-8198, 8201, 8211, 10982-10984 from locality 100/30; CPC 8196, 8210 from locality 117/1; CPC 8199-8200, 8202, 10985 from locality 100/24; CPC 10986-10987 from locality 101/10; CPC 11117 from locality 109/2. More than 50 specimens.

Age: Tournaisian.

Genus TANGSHANELLA Chao, 1929

Type species: Tangshanella kaipingensis Chao, 1929, from the Upper Carboniferous Tangshan Limestone, Penchi Series, of the Kaiping coal basin, Chihli, China, by original designation of Chao.

Description: Shell ovate, with greatest width near midlength, and with an angular sinus and a well-defined fold; both fold and sinus ornamented with accessory costae; sinus having a prominent median groove; costae on lateral slopes bifurcating, but evenly distributed; micro-ornamented of fine radial lirae; ventral interarea triangular, having vertical denticle grooves; dental ridges extending down inner margins of delthyrium; ventral adminicula absent.

Remarks: Waterhouse (1968), in a revision of the classification of the Spiriferida, suggested that *Tangshanella* belonged to the Brachythyridinae, presumably because it lacked ventral adminicula. *Tangshanella,* however, is characterized by the possession of rounded costae and a micro-ornament of well-defined radial lirae, both of which suggest an affinity with *Choristites* Fischer, and hence with the Choristitinae. External characters such as these may give a more natural grouping of the Brachythyrididae than the degree of development of the ventral adminicula advocated in Waterhouse's classification.

Tangshanella is recorded from the Upper Carboniferous of China, and a species tentatively referred to the genus is from the upper Tournaisian Septimus Limestone in the Bonaparte Gulf Basin. This species, *Tangshanella? fasciculata* sp. nov., is tentatively placed in *Tangshanella* because the characters of the brachial valve are poorly known, and there are a number of morphological differences in the pedicle valve. It differs from *T. kaipingensis,* the type species (Chao, 1929, pp. 57-58, pl. 8, figs 12-14) in having the inner costae on the lateral slopes arranged in bifurcating or trifurcating bundles instead of being evenly distributed; and on all except the posterior of the shell, the costae are wider and flatter. The fold and sinus are weaker, and are ornamented with fewer accessory costae, the floor of the sinus is wider and less angular, and the fold is outlined by deep furrows. The delthyrium is covered by a convex pseudodeltidium which is nearly always preserved, and is apparently absent in *T. kaipingensis*. There does not appear to be any radial micro-ornament. In contrast to the affinities of *Tangshanella,* these characters, especially the flat strap-like costae, and the absence of radial lirae, suggest an affinity with the Brachythyridinae. If better preserved material becomes available for description, *T? fasciculata* may require assignment to a new genus.

TANGSHANELLA? FASCICULATA sp. nov.

(Pl. 51, figs 1-16)

Diagnosis: Shell unequally biconvex, having a moderately convex pedicle valve and a less convex brachial valve; posterolateral margins possibly pointed in adult specimens; commissure weakly uniplicate; costae wide and strap-like, especially at the anterior, and numbering fifteen to seventeen around the margins of each lateral slope; the first three to five costae on either side of the fold and sinus form bifurcating or trifurcating bundles; micro-ornament of concentric growth lines; fold and sinus weakly developed; sinus having a flat floor bordered by bifurcating or trifurcating primary costae on the flanks; apical angle 100-110°; delthyrium covered with a convex pseudodeltidium; dental ridges flanged; diductor muscle scars deeply impressed posteriorly.

Description: External. Shell subovate to transversely oval, usually slightly wider than long, and unequally biconvex; posterolateral margins rounded in young and intermediate growth stages, to possibly pointed in adult; greatest width at or immediately in front of hinge; commissure weakly uniplicate; costae relatively high, rounded, and separated by deep furrows posteriorly, anteriorly becoming broader and flatter, and separated by narrow intercostal grooves; first three to five costae on either side of fold and sinus, including primary costae bordering sinus, form bifurcating or trifurcating bundles, the multiplication taking place usually towards posterior of shell; remaining costae on lateral slopes simple, becoming obsolete towards lateral margins; costae numbering fifteen to seventeen around margins of each lateral slope; micro-ornament of closely spaced concentric growth lines.

Pedicle valve most strongly convex at umbo, becoming less convex anteriorly, greatest height near midlength; umbo moderately incurved, extending behind hinge, having gently concave umbonal shoulders, and forming an apical angle of 100-110°; lateral slopes evenly convex, in larger individuals becoming flattened on posterolateral extremities; sinus commencing at tip of umbo, having an extremely narrow flat floor in some instances appearing as a single median costation, and bordered by bifurcating or trifurcating primary costae; each flank of sinus ornamented with a costa branching from primary costae; sinal angle 15-20°; ventral interarea apsacline to orthocline, gently concave, triangular, ornamented with horizontal growth lines and vertical denticle grooves; delthyrium covered by a large convex pseudodeltidium.

Brachial valve much lower than pedicle valve; greatest height a short distance in front of umbo; umbo small, blunt, extending a short distance behind hinge; umbonal shoulders gently concave; lateral slopes moderately convex medially, becoming flat on posterolateral extremities; fold commencing at umbo, smoothly rounded, outlined by two deep furrows, becoming wide at anterior of large individuals.

Internal. Pedicle valve (Fig. 76). Posterior parts of pedicle valve thickened with columnar shell material, apical callus extending even into proximal tip of pseudodeltidium; dental ridges extending down inner margins of delthyrium, having small internal flanges, and distally bearing pointed teeth; ventral adminicula absent; muscle field deeply impressed into thick columnar calcite; diductor muscle scars in deep narrow troughs separated by a convex callus bearing adductor muscle scars posteriorly, and broader and not as clearly distinguishable from adductor muscle scars anteriorly.

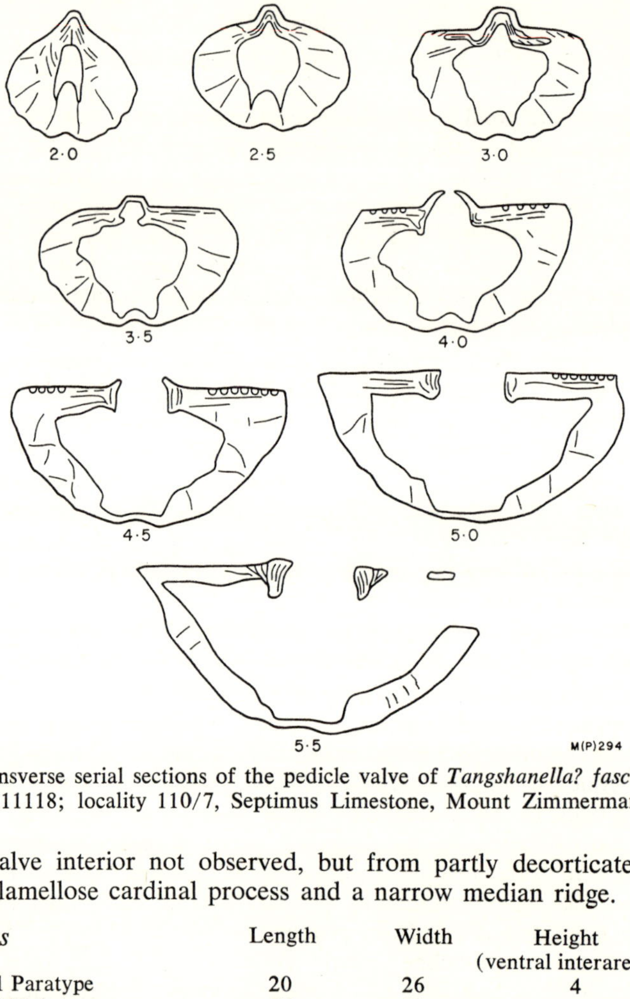

Figure 76. Transverse serial sections of the pedicle valve of *Tangshanella? fasciculata* sp. nov. CPC 11118; locality 110/7, Septimus Limestone, Mount Zimmerman. × 2.5.

Brachial valve interior not observed, but from partly decorticated specimens, possessing a lamellose cardinal process and a narrow median ridge.

Measurements	Length	Width	Height (ventral interarea)	Height
CPC 11021 Paratype	20	26	4	9.5
CPC 11022 Holotype	25	31 est.	—	10
CPC 11024 Paratype	24.5	32 est.	3	11.5
CPC 11025 Paratype	25	33 est.	4.5	10.5
CPC 11026 Paratype	21	27 est.	3.5	10
CPC 11027	24	—	—	5.5

Remarks: A comparison with *Tangshanella kaipingensis* Chao (1929), the type species, is given in the remarks on the genus (p. 234). I know of no other comparable species.

The species name is taken from the Latin *fasciculus,* a little bundle, and refers to the morphology of the innermost costae on the lateral slopes.

Occurrence: Localities 110/6 and 110/7 (the type locality), Septimus Limestone, Mount Zimmermann.

Material: CPC 11021-11027, CPC 11118. Holotype CPC 11022; paratypes CPC 11021, 11024, 11025, 11026. CPC 11021-11026, 11118 from locality 110/7; CPC 11027 from locality 110/6. Forty specimens preserved in calcarenite.

Age: Tournaisian.

Subfamily BRACHYTHYRIDINAE Fredericks, 1919 (1924)

Genus LITOTHYRIS nov.

Type species: Litothyris alticostata gen. et sp. nov. from the Famennian part of the Ningbing Limestone, Jeremiah Hills, Bonaparte Gulf Basin.

Description: Shell brachythyrid in shape, with rounded lateral margins, and a hinge less than maximum width; lateral slopes, fold, and sinus ornamented with costae having low broadly arched surfaces separated by narrow intercostal furrows; costae mainly simple, but those on and adjacent to fold and sinus sometimes increasing by bifurcation near front of shell; micro-ornament of concentric growth lines; pedicle valve with a strongly incurved umbo, and a triangular interarea without denticle grooves; delthyrium possibly closed by a thin pseudodeltidium; dental plates extending down inner margins of delthyrium; adminicula absent; brachial valve with a short swollen umbo, and a high fold; crura long, arising from inner margins of thickened inner socket ridges.

Remarks: Litothyris, found in the Famennian part of the Ningbing Limestone, is one of the earliest members of the Brachythyridinae and may represent the forerunner to the Lower Carboniferous genus *Brachythyris* M'Coy. Stainbrook (1947) referred two species from the Famennian Percha Shale in Arizona and New Mexico to the genus *Brachythyris;* one of these, *B. putilla* Stainbrook, has ventral adminicula and should not be referred to *Brachythyris,* and the other, *B. bisbeensis* Stainbrook, does not resemble *Brachythyris* externally, and is unknown internally.

In external features, *Litothyris* resembles *Choristites* Fischer, 1825, and *Eochoristites* Chu, 1933. *Choristites* has a convex pedicle valve and an incurved ventral umbo comparable to those on *Litothyris,* but is distinguished by long ventral adminicula which intersect the muscle field, a denticulate hinge, a trapezoidal ventral interarea, numerous bifurcating costae in the sinus, and the lack of crural plates in the brachial valve. *Eochoristites* has a slightly coarser costate ornament, a weaker fold and sinus, adminicula in the pedicle valve, and shorter crura in the brachial valve.

The Upper Carboniferous genus *Tangshanella* Chao, 1929, has a similar biconvex shape and brachythyrid outline, and also lacks ventral adminicula. It is distinguished from *Litothyris* by many bifurcating costae on the lateral slopes and in the sinus, and a striate micro-ornament.

The name of the genus is derived from the Greek *litos* — simple.

LITOTHYRIS ALTICOSTATA gen. et sp. nov.

(Pl. 52, figs 1-17)

Diagnosis: Shell unequally biconvex, having twelve to fourteen costae on each lateral slope; costae obsolete on the posterolateral parts of the shell; ventral interarea apsacline; costae in the sinus numbering from eight to thirteen, frequently simple, but sometimes bifurcating towards the front; sinus extending into an anterior lingual extension; fold obsolete posteriorly, high anteriorly.

Description: External. Shell brachythyrid in outline, approximately as wide as long, widest at midlength; unequally biconvex, with evenly rounded lateral margins; hinge straight, two-thirds the width of shell; commissure uniplicate, with an abrupt median dorsal deflection; costae commencing at umbo, ornamenting fold and sinus as well as lateral slopes, having low broadly arched surfaces, and separated by narrow intercostal furrows; costae mainly simple, but those on and adjacent to fold and sinus sometimes increasing by bifurcation anteriorly; twelve to fourteen costae on each lateral slope, becoming obsolete on posterolateral margins; micro-ornament of closely spaced growth lines.

Pedicle valve most strongly convex at umbo, becoming less incurved towards front; greatest height between one-third and one-half length of valve; umbo broad near body of valve, tapering towards tip, extending well behind hinge (5 mm on a valve 30 mm long and 33 mm wide), and strongly incurved; umbonal shoulders gently concave; lateral slopes running steeply to margins; sinus originating at umbo, extending into a rounded or sometimes pointed lingual extension at front of valve, having a sinal angle of 20-25°, and ornamented by from eight to thirteen costae similar to those on lateral slopes; costae in sinus simple posteriorly, some increasing by bifurcation towards front (Fig. 77); interarea low (3.5 mm high

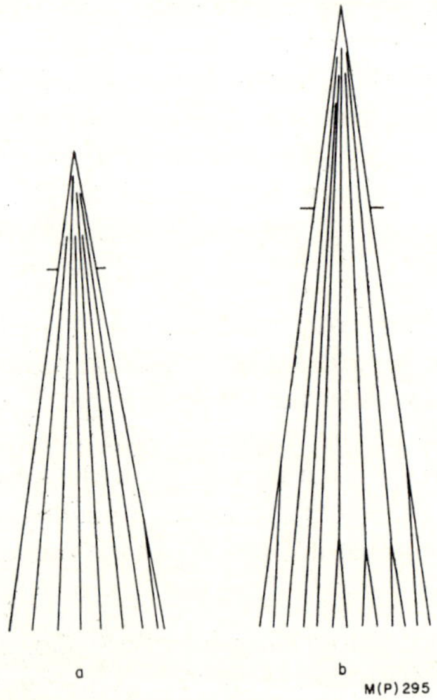

Figure 77. Diagrammatic representation of the sinal costae of *Litothyris alticostata* gen. et sp. nov. The posterior parts of the diagrams are hypothetical. a-CPC 8220, b-CPC 8221; locality 443/22, Ningbing Limestone (Famennian part), Jeremiah Hills. × 1.5.

on a valve 19 mm long and 24 mm wide), straight near hinge, but concave beneath umbo; delthyrium open on all specimens examined except sectioned specimen, but possibly closed by a thin convex pseudodeltidium (Fig. 78, 3.3 and 3.7); delthyrial angle 50-60°.

Brachial valve with a short swollen umbo extending 0.5 mm behind hinge on a valve 23 mm long and 29 mm wide; umbonal shoulders concave; lateral slopes evenly convex, or convex towards posterior of shell, becoming flatter towards front of valve adjacent to strongly elevated front of fold; fold narrow, low, almost obsolete at posterior of valve, becoming much higher and wider at front, and frequently projecting a short distance in front of remainder of valve; fold ornamented by both simple and bifurcating costae similar to those on lateral slopes; highest part of valve along apex, or frequently at front of fold.

Internal (Fig. 78). Pedicle valve thickened apically, with unsupported dental plates extending down inside margins of delthyrium; dental plates short and thickened near apex of valve, becoming longer and more slender towards hinge, and subtending small teeth; in transverse section, dental plates diverging dorsolaterally, and slightly hook-shaped near hinge; muscle field impressed into posterior part of valve; diductor muscle scars deeply impressed, pointed, and separated by a small callus posteriorly, becoming less well impressed and wider towards front of valve; adductor muscle scars not observed.

Brachial valve. Sockets wide, and having abruptly upturned thickened inner socket ridges; cardinal process (poorly preserved in the sectioned specimen) apparently supported by inner socket ridges and crura; crura originating from inner socket ridges, short and thickened near umbo, longer and more slender anteriorly, slightly convex outwards, and converging dorsally in transverse section; spires projecting laterally, having from twelve to fourteen volutions.

Measurements	Length (pedicle valve)	Length (brachial valve)	Width	Height
CPC 8220 Holotype	25	20	25	17.5
CPC 8221 Paratype	32 est.	25.5	30.5	20.5
CPC 8222 Paratype	30	23.5	33 est.	21.5
CPC 8224	18.5	15	23 est.	13

Remarks: No comparable species are known.

The specific name *alticostata* is taken from the Latin *altus* — high, and *costa* — rib, and refers to the high fold on the brachial valve.

Occurrence: Localities 443/13, 443/22 (the type locality), and 443/23, Ningbing Limestone (Famennian back-reef), Jeremiah Hills.

Material: CPC 8220-8225, CPC 11119. Holotype CPC 8220, paratypes CPC 8221 and 8222. CPC 8220-8222, 8224-8225, 11119 from locality 443/22; CPC 8223 from locality 443/13.

Age: Famennian.

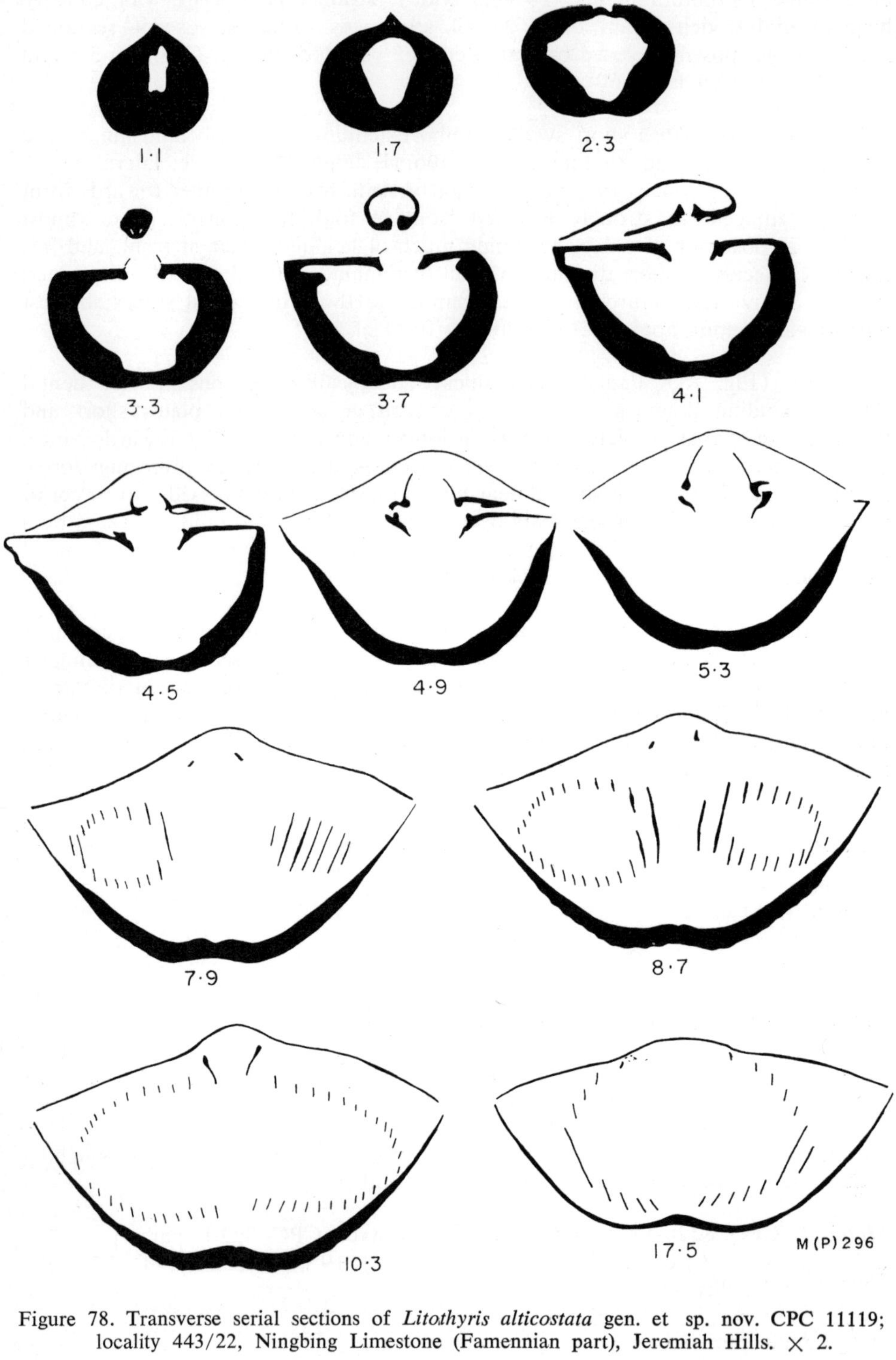

Figure 78. Transverse serial sections of *Litothyris alticostata* gen. et sp. nov. CPC 11119; locality 443/22, Ningbing Limestone (Famennian part), Jeremiah Hills. × 2.

Genus Brachythyris M'Coy, 1844

Type species: Spirifer ovalis Phillips, 1836, from the Visean of Yorkshire, England, by original designation of M'Coy, 1844.

Brachythyris planulata sp. nov.

(Pl. 52, figs 18-29)

Diagnosis: Shell of moderate size for genus, subovate to subelliptical, and subequally biconvex; from eight to twelve plicae on each lateral slope of a mature shell, increasing occasionally by bifurcation; pedicle valve umbo short; sinus having a broad flat to gently concave floor, and lateral margins with generally one or two accessory plications; apex of fold bearing a median furrow.

Description: External. Shell of moderate size for genus, subovate to subelliptical, subequally biconvex; posterolateral margins well rounded; greatest width at one-third length of shell; commissure uniplicate; hinge about two-thirds width of shell; ornament of wide broadly rounded plicae separated by very narrow interplical furrows; plicae commencing at umbo, increasing occasionally by bifurcation, becoming obsolete towards posterolateral margins, from eight to twelve on each lateral slope of a mature shell; posterolateral margins almost completely smooth; micro-ornament of fine evenly spaced concentric growth lines having a density of about ten per 1 mm towards median anterior part of an adult shell; shape of shell not altering appreciably during ontogeny.

Pedicle valve. Umbo sharply pointed, short, extending 4 mm behind hinge on a valve 30 mm long and 34.5 mm wide, and strongly incurved over delthyrium, umbonal tip facing anteriorly; umbonal shoulders moderately concave, separated from ventral interarea by beak ridges; valve highest at midpoint, evenly convex in lateral profile except for a slightly greater convexity at umbo; lateral slopes steep except on posterolateral extremities; ventral interarea apsacline, narrowly

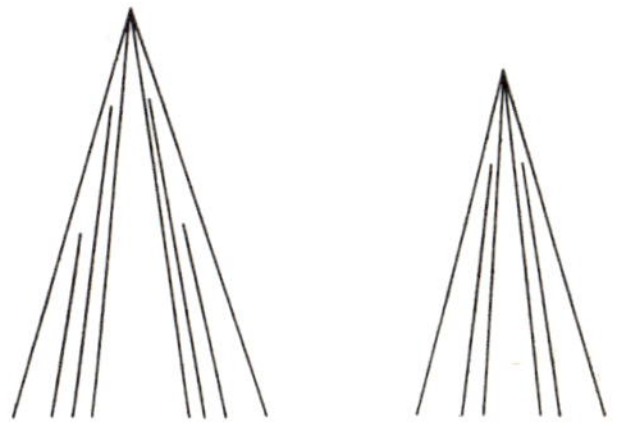

Figure 79. Diagrammatic representations of the bifurcation of the plicae on the flanks of the sinus of *Brachythyris planulata* sp. nov. a-CPC 8233, b-CPC 8231; locality 7/1, Ningbing Limestone, Ningbing Range; the solid lines represent the furrows between the plicae, and the median V-shaped portion the nearly flat floor of the sinus. × 1.

triangular in outline, concave immediately beneath umbo but flat towards hinge and lateral margins, 2.5 mm high on a valve 30 mm long and 34.5 mm wide; delthyrium open, delthyrial angle 40-50°; sinus commencing at tip of umbo, becoming broader and deeper anteriorly, forming a small rounded linguiform projection at front of valve; sinal angle 20-25°; floor of sinus broad and flat to slightly concave; flanks of sinus ornamented by up to three pairs of bifurcating plicae (Fig. 79), but some individuals with smooth flanks.

Brachial valve subovate, moderately convex with greatest convexity at umbo, highest at about one-third length of valve; umbo broadly rounded, incurved, projecting a short distance behind hinge; umbonal shoulders moderately concave;

lateral slopes evenly convex except for a flattening at posterolateral margins; fold commencing at tip of umbo, becoming broader anteriorly, and truncated by dorsal deflection of sinus; median furrow on crest of fold narrow and rounded near umbo, becoming broader and lower towards front; lateral margins of fold rounded, bearing one or two low ribs.

Internal (Fig. 80). Pedicle valve. Muscle field strongly impressed posteriorly, shallower anteriorly, consisting of elongate ovate diductor muscle scars flanking narrow linear adductor muscle scars; mantle canals of vascula genitalia linear, slightly irregular, and radiating from flanks of muscle field; vascula media canals

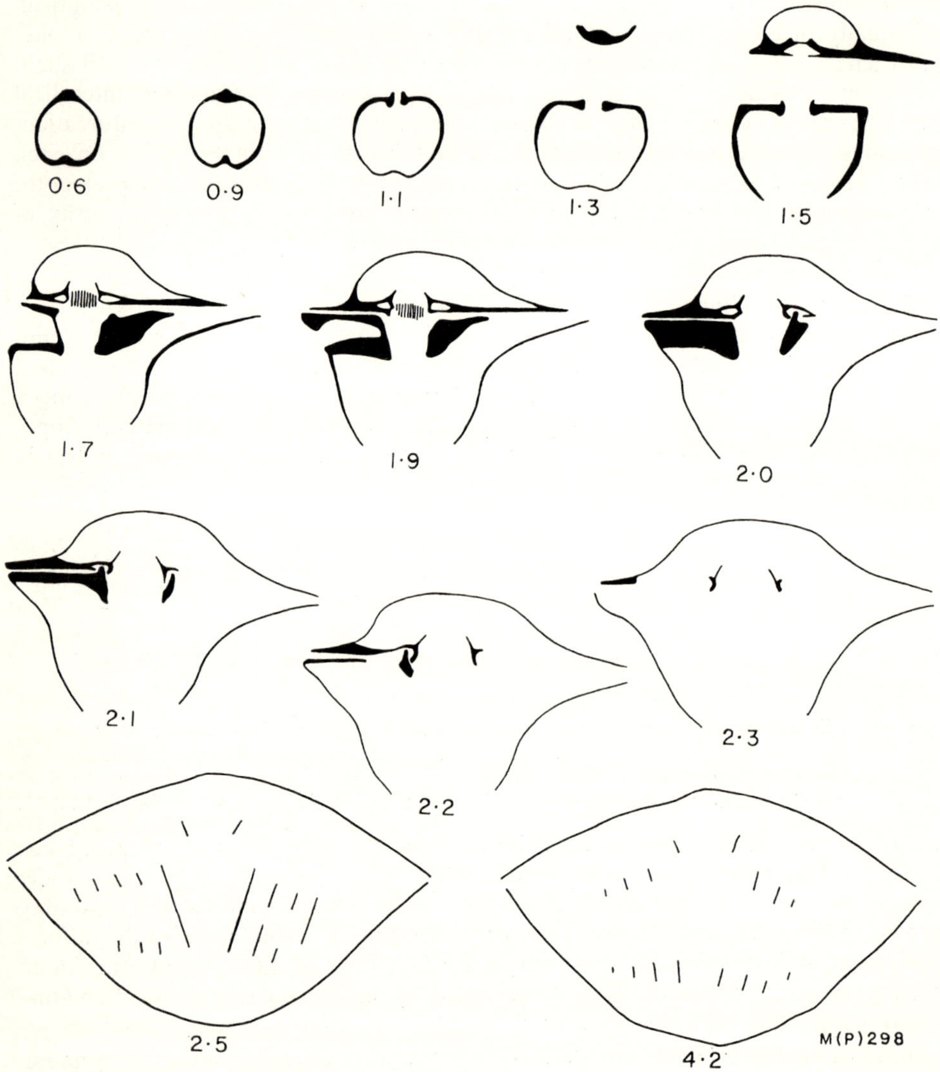

Figure 80. Transverse serial sections of *Brachythyris planulata* sp. nov. CPC 11120; locality 7/1, Ningbing Limestone (Tournaisian part), Ningbing Range. × 2.

arising from front of muscle field, extending along margins of sinus; dental plates small, confined to margins of delthyrium, and bearing small pointed teeth; tip of umbo thickened, but having a median longitudinal furrow.

Brachial valve. Adductor muscle scars narrowly triangular, situated on inner flanks of fold, and probably separated by a median ridge; cardinal process supported on a thin transverse plate, consisting of about ten lamellar platelets; crura arising from inner margins of inner socket ridges, robust posteriorly, becoming more slender anteriorly; inner socket ridges narrow, enclosing deep rounded sockets.

Measurements	Length (pedicle valve)	Length (brachial valve)	Width	Height
CPC 8233 Holotype	28	25	35	20
CPC 8229 Paratype	—	16	25	—
CPC 8231 Paratype	24.5	—	31	—

Remarks: Brachythyris planulata belongs to George's (1927) group A of *Brachythyris* species in which the fold is divided by a median groove.

B. planulata is closest morphologically to specimens of *B. chouteauensis* (Weller) from the Chappel Limestone of Texas, figured by Carter (1967, pl. 39, figs 1-9). *B. chouteauensis* has a comparable shape and a similar plicate ornament on the lateral slopes and on the flanks of the sinus (*see* Carter, 1967, pl. 39, fig. 1a); externally it is distinguished from *B. planulata* by its slightly higher pedicle valve and flatter brachial valve. Serial sections of *B. chouteauensis* (Carter, 1967, fig. 37) show the internal structures to be very close to those of this species. The specimens of *B. chouteauensis* figured by Weller (1914, pl. 52, figs 4-11) are smaller and less transverse than most specimens of *B. planulata,* and are probably younger individuals.

Specimens referred to *Spirifer subrotundatus* M'Coy by de Koninck (1887, pl. 31, figs 16-18) from the Visean (stage III) of Belgium are morphologically close externally to this species. The Belgian specimens have a similar shape, a sinus with a flat floor and ribbed flanks, and a median furrow on the fold. *B. planulata* is distinguished by its smaller size, and finer plicae on the lateral slopes.

The species referred to *Spirifer* aff. *integricostatus* Phillips by Nalivkin (1937, pl. 27, figs 6, 10) from the Ishim and Yagovkin Beds (Visean) of Kazakhstan should be referred to *Brachythyris* (Muir-Wood, 1948, p. 47). It differs from *B. planulata* in having a longer hinge, and a narrower median sinus having numerous accessory plicae.

The pedicle valve of *B. krapivnensis* Besnosova (1959, pp. 123-25, pl. 10, fig. 6), from the Tournaisian of the Kuznetsk Basin has an outline and ribbed flanks on the sinus comparable with that in *B. planulata.* However, because the brachial valve of *B. krapivnensis* has not been figured a closer comparison cannot be made.

Two species of *Brachythyris* from the Septimus Limestone in the Bonaparte Gulf Basin have been described by Thomas (1970). *B. laticardinalis* Thomas (1970, pl. 6, figs 3, 6, 7) has a wide hinge and is more trigonal than *B. planulata,* and is characterized by a weaker narrower median sinus which has very faint accessory costae on the flanks. The brachial valve is poorly preserved and is not adequately known for comparison. *B.* sp. cf. *B. peculiaris* (Shumard), figured by Thomas (1970, pl. 6, figs 4, 5), is distinguished from *B. planulata* by having a more subcircular outline, narrower umbo, and fewer plicae on the lateral slopes.

The specific name, from the Latin *planus* — level, refers to the flattened floor of the median sinus.

Occurrence: Locality 7/1, Ningbing Limestone (Tournaisian part), Ningbing Range.

Material: CPC 8226-8233, CPC 11120. Holotype CPC 8233, paratypes CPC 8229, 8231, 8232. All from locality 7/1. About 30 specimens obtained by calcining reef limestone.

Age: Tournaisian.

Family MARTINIIDAE Waagen, 1883

Subfamily MARTINIINAE Waagen, 1883

Genus EOMARTINIOPSIS Sokolskaya, 1941

Type species: Eomartiniopsis elongata Sokolskaya, 1941, from the Tournaisian of the Moscow Basin, USSR, by original designation of Sokolskaya, 1941.

Remarks: A detailed description, and a discussion on the affinities of *Eomartiniopsis* is given by Carter (1967, pp. 418-19).

Eomartiniopsis may be a junior synonym of *Martiniella* Grabau, 1931 (Sokolskaya, 1941; Havlicek, 1959). However, because its type species, *Martiniella nasuta* Grabau, has never been described, *Martiniella* is presently unavailable. Carter (1967) considered *Martiniella* an invalid name. The genus will become valid, and probably a senior synonym of *Eomartiniopsis,* when *M. nasuta* is formally described (International Code of Zoological Nomenclature, Article 68).

EOMARTINIOPSIS COSTATA sp. nov.

(Pl. 53, figs 1-15)

Diagnosis: Shell large, subovate to subquadrate, and unequally biconvex; hinge wide, and slightly less than greatest width; commissure weakly uniplicate; shell material thick, especially at the apex of mature individuals; shell usually ornamented with from ten to fifteen broad low costae on each lateral slope, but occasionally smooth; micro-ornament of radial lirae and concentric growth lines; ventral sinus narrow and shallow; in costate individuals the flanks of the sinus bear one or two weak costae branching from the primary costae bordering the sinus; apex of delthyrium covered by an arched pseudodeltidium; pedicle valve interior with strong dental plates supported by short slender adminicula; adminicula buried by apical callus in mature individuals; brachial valve less convex than pedicle valve, and having a low median fold.

Description: External. Shell large, subovate to subquadrate (Fig. 81), unequally biconvex, and widest at midlength; hinge straight, slightly less than greatest width; cardinal extremities and lateral margins well rounded; commissure weakly uniplicate; shell material thick, especially at apex of mature individuals; shell generally ornamented with low broad costae on lateral slopes, but sometimes completely smooth; ten to fifteen costae on each lateral slope, varying from moderately developed to almost completely obsolete, becoming lower towards margins, and each costa usually bifurcating towards front; micro-ornament of radial lirae and concentric growth lines.

Pedicle valve moderately convex, with greatest curvature at umbo; umbo well defined, projecting 5 mm behind hinge on a valve 33.5 mm wide and 31 mm long, and in some instances incurving over and abutting against umbo of brachial valve; umbonal angle 90°; umbonal shoulders sloping steeply to cardinal margins; lateral slopes moderately steep, becoming flattened on posterolateral extremities; cardinal margins straight, well differentiated from ventral interarea; interarea flat, orthocline to apsacline, triangular, marked with vertical denticle grooves having a density of approximately four per 1 mm; uppermost third of delthyrium covered by an arched pseudodeltidium; delthyrial angle approximately 80°; median sinus originating at tip of umbo, narrow, shallow, having a slightly rounded floor, becoming wider anteriorly; sinal angle 15-20°; on costate individuals each flank of sinus ornamented with one or two weak costae branching from primary costae bordering sinus.

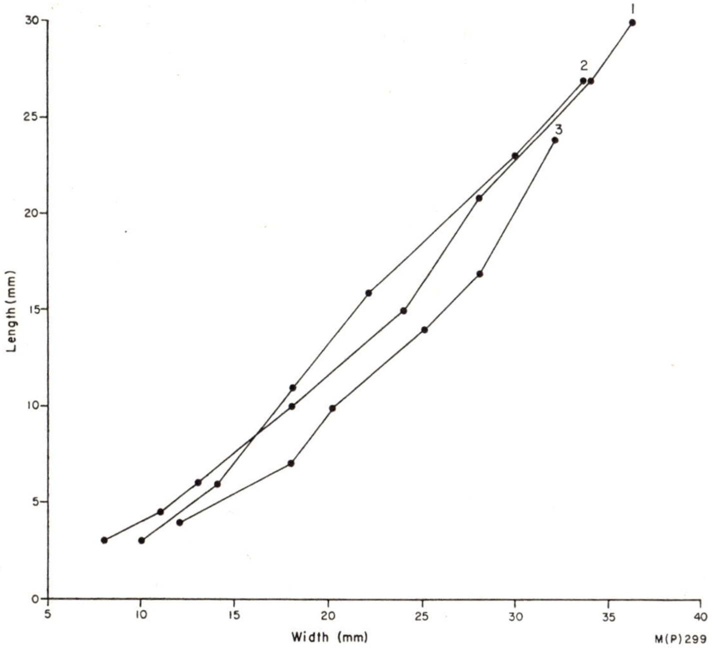

Figure 81. Length/width ontogenies of the pedicle valves of three specimens of *Eomartiniopsis costata* sp. nov. Locality 102/4, Enga Sandstone, Enga Ridge.

Brachial valve less convex than pedicle valve, highest towards umbo, becoming flat on posterolateral and lateral margins; umbo small, barely projecting over hinge, and not incurved; interarea 2 mm high on a valve approximately 28 mm long, orthocline to anacline; lateral slopes gently convex; fold smooth or with a shallow median furrow, commencing at umbo, becoming wider anteriorly.

Internal (Figs 82, 83). Pedicle valve. Dental plates strong, curved, situated on inner margins of delthyrium, supported by ventral adminicula; adminicula slender, divergent, enclosing posterior end of muscle field, 5 mm long in a valve 15 mm long, but often much shorter, and obscured in older individuals by an extensive apical callus; teeth not observed; muscle field elongate, extending to midlength of valve, bluntly rounded posteriorly, broadening to approximately two-thirds of its length and then tapering anteriorly, very deeply impressed posteriorly in older specimens; adductor muscle scars narrow, situated on linear rounded ridges, divided medially by a shallow furrow; diductor muscle scars much broader, flat to gently convex, having occasional longitudinal striations, and separated by furrows from adductor muscle scars.

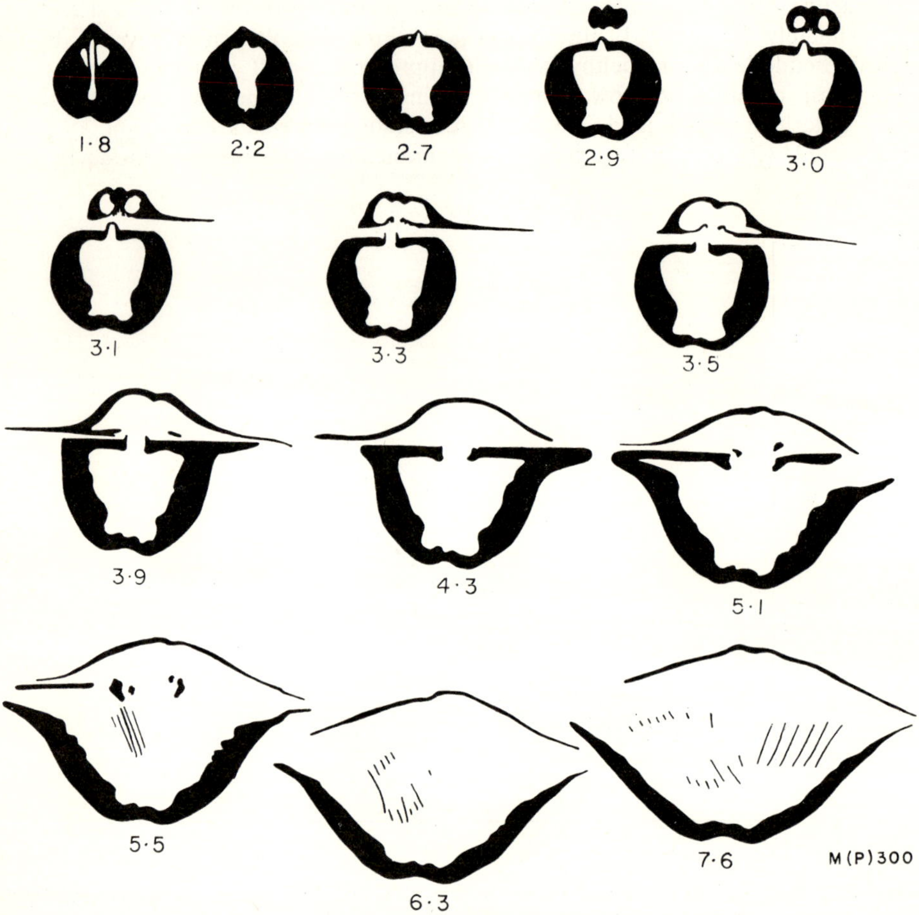

Figure 82. Transverse serial sections of *Eomartiniopsis costata* sp. nov. Note the short ventral adminicula in 1.8 and the absence of dorsal adminicula. CPC 11121; locality 102/4, Enga Sandstone, northern end of Enga Ridge. × 2.

Brachial valve. Cardinal process with approximately eighteen thin vertical lamellar plates, the longest more than 1 mm long; socket plates diverging at 85-90°, flanking cardinal process; inner socket ridges very strong distally, bordering deep rounded sockets, and giving rise from their median anterior surface to crura; crura flat, triangular posteriorly, more than 1 mm long; adductor muscle scars ovoid, extending approximately one-quarter length of valve, bluntly rounded at either end; median ridge dividing adductor muscle scars, and extending about three-quarters length of valve.

Remarks: Eomartiniopsis costata sp. nov. is characterized by an extremely thick shell which buries the ventral adminicula in adult specimens, a weakly costate ornament on most individuals, and a transverse outline. Dorsal adminicula, or basal plates, which characterize the type species *E. elongata* Sokolskaya, have not been observed in *costata,* but they too may have been obscured by the apical thickening of the shell. Additional differences distinguishing *costata* from the type species include its more transverse outline, much wider hinge, more strongly costate ornament on most individuals, weaker fold and sinus, and much thicker shell.

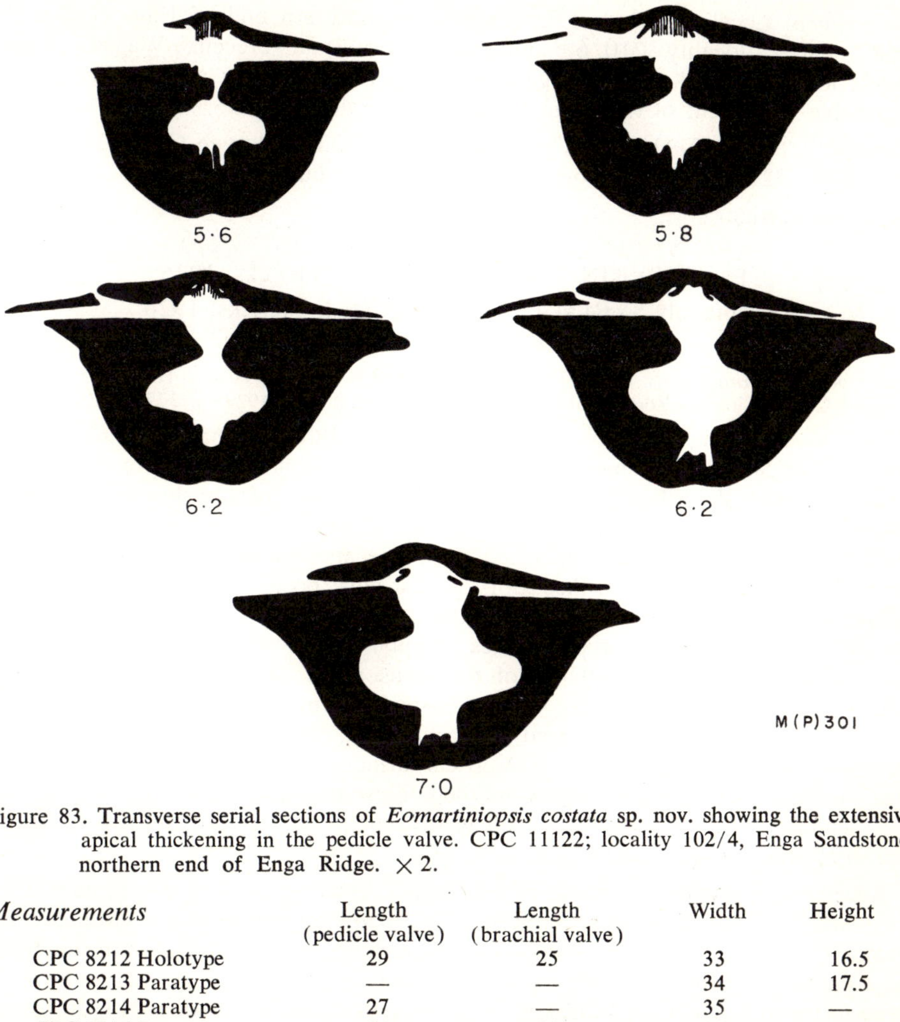

Figure 83. Transverse serial sections of *Eomartiniopsis costata* sp. nov. showing the extensive apical thickening in the pedicle valve. CPC 11122; locality 102/4, Enga Sandstone, northern end of Enga Ridge. × 2.

Measurements	Length (pedicle valve)	Length (brachial valve)	Width	Height
CPC 8212 Holotype	29	25	33	16.5
CPC 8213 Paratype	—	—	34	17.5
CPC 8214 Paratype	27	—	35	—
CPC 8216 Paratype	28 est.	21 est.	35	—

The morphologically closest species to *costata* is *E. girtyi* (Branson), described by Carter (1967, pp. 419-25, pl. 43, figs 1-7), which has a weakly costate ornament and a subovate shape. This species is distinguished from *girtyi* by its much thicker shell in adult individuals, slightly more quadrate outline, wider hinge, flatter brachial valve, lower dorsal fold and anterior dorsal deflection, narrower and shallower sinus, the possession of a pseudodeltidium at the apex of the delthyrium, and possibly by lacking well-defined dorsal adminicula. *E. girtyi* is recorded from the Chouteau Limestone of Missouri (Branson, 1938) and the Chappel Limestone of Texas (Carter, 1967).

The closest Russian species is *E. tscherepeti* Sokolskaya (1941, pl. 7, figs 7-9) from the late Tournaisian of the Moscow Basin. *E. tscherepeti* has a pedicle valve of similar outline to that of *costata,* but lacks a costate ornament, and has a thin shell; the brachial valve of *tscherepeti* is unknown.

Occurrence: Locality 102/4, Enga Sandstone, northern end of Enga Ridge, the type locality; locality 210/6, unnamed Tournaisian breccia, Waggon Creek valley; locality 124/1, Burt Range Formation, Flapper Hills; and locality 136/1, a possible equivalent of the Enga Sandstone at Spirit Hill.

Material: CPC 8212-8219, CPC 11121, CPC 11122. Holotype CPC 8212; paratypes CPC 8213-8219. All from locality 102/4. About 70 specimens.

Age: Tournaisian.

Family SYRINGOTHYRIDIDAE Fredericks, 1926
Subfamily SYRINGOTHYRIDINAE Fredericks, 1926
Genus SYRINGOTHYRIS Winchell, 1863

Type species: Syringothyris typa Winchell, 1863, equals *Spirifer carteri* Hall, 1857 (International Commission on Zoological Nomenclature Opinion 100, 1928, p. 377).

Remarks: According to North (1920, p. 171) and Pitrat (1965, p. H692), *Syringothyris* typically has a smooth fold and sinus. Specimens of *Syringothyris fontanalis* sp. nov., described below, have up to three pairs of very low broad costae on the flanks of the sinus. The costae appear to branch from the primary costae bordering the sinus, and are separated from one another by narrow grooves. The floor of the sinus is smooth. Accessory costae have been observed on the fold of one brachial valve interior also. All other characters of *S. fontanalis* compare closely with those of the type species. Similar low costae outlined by narrow furrows are present on the flanks of the sinus and fold of *?Syringothyris bifida* Campbell (1961, pl. 62, fig. 6) from the Isaacs Formation (Stephanian) of New South Wales, and Campbell used this character, among others, to query the generic assignment of his material. Vertical striation over the entire ventral interarea, another character cited by Campbell, may also occur in *S. fontanalis,* but my specimens are not sufficiently well preserved to see whether the perideltidium extends farther than two-thirds the width of the ventral interarea.

Accessory costae on the flanks of the sinus are also present in a specimen from the Carboniferous of Kashmir, referred to *Syringothyris cuspidata* (Martin) by Diener (1915, pl. 1, fig. 7), and in some specimens of *S. angulata* Simpson from the Conewango Formation, Pennsylvania (Caster, 1930, pl. 5, fig. 2; and specimens in the U.S. National Museum collection). In *S. distans* (Sowerby), from the Carboniferous of Ireland (Davidson, 1858, pl. 8, figs 2, 7, 13) there appear to be costae on the floor as well as on the flanks of the sinus.

Syringothyris angulata Simpson is the senior synonym of *S. randalli* Simpson (1889), the type species of *Syringopleura* Schuchert (1910) (Caster, 1930). *Syringopleura* embraced *Syringothyris*-like specimens which possessed a syrinx, and a fully ribbed fold and sinus. Both Girty (1911) and Caster (1930) pointed out that no specimens of *S. randalli* have a medially costate fold and sinus, and suggested that the specimen showing these costae and illustrated as *S. randalli* by Simpson (1889) was in fact a *Cyrtospirifer*. North (1920) and Pitrat (1965) placed *Syringopleura* in synonymy with *Syringothyris*. I agree with their decision, but emphasize that contrary to Pitrat's diagnosis (1965, p. H692), species of *Syringothyris* may have weak accessory costae on the flanks of the sinus and fold.

Accessory costae comparable with those in *S. fontanalis* are found on the flanks of the sinus and fold of *Asyrinxia* Campbell (1957, pp. 80-81), a syringothyridinid genus which lacks a high ventral interarea and a delthyrial plate bearing a syrinx.

Syringothyris langfieldensis sp. nov., from the Septimus Limestone (Pl. 55, figs 1-19) has a smooth sinus, but the primary costae bordering the sinus are divided by narrow median furrows. This may be a fore-runner to the accessory costae on the flanks of the sinus in the younger species *S. fontanalis*.

SYRINGOTHYRIS FONTANALIS sp. nov.

(Pl. 54, figs 5-18)

1970 *Syringothyris* sp. A Thomas, pp. 134-35, p. 9, figs 8, 9.

Diagnosis: Shell pyramidal, twice as wide as long, with the greatest width at hinge; eighteen to twenty-four costae on each lateral slope; pedicle valve with a pointed, slightly incurved umbo, and an apical angle of 120°; ventral interarea steeply apsacline, and moderately to strongly concave beneath the beak; delthyrial angle 40°; perideltidium extremely wide; sinus moderately deep, with a short anterior tongue, and with up to three pairs of accessory costae on the lateral margins; sinal angle 25-30°; delthyrial plate short; syrinx of small diameter and with a narrow anterior opening; brachial valve convex posteriorly, with a moderately high fold bearing obsolete costae on the flanks.

Description: External. Shell pyramidal, twice as wide as long, with greatest width at hinge; anterior and lateral margins rounded, with lateral margins meeting hinge nearly at right angles; commissure moderately uniplicate; costae broadly rounded, becoming wider anteriorly, separated by narrow intercostal grooves; eighteen to twenty-four costae on each lateral slope; micro-ornament not preserved.

Pedicle valve trigonal; ventral interarea steeply apsacline, moderately to strongly concave beneath beak, becoming flatter towards hinge; interarea 16.5 mm high on a valve 45 mm wide and 23 mm long; delthyrium with a small apical delthyrial plate, apparently lacking a covering plate, and a delthyrial angle of 40°; perideltidium poorly preserved, but on one valve extremely wide, extending at least two-thirds width of hinge, ornamented by deep grooves having a density of four to five per 1 mm; umbo sharply pointed, extending behind and slightly incurved over interarea; apical angle 120°; lateral slopes steep medially, becoming a little flatter posterolaterally; sinus moderately deep, with a U-shaped floor, steep lateral margins, and a short anterior tongue; lateral margins of sinus ribbed by up to three pairs of broad low costae apparently branching from marginal costae; sinal angle 25-30°; sinus crossed by concentric growth lines.

Brachial valve moderately convex, with greatest convexity at umbo, and greatest height a short distance behind umbo; umbo incurved, with concave umbonal shoulders; lateral slopes convex medially, becoming flat towards posterolateral margins; interarea short, orthocline, with a wide notothyrium; fold moderately high, narrow at umbo, wider anteriorly, with straight lateral margins, a flat crest, and steep sides bearing obsolete ribbing (one specimen only); fold truncated by anterior extension of sinus.

Internal. Pedicle valve. Delthyrial plate short, with a concave outer surface; syrinx of moderate diameter, with a narrow anterior opening; dental plates inclined anteromedianly from inner margins of delthyrium, diverging at about 40°, supported by ventral adminicula; internal surfaces of delthyrial plate, dental plates, and posterior parts of adminicula thickened by callus; adminicula diverging posteriorly, but curving inwards around anterior of muscle field; muscle field subovate, divided posteriorly by a pointed convex callus emerging from beneath delthyrial plate; diductor muscle scars narrowly elliptical, pointed posteriorly, and bearing fine striae; adductor muscle scars poorly defined, situated on ridge of median sinus in front of pointed apical callus.

Brachial valve. Cardinal process triangular, having approximately sixteen lamellar plates, and supported on a high apical callus; front of callus with its anterior margin projecting into a short septum-like ridge, but not joining median ridge separating adductor muscle scars; inner socket ridges high, narrow, swollen distally, and on their inner margins giving rise to wide flat to slightly dorsally inclined crura; sockets narrow posteriorly, deeper and wider distally, with a narrow longitudinally grooved floor; adductor muscle scars narrow posteriorly, slightly wider anteriorly, divided by a low median ridge, and situated within fold.

Measurements

	Length (pedicle valve)	Width	Height (ventral interarea)
CPC 11041 Holotype	22.5	45	13.5
CPC 11042 Paratype	25.5	57 est.	18
CPC 11043	27 est.	69	17

Remarks: Syringothyris fontanalis is distinguished from most species of *Syringothyris* by the possession of weak accessory costae on the flanks of the sinus (*see* remarks on the genus, p. 248). Specimens referred by Diener (1915, pp. 9-10, pl. 1, figs 6-8) to *S. cuspidata* (Martin), from the Carboniferous of Kashmir, are externally close to *S. fontanalis.* They have a similar shape, a concave ventral interarea, a comparable costate ornament on the lateral slopes, and broad costae on the flanks of the median sinus (Diener, 1915, pl. 1, fig. 7). Diener's specimens are not conspecific with *S. cuspidata* (Martin), having a concave ventral interarea, a greater number of costae on the lateral slopes, and ribbed flanks on the sinus and fold.

Ribbing on the sides of the sinus is also present in a specimen from the Lower Carboniferous of Siberia, referred to *S. texta* (Hall) by Besnosova et al. (1962, pl. C22, fig. 1). It is distinguished from *S. fontanalis* by slightly coarser costae on the flanks. American specimens of *S. texta* are distinguished from *S. fontanalis* by their flatter ventral interarea, coarser costae, and smooth sinus (Weller, 1914, pl. 69, figs 6-9; pl. 70, figs 1-4).

S. hanibalensis (Swallow) described by Weller (1914, pp. 388-90, pl. 68, figs 1-7) from the Louisiana Limestone of Missouri, has a transverse outline and apical angle similar to *S. fontanalis. S. hanibalensis* has a less concave ventral interarea, coarser costae, a smooth sinus, a shorter and larger diameter syrinx, and a higher delthyrial plate. Hyde (1953, p. 270) considered the specimen illustrated by Weller (pl. 68, figs 1-5), which I compare with *S. fontanalis,* to be atypical of *S. hanibalensis* (Swallow) and more like *S. bedfordensis* Hyde.

S. bedfordensis Hyde (1953, pp. 268-70, pl. 27, figs 1-15; pl. 28, figs 1-11), from the Bedford Formation near Cleveland, Ohio, is morphologically close to *S. fontanalis,* having a transverse outline, a concave ventral interarea, a short delthyrial plate, a similar costate ornament on the lateral slopes, and a comparable fold and sinus. *S. bedfordensis* is distinguished by its wider lateral slopes, smooth sinus, and larger diameter syrinx which is open anteriorly and bears muscle impressions.

S. distans (Sowerby) from the Carboniferous of Ireland has a median sinus with accessory costae on the floor as well as on the margins (Davidson, 1858, pl. 8, figs 2, 7, 13). The holotype (Davidson, 1858, pl. 8, figs 1-4) has an incurved umbo similar to that of *S. fontanalis,* but in other specimens the umbo does not extend beyond the ventral interarea (pl. 8, figs 9, 12). *S. distans* is distinguished from *S. fontanalis* by its less transverse outline, costate floor on the sinus, and coarser costae on the lateral slopes.

S. lydekkeri (Diener), from the Ladakh and Kashmir Valleys (Diener, 1899, pl. 3, figs 1-11) and the Marbal Valley (Bion, 1928, pl. 3, figs 9-14) of the Himalaya region, is distinguished from *S. fontanalis* by its lower ventral interarea, alate outline, and smooth sinus. Specimens from the *Fenestella* Beds of Kashmir, figured by Diener (1915, pl. 4, figs 3-6), however, have a higher ventral interarea, and appear to be closer in shape to *S. fontanalis.*

The name of this species, *fontanalis,* is Latin for 'of the spring'. It refers to springs close to localities 25 and 435/0 at the foot of the Weaber Range escarpment.

Occurrence: Locality 25 (the type locality), 2.5 miles east-northeast of Point Spring, locality 435/0, 5 miles northeast of Point Spring, both at the foot of the Weaber Range escarpment; and localities 453/5 and 453/7, Burvill Point, Burvill Beds.

Material: CPC 11038-11043. Holotype CPC 11041; paratypes CPC 11039, 11042. CPC 11038 from locality 453/7; CPC 11039, 11041-11042 from locality 25; CPC 11040 from locality 453/5; CPC 11043 from locality 435/0. Ten specimens.

Age: Visean.

Syringothyris langfieldensis sp. nov.

(Pl. 55, figs 1-19)

1970 *Syringothyris?* Thomas, pl. 28, fig. 7.

Diagnosis: Shell transverse, and moderately to weakly uniplicate; fifteen to twenty-five (usually nineteen) costae on each lateral slope; pedicle valve with an apical angle of 160-175°, and a straight umbo; ventral interarea catacline, and flat to very gently concave; sinus shallow; costae bordering the sinus bearing a narrow median groove; syrinx channel-like, and open anteriorly; delthyrial plate deeply submerged dorsally, and covered by a convex plate; apex of pedicle valve interior unthickened, having a low median ridge, and short adminicula diverging at 40-50°; ventral muscle field in front of the adminicula in the midpart of valve; brachial valve convex posteriorly, with a low median fold.

Description: External. Shell of moderate size for genus, pyramidal in profile, transverse in outline; greatest width at hinge, lateral margins meeting hinge at acute angles; commissure moderately to weakly uniplicate; costae low, broadly rounded, separated by narrow intercostal furrows; costae numbering from fifteen to twenty-five (usually nineteen) on each lateral slope; some growth stages lamellose.

Pedicle valve. Interarea catacline, measuring 24 mm high on a valve approximately 60 mm wide, and ranging from entirely flat to mainly flat with a slight concavity beneath umbo; delthyrial angle 30-40°; delthyrium almost entirely covered by a thick triangular irregularly lamellose plate extending from a few mm beneath apex to near hinge; perideltidium wide, ornamented with coarse vertical striations having a density of five per 1 mm, and finer horizontal growth lines; remainder of interarea ornamented with horizontal growth lines only; umbo forming highest part of valve, pointed, usually straight, occasionally slightly incurved, and level with posterior of lateral slopes; lateral slopes high medially, becoming gradually lower laterally; apical angle 160-175°; sinus shallow, deepest at anterior margin, having a smoothly rounded floor and steep sides; costae bounding sinus bearing a shallow median groove; sinal angle 30-40°.

Brachial valve moderately convex, especially posteriorly, highest at one-third length of valve; umbo convex, abruptly incurved posteriorly, projecting a short distance behind hinge, flanked by gently concave umbonal shoulders; lateral slopes moderately convex medially, grading into flat posterolateral margins; fold low, narrow posteriorly, becoming much wider anteriorly and having steep sides, a wide flat crest, and curved lateral margins; interarea short, apsacline, ornamented with parallel growth lines, and near notothyrium with vertical lirae, possibly forming a perideltidium; notothyrium wide.

Internal. Pedicle valve. Dental plates long, joined posteriorly by a sunken delthyrial plate, and supported by low ventral adminicula; delthyrial plate flat apically, slightly keeled, becoming deeper below the interarea distally, and extending one-third to one-half height of interarea; adminicula extending approximately one-third length of valve, and diverging at 40-50°; syrinx open anteriorly, extending a short distance in front of delthyrial plate; median ridge low, and commencing at tip of unthickened umbo; muscle field in front of adminicula near midlength of valve; diductor muscle scars (one specimen only) subovate, and dendritic; adductor muscle scars on a narrow elongate ridge between diductor muscle scars.

Brachial valve. Cardinal process resting on a massive umbonal callus, consisting of thirty lamellar plates; lateral plates of process extending on to posterior parts of crura; sockets long, narrow, with rounded floors, expanding abruptly at their distal extremities; inner socket ridges high and rounded ventrally, with their inner margins subtending crura; crura round and thickened near cardinal process, becoming blade-like and dorsally inclined distally; adductor muscle scars in two pairs; inner scars situated in fold, elongate, with rounded extremities, expanding slightly anteriorly, separated from one another by a narrow median furrow; outer scars situated outside fold at posterior of muscle field, narrow, and elongate.

Measurements	Length (pedicle valve)	Length (brachial valve)	Width	Height (of ventral interarea)
CPC 11044	17	—	39 est.	18
CPC 11045	—	22 est.	50	—
CPC 11051 Holotype	25	—	60 est.	24

Remarks: Syringothyris langfieldensis is comparable with *S. spissa* Glenister from the Moogooree Limestone of the Carnarvon Basin (Glenister, 1955, pp. 70-71, pl. 7, fig. 15; pl. 8, fig. 9; Thomas, 1970, pp. 131-34, pl. 9, figs 6, 7; pl. 30, figs 1-5; pl. 31, fig. 3), having a similar external ornament, moderately developed fold and sinus, and morphologically close internal features in the brachial valve. Both *S. langfieldensis* and *S. spissa* are known from fragmentary material. *S. langfieldensis* is probably more transverse, has a ventral interarea which is catacline rather than apsacline, and a wider apical angle on the pedicle valve. In the pedicle valve interior it lacks the apical thickening typical of *S. spissa,* and the median ridge is weaker and does not arise from a boss of callus beneath the syrinx. The syrinx, observed on one silicified specimen only, is an open channel, whereas that in *S. spissa* is tubular with a narrow anterior slit. Adminicula are less divergent, and the muscle field is located near the midpart of the valve instead of between the adminicula as in *S. spissa.*

S. australis Maxwell (1954, pp. 41-42, pl. 5, figs 1-4; *see also* Thomas, 1970, pl. 31, figs 1, 2), from the Neil's Creek Clastics of the Mount Morgan district, Queensland, has a lower ventral interarea, an apically thickened pedicle valve, longer ventral adminicula, and an apparent two-fold division of the cardinal process; Thomas's figures 2a and 2b show that the inner margins of the inner socket ridges lack the flange-like crura found in *S. langfieldensis.*

S. elongata North (1920, pp. 183-84, pl. 12, figs 3-5) from the Z and C Zones of Ireland, and the D_2 Zone of Yorkshire, England, is comparable with *S. langfieldensis* in outline, number of costae, and in the possession of an open channel-like syrinx (North, 1920, figs 1e, 1f). *S. elongata* is distinguished by a shallower delthyrial plate and a higher fold.

The closest North American species is a Devonian form referred to *S. chemungensis* Cushing, by Hyde (1953, pl. 33, figs 1-16; pl. 34, figs 1-8), which has a comparable shape, ventral interarea, and ornament. *S. chemungensis* is distinguished from this species by its longer and larger syrinx, more strongly keeled delthyrial plate, and more robust fold and sinus.

This species is named after Langfield Point in the Central Burt Range.

Occurrence: Localities 104/11, 104/11A, 104/15, 150/5, and 150/6, Mount Septimus; localities 110/5 (the type locality), 110/6, and 110/7, Mount Zimmermann; locality 112/2, Langfield Point; and locality 138/2, Burt Range Amphitheatre. Septimus Limestone.

Material: CPC 11044-11051. Holotype CPC 11051; paratype CPC 11049. CPC 11044 from locality 104/11; CPC 11045 from locality 138/2; CPC 11046-11048 from locality 112/2; CPC 11049, 11051 from 110/5; CPC 11050 from locality 110/6. Thirty fragmentary specimens.

Age: Tournaisian.

SYRINGOTHYRIS sp. cf. S. SPISSA Glenister

(Pl. 54, figs 1-4)

Description: External. Pedicle valve pyramidal, wider than long; umbo blunt, almost straight, not differentiated from lateral slopes, projecting a short distance behind interarea; apical angle 170°; interarea apsacline to catacline, 16.5 mm high on a valve approximately 45 mm wide, slightly concave beneath umbo, but flat near hinge; delthyrium closed by a concave delthyrial plate, delthyrial angle 35°; lateral slopes steep; sinus shallow; costate ornament not observed.

Internal. Pedicle valve. Dental plates long, supported anteriorly by thickened ventral adminicula; adminicula strongly divergent, with their anterior extremities curving around muscle field; diductor muscle scars dendritic, subovate, pointed posteriorly, and deeply impressed into floor; median ridge dividing diductor muscle scars arising from an elongate boss of callus at posterior of muscle field; adductor muscle scars not visible; delthyrial plate concave posteriorly, becoming more deeply submerged below level of interarea dorsally, and extending two-thirds the height of delthyrium; syrinx tubular, with a narrow anterior slit, and projecting well above delthyrial plate; apex of valve thickened.

Remarks: The few specimens of *Syringothyris* from the Utting Calcarenite resemble *S. spissa* Glenister (1955, pp. 70-71, pl. 7, fig. 15; pl. 8, fig. 9; Thomas, 1970, pp. 131-34, pl. 9, figs 6, 7; pl. 30, figs 1-5; pl. 31, fig. 3) from the Moogooree Limestone of the Carnarvon Basin. They have an apically thickened pedicle valve in which the muscle field is situated between the ventral adminicula; a moderately well-developed median ridge arises from a boss of callus near the apex of the valve, and divides the diductor muscle scars. A single partially complete pedicle valve differs from *S. spissa* Glenister in having a larger apical angle.

Occurrence: Localities 108/3 and 108/4, Utting Calcarenite, Utting Gap.

Material: CPC 11037. From locality 108/3. One partially complete and 3 fragmentary pedicle valves.

Age: Visean.

Superfamily RETICULARIACEA Waagen, 1883
Family RETICULARIIDAE Waagen, 1883
Genus CRASSUMBO Carter, 1967

Type species: Crassumbo inornatus Carter (1967, pp. 410-13, pl. 23, fig. 14; pl. 41, figs 1-12) from the Chappel Limestone of central Texas, by original designation of Carter, 1967.

CRASSUMBO? JONESI sp. nov.
(Pl. 56, figs 1-15)

Diagnosis: Shell smooth posteriorly, but with six to ten weak costae on the anterior two-thirds of lateral slopes, and lacking a micro-ornament of capillae and spine bases; sinus with a narrow median trough, and wide lateral margins; fold high; interior of shell mainly unthickened; pedicle valve with slender adminicula, and slightly thickened dental plates; brachial valve with robust margins on notothyrium, and with long crura arising from the dorsal margins of inner socket ridges, but not reaching the floor.

Description: External. Shell of moderate size for genus, semi-elliptical, and equally biconvex; lateral margins rounded; hinge five-eighths the width of shell, greatest width near midlength; commissure strongly uniplicate; umbonal region smooth; costae commencing at about one-third length of shell, low, broad, rounded, and separated by narrow intercostal furrows, numbering six to ten on each lateral slope; micro-ornament not observed; growth lamellae crowded at margins.

Pedicle valve. Umbo narrow, strongly incurved, extending 1.5 mm behind hinge on a valve 16 mm long and 17 mm wide; apical angle approximately 80°; umbonal shoulders strongly concave, poorly differentiated from ventral interarea; lateral slopes steep except for a slight flattening at posterolateral margins; greatest convexity at umbo; valve highest at about one-third its length; median sinus narrow posteriorly, becoming much wider towards front, and produced into a broad flat to concave dorsally directed trough; floor of sinus having a narrow rounded median furrow extending from tip of umbo to anterior margin; lateral margins of sinus flat to gently convex, becoming wider and in some individuals steeper towards front, and smooth or possibly with one or two very obsolete costae at anterior margin; sinal angle 40°; ventral interarea triagular, gently concave, apsacline to slightly catacline, and 2 mm high beneath umbo on a valve 15 mm wide and 12 mm long; delthyrium open, almost twice as wide as high, subtending an angle of 70°.

Brachial valve. Umbo almost equal in size to that on pedicle valve, bluntly rounded, with short concave umbonal shoulders; lateral slopes moderately convex, and, except for a slight flattening at posterolateral extremities, sloping evenly to margins; fold evenly rounded, commencing at one-third length of valve, becoming broader and higher anteriorly; fold ornamented by concentric growth lines only; valve highest at front of fold, most convex at umbo; dorsal interarea low, orthocline, bearing a small open notothyrium.

Internal (Fig. 84). Pedicle valve. Adminicula slender, subparallel both along floor and in transverse section, extending almost one-third length of valve; dental plates slightly thickened near delthyrium, widely divergent in transverse section, and supported by adminicula; teeth and muscle scars not observed.

Brachial valve. Cardinal process consisting of small lamellar platelets in apex of valve; margins of notothyrium thickened, and prominent in transverse section (Fig. 84b, 0.4 and 0.6); sockets small, rounded, and bounded by thickened inner socket ridges; crura long, arising from anterior edge of thickened region alongside cardinal process, and from dorsal surfaces of inner socket ridges; in transverse section, crura having concave inner faces, and converging distal extremities; spires and muscle scars not observed.

Measurements	Length (pedicle valve)	Length (brachial valve)	Width	Height
CPC 8255 Holotype	12.5	12.5	17.5	11
CPC 8254 Paratype	13	12.5	16.5	15
CPC 8256 Paratype	8	—	13.5	—
CPC 8257 Paratype	11	—	15	—
CPC 8258 Paratype	13	—	17.5	—
CPC 8259 Paratype	—	12.5	20 est.	—
CPC 8260 Paratype	—	12.5 est.	18 est.	—

Remarks: The type species of *Crassumbo, C. inornatus* Carter (1967), is characterized by an external ornament of numerous growth halts, fine radial capillae, and scattered nodes or spine bases; on some individuals there are faint ribs on the flanks. *C? jonesi* does not appear to bear radial capillae or spine bases, but these could be absent by decortication, and it has low rounded costae on the anterior two-thirds of the lateral slopes. The costae affect the entire thickness of the shell, and are not confined to the inner layers, as with radial ornament in many reticulariid genera. When compared with *C? jonesi*, *C. inornatus* is less inflated, has a narrower sinus, a smaller dorsal anterior deflection, and a lower fold. Internally *inornatus* is distinguished by the thick apical callus which in the brachial valve buries the crural plates.

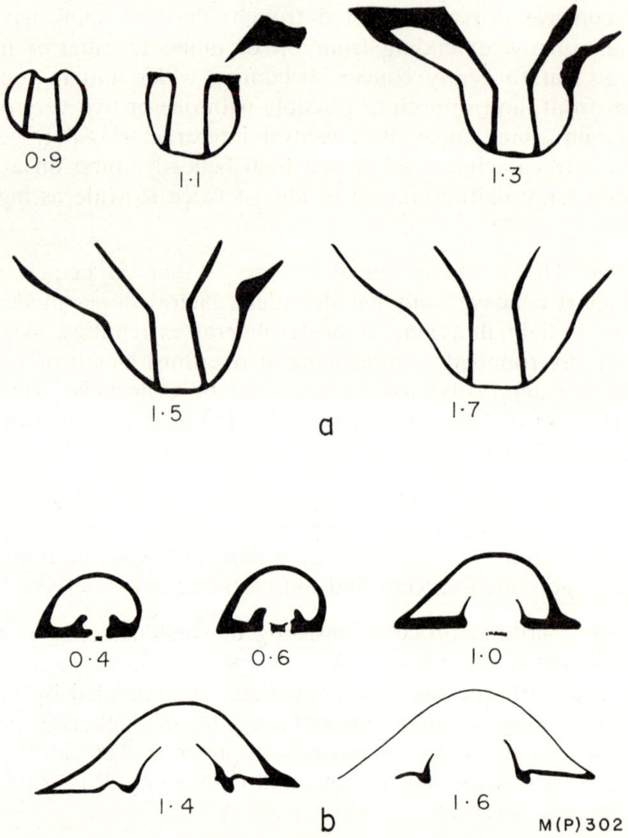

Figure 84. Transverse serial sections of a pedicle valve (a) and a brachial valve (b) of *Crassumbo? jonesi* sp. nov. a-CPC 11123, b-CPC 11124; locality 7/1, Ningbing Limestone (Tournaisian part), Ningbing Range. × 2.

C? jonesi is externally morphologically close to *C. turgidus* Carter (1967, pp. 413-18, pl. 42, figs 1-6) from the Chappel Limestone of central Texas. *C. turgidus*. however, lacks costae on the anterior of the shell except in partially decorticated specimens, has a lower fold, and has a capillate micro-ornament. Internally *C. turgidus* differs from *C? jonesi* in the possession of a thick apical callus, and, in juveniles, crural plates which reach the floor at the posterior of the brachial valve, and a median ridge in the pedicle valve (Carter, 1967, fig. 44); in adult specimens these latter structures are buried by the apical callus.

Thomasaria? voiseyi Roberts (1963, pp. 21-22, pl. 4, figs 5-11), from the Lower Carboniferous of New South Wales, is characterized by the possession of fine radial lirae, small spine bases (possibly paired), growth halts, subparallel ventral adminicula, and crural plates which reach the floor at the posterior of the dorsal valve, and should be referred to *Crassumbo.* Small specimens of *C. voiseyi* are smooth, but mature individuals have indistinct costae at the front of the lateral slopes; the latter specimens lack the apical thickening which characterizes the type species. *C. voiseyi* is distinguished from *C? jonesi* by its more subcircular outline, flatter brachial valve, lower fold, and crural plates which reach the floor at the posterior of the brachial valve.

Spirifer rhomboidea Phillips, figured by M'Coy (1844, pl. 22, fig. 11) and Davidson (1858, pl. 12, figs 6, 7) from Cork and Millecent, Ireland, is externally morphologically close to *C? jonesi,* having a similar outline, fold and sinus, and an obsolete costate ornament on the lateral slopes. The internal structures of both valves are unknown, and hence a detailed comparison cannot be made. Phillips's original illustrations (1836, pl. 9, figs 8-9) do not resemble those of either M'Coy or Davidson, so that the name of the material with which *C? jonesi* is compared is of doubtful validity.

This species is named in honour of Mr P. J. Jones.

Occurrence: Locality 7/1, Ningbing Limestone (Tournaisian part), Ningbing Range.

Material: CPC 8254-8260, CPC 11123, CPC 11124. Holotype CPC 8255; paratypes CPC 8254, 8256-8260. All from locality 7/1. About 35 specimens extracted by calcining massive reef limestone.

Age: Tournaisian.

Family ELYTHIDAE Fredericks, 1924

Genus KITAKAMITHYRIS Minato, 1951

Type species: Torynifer (Kitakamithyris) tyoanjiensis Minato, 1951, from the Hikoroiti Series, North East Honsyu, Japan, by original designation of Minato, 1951.

Remarks: The genus *Kitakamithyris* was designated by Minato (1951) to include forms like *Torynifer* Hall & Clarke having ventral adminicula and a median septum, but lacking a dorsal apical plate. He did not state whether the term apical plate referred to the cardinal plate and the dorsal median septum of *Torynifer,* but in a later publication (1952) he stated that the brachial valve of *Kitakamithyris* lacked all kinds of apical plates, in contrast to that of *Torynifer* which had a median septum. As already pointed out by Campbell (1955), these remarks are difficult to follow because a specimen of *K. tyoanjiensis* figured by Minato (1952, pl. 6, fig. 5b) shows definite plates, probably socket plates and a median ridge, in a brachial valve. Campbell suggested that Minato intended to distinguish the genus on 'the absence of an entire concave hinge-plate supported on a strong median septum'.

Species of *Kitakamithyris* have been recorded from Australia (Roberts, 1964; 1965), Japan (Minato, 1951; 1952), western Europe, and Kazakhstan (Weyer, 1967). The Australian species generally possess a low median ridge in the brachial valve (*see K. uniplicata* (Campbell) illustrated by Roberts, 1964, pl. 4, fig. 1b), and were previously referred to the genus *Phricodothyris* George (Campbell, 1955; 1961; Cvancara, 1958).

The type species of *Phricodothyris, P. lucerna* George, lacks adminicula and a median septum in the pedicle valve (George, 1932, fig. 8). Other European species referred to *Phricodothyris,* for example *P. lineata* (Martin) figured by George

(1932, fig. 6), have variably developed adminicula, but only one species so far described (*P. tripustulosa* Demanet, 1938, p. 96) has a median septum in the pedicle valve. Because of this variability, it is difficult to assess the relationship between *Kitakamithyris* and *Phricodothyris;* until the extent of the variation of the ventral adminicula and median septum in European species is ascertained, it is proposed to assign species which consistently have adminicula and a median septum in the pedicle valve to *Kitakamithyris.*

KITAKAMITHYRIS OCCIDUA sp. nov.

(Pl. 57, figs 1-24)

Diagnosis: Shell subovate to ovate, and weakly to moderately uniplicate; concentric lamellae with a density of four to six per 5 mm at 10 mm from umbo; lamellae at front of shell bear four to five narrow biramous spine bases per 1 mm; ventral median sinus usually shallow; adminicula divergent, extending one-third the length of pedicle valve; ventral median septum broad, low, and extending to near the midpoint of valve; brachial valve with a fold of variable height; some brachial valves with a weak median ridge.

Description: External. Shell transverse, subequally biconvex, subovate to ovate, widest at midlength, and having rounded lateral margins; hinge two-thirds width of valve; commissure weakly to moderately uniplicate; concentric lamellae evenly spaced posteriorly, becoming crowded anteriorly; lamellae having a density of four to six per 5 mm at 10 mm from umbo, and near front of shell bearing four to five spine bases per 1 mm; spine bases narrow, pointed posteriorly, having an elevated subovate probably biramous base at front; inner layers of shell having a well-defined radially costellate ornament.

Pedicle valve more strongly convex than brachial valve, most convex at umbo; valve highest between one-third and one-half its length; umbo pointed, incurved, and overhanging delthyrium, apical angle 75-80°; umbonal shoulders concave, separated from interarea by prominent beak ridges; ventral interarea apsacline, concave near umbo, becoming flatter laterally and towards the hinge, ornamented with denticle grooves having a density of about eight per 1 mm, 3 mm high at umbo on a valve 24 mm wide and 21 mm long; delthyrium open, delthyrial angle 65-70°, bordered on its inner margins by wide dental flanges; lateral slopes convex, sloping evenly to lateral margins; sinus of variable strength, usually shallow, rounded, commencing near tip of umbo, and broadening anteriorly; some specimens lacking anterior part of sinus, others with sinus deep and narrow anteriorly.

Brachial valve moderately convex, with greatest height at between one-third and one-half its length; umbo blunt, gently incurved, projecting a short distance behind hinge; interarea low, orthocline; notothyrium broadly triangular; lateral slopes running evenly to margins; fold variable in height, almost entirely obsolete in some valves, and when present low near umbo becoming higher and wider anteriorly.

Internal (Fig. 85). Pedicle valve. Dental plates stout, forming wide flanges along inner margins of delthyrium, and bearing small peg-like teeth; dental plates supported by slender adminicula extending about one-third length of valve, diverging between 25° and 50°; muscle field divided by a broad low median septum extending to about midpoint of valve; adductor muscle scars elongate, subrectangular, situated along sides of median septum; diductor muscle scars striated, pointed posteriorly, bordered posterolaterally by adminicula, and expanding anteriorly; interior of valve bearing narrow linear radially arranged mantle canals.

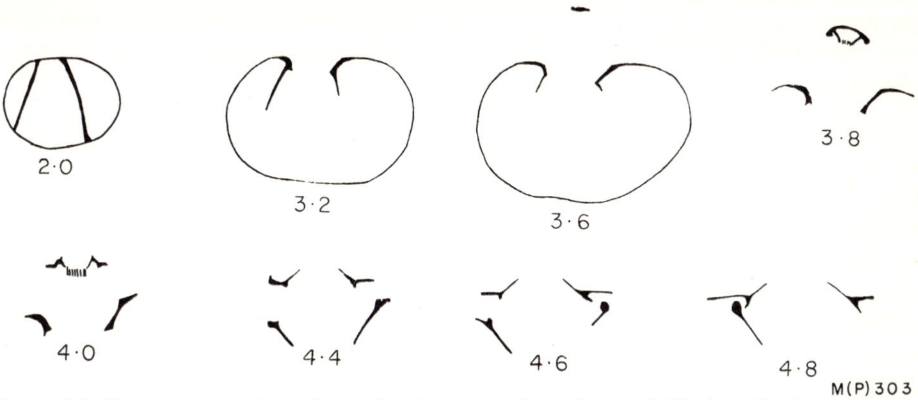

Figure 85. Transverse serial sections of a decorticated specimen of *Kitakamithyris occidua* sp. nov. showing the morphology of the apical plates. CPC 11125; locality 7/1, Ningbing Limestone. × 2.

Brachial valve. Cardinal process small, rectangular, having a lamellose myophore; sockets narrow, bordered by slender inner socket ridges; crura originating from inner margins of inner socket ridges, and not extending to floor of valve; some specimens with an extremely obsolete median ridge extending one-third length of valve; diductor muscle scars narrow, elongate, and situated on median posterior part of valve; interior of valve bearing narrow linear radially arranged mantle canals.

Measurements	Length (pedicle valve)	Length (brachial valve)	Width	Height
CPC 8509 Holotype	23	21	29	15.5
CPC 8510 Paratype	20.5	17.5	25 est.	15
CPC 8511 Paratype	—	18	25	—
CPC 8512	—	17.5	28	—
CPC 8513 Paratype	—	11.5	18	—
CPC 8514	—	20.5	34 est.	—
CPC 8517 Paratype	23	—	32	—
CPC 8518	—	15	—	—
CPC 8520	20 est.	—	28 est.	—
CPC 8521	25	—	37	—

Remarks: Specimens of *Kitakamithyris occidua* sp. nov. from the Burt Range Formation, the Tournaisian part of the Ningbing Limestone, and the Enga Sandstone, are all at least partially abraded externally, and hence it is difficult to determine precisely the morphology of the spine bases. The description of the spine bases is based on a single specimen, from locality 102/4 in the Enga Sandstone, in which the spines appear to be biramous; in all other specimens decortication of the shell has destroyed the spinose micro-ornament.

The ventral median septum in *K. occidua* is lower than those in species of *Kitakamithyris* described from eastern Australia (viz. *K. uniplicata* (Campbell) figured by Campbell, 1955, pl. 18, figs 1-9, and Roberts, 1964, pl. 4, figs 1-4;

and *K. triseptata* figured by Campbell, 1955, pl. 18, figs 10-15, and Roberts, 1965, pl. 11, figs 1-3). The strength of the septum is, however, comparable with that in *K. moogooriensis* described by Thomas (1970, pp. 159-62, pl. 20, figs 10, 11) from the Moogooree Limestone in the Carnarvon Basin, Western Australia. *K. occidua* is distinguished from the Moogooree species by slightly finer and more elongate spine bases, a more transverse shell, and shorter and more divergent ventral adminicula.

Two eastern Australian species with comparable densities of the spine bases, *Kitakamithyris* sp. (Cvancara, 1958, pp. 872-73, pl. 111, figs 21, 22, 24, 26) and *K. uniplicata* (Campbell, 1955, pp. 377-79, pl. 18, figs 1-9), are distinguished by the possession of a stronger median septum in the pedicle valve; *K. uniplicata* also has a stronger fold and sinus.

Torynifer? dorsiseptatus Thomas (1970, pp.163-66, pl.20, figs 1-9, 13) from the Septimus Limestone in the Bonaparte Gulf Basin resembles *Kitakamithyris occidua* in outline, convexity of the valves, and in the density of the concentric lamellae. *T? dorsiseptatus* is distinguished by the possession of a well-defined median septum supporting a concave cardinal plate in the brachial valve. Another feature which at first appears to be a specific difference is the manner of imbrication of the concentric lamellae. Those in well-preserved specimens of *K. occidua* from locality 7/1 are imbricating, and overlap one another from the posterior towards the anterior. Lamellae in silicified specimens of *T? dorsiseptatus* from the Septimus Limestone, however, have their posterior margins overlapping the preceeding lamella — the reverse of well-preserved specimens of *K. occidua.* However, because some partially decorticated specimens of *K. occidua* show a reverse imbrication also, it is likely, as already suggested by Thomas (1970), that the outer surfaces of the silicified specimens from the Septimus Limestone are not preserved and that we are observing partially decorticated shells: this would account for the poor state of preservation of the spine bases in *T? dorsiseptatus* (Thomas, 1970, p. 163).

The name *occidua* is the Latin for western, and refers to the occurrence of the species in Western Australia.

Occurrence: Locality 7/1 (the type locality), Ningbing Limestone (Tournaisian part), Ningbing Range. Localities 100/21 and 100/22, 7 miles northwest of Mount Septimus; locality 101/7C, 5.5 miles west of Mount Septimus; locality 117/1, Spirit Hill; Burt Range Formation. Locality 102/4, Enga Sandstone, northern end of Enga Ridge.

Material: CPC 8509-8522, CPC 11125. Holotype CPC 8509; paratypes CPC 8510, 8511, 8513, 8516, and 8517. CPC 8509-8510, 8513-8517, 11125 from locality 7/1; CPC 8512, 8519 from locality 100/21; CPC 8518, 8522 from locality 102/4. Twenty specimens prepared by calcining reef limestone from the Ningbing Limestone, several specimens preserved in limestone and as internal moulds from the Burt Range Formation, and two loose specimens from the Enga Sandstone.

Age: Tournaisian.

Genus TORYNIFER Hall & Clarke, 1894

Type species: Tornifer criticus Hall & Clarke, 1894, equals *Spirifer pseudolineatus* Hall, 1858, by original designation of Hall & Clarke, 1894.

TORYNIFER? sp.

(Pl. 51, figs 24-33)

Description: External. Shell subelliptical, unequally biconvex, with pedicle valve much more convex than brachial valve; greatest width at midlength; hinge approximately half as wide as shell; commissure rectimarginate; concentric lamellae evenly spaced over greater part of shell, crowded around margins of mature shells, and on posterolateral extremities, numbering from seven to nine per 5 mm at between 10 and 15 mm from umbo; front of lamellae bearing fine biramous spines, and a single anteriorly projecting interspinous pustule between each biramous spine base; spine bases narrow, numbering four to five per 1 mm on midanterior part of a mature shell.

Pedicle valve most strongly convex at umbo, becoming gradually less convex anteriorly; greatest height at about one-third length of valve; umbo pointed, strongly incurved over apex of delthyrium; apical angle 70-80°; umbonal shoulders concave, poorly differentiated from interarea; lateral slopes moderately steep, extending evenly to margins; median sinus usually absent, but large specimens with an extremely shallow sinus at front; ventral interarea apsacline, narrowly triangular, concave beneath umbo to flat near hinge, marked by horizontal growth lines and fine vertical denticle grooves having a density of approximately thirteen per 1 mm, and having a height of 2.5 mm on a valve 16 mm wide and 14 mm long; delthyrial angle 70-80° in large individuals, and approximately 60° in small specimens; lateral margins of delthyrium bordered by raised flanges (flanges separate from posterior of dental plate).

Brachial valve transversely elliptical in outline, and much lower than pedicle valve; umbo small, bluntly rounded, with concave umbonal shoulders; lateral slopes becoming flat on posterolateral margins; fold obsolete to absent.

Internal. Pedicle valve. Dental plates extending down inner margin of delthyrium, flanged along their inner margins at junction with ventral adminicula; in some specimens flange plugs apex of delthyrium, teeth small and elongate; ventral adminicula short in small specimens, but extending farther anteriorly along floor of valve in large individuals; adminicula unthickened, and diverging at approximately 50°; median septum commencing a short distance in front of umbo, extending to nearly the length of valve, low and rounded posteriorly, narrower and ridgelike at its midlength, and again rounded towards front; adductor muscle scars longitudinally striated, slightly impressed, elongate, pointed posteriorly, bounded on their posterolateral margins by ventral adminicula, and wider and rounded posteriorly; diductor muscle scars narrow, elongate, and situated along either side of median septum; interior of valve marked by many narrow radially arranged, occasionally branching mantle canals.

Brachial valve (Fig. 86). Sockets narrow, having a rounded floor, and bordered by moderately high curving inner socket ridges; crura forming two discrete flange-like platforms on inner margins of inner socket ridges; inner margins of platforms supported by two short inwardly curving plates extending to floor of valve, and forming a concave septalium-like structure at posterior extremity of median septum; cardinal process on a small convex callus at apex of valve; median septum blade-like, 3 mm long in a valve approximately 15 mm long, and dividing posterior of muscle field; adductor muscle scars subrectangular, slightly impressed, and separated anteriorly by a low rounded ridge probably coming off anterior extremity of median septum.

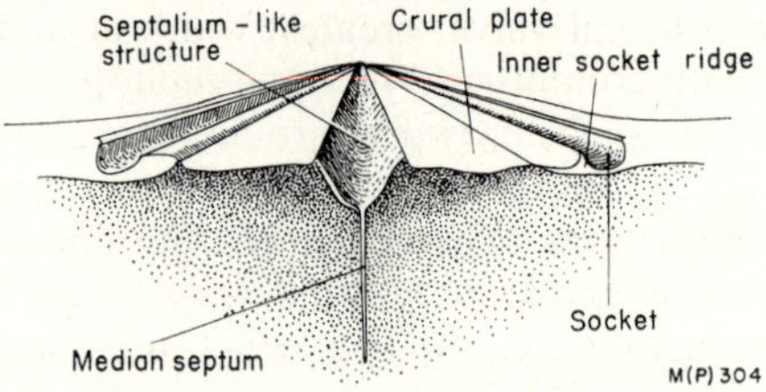

Figure 86. Apical plates in the brachial valve of *Torynifer?* sp. CPC 11035, locality 108/0, Utting Calcarenite, Utting Gap. × 6.

Remarks: The morphology of the dorsal apical plates in this material (Fig. 86) is comparable with that of one elythid genus only, *Torynifer* Hall & Clarke, 1894. The crura form wide flange-like platforms which are almost level with the ventral margins of the inner socket ridges, and are in the same plane as the commissure; the platforms or cardinal plates are discrete, and are supported medially by two short plates extending to the floor of the valve and forming a convex septalium-like structure at the posterior of the median septum. *Torynifer* has a concave cardinal plate formed by the crura joining with the apex of the median septum. The latter plate is well above the floor of the valve, has inclined rather than flat lateral regions, lacks the two nearly vertical plates extending to the floor, and except for a small apical foramen is undivided medially. These differences could, however, be caused by variation in the curvature of the crura and the height of the median septum, both structures being septalium-like, and both presumably performing similar functions. Because the description of the apical plates in the present material is based on two specimens only, I consider that until further material becomes available it is best placed tentatively in *Torynifer* Hall & Clarke.

Martinothyris Minato, 1953, with *M. lineata* (Sowerby) as the type species, is defined as having short crural plates which reach the floor at the apex of the brachial valve, and an incipient dorsal median septum (Minato, 1953). In *Martinothyris,* the short crural plates diverge towards the floor in contrast to the convergent nature of those in the present material, and there are no platform-like extensions on the inner margins of the inner socket ridges.

Torynifer dorsiseptatus Thomas (1970, pp. 163-65, pl. 20, figs 1-5, 7-9, 13) from the Septimus Limestone in the Bonaparte Gulf Basin has a cardinal plate which is typical for the genus, and hence differs from *Torynifer?* sp. Additional differences in *T. dorsiseptatus* are a more massive dorsal median septum, longer and usually less divergent ventral adminicula, and a higher ventral median septum. The spinose ornament on *T. dorsiseptatus* has not been described.

Occurrence: Localities 108/0, 108/4, and 108/5A, Utting Calcarenite at Utting Gap, and possibly at locality 446/1, Waggon Creek Breccia at Waggon Creek.

Material: CPC 11032-11036. CPC 11032 from locality 108/4; CPC 11033-11036 from locality 108/0. Seventeen silicified specimens.

Age: Visean.

Superfamily CYRTIACEA Fredericks, 1919 (1924)

Family AMBOCOELIIDAE George, 1931

Genus CRURITHYRIS George, 1931

Type species: Spirifer urei Fleming, 1828, from the Lower Carboniferous of the British Isles, by original designation of George, 1931.

CRURITHYRIS APENA Veevers, 1959

(Pl. 56, figs 16-26)

1959 *Crurithyris apena* Veevers, pp. 133-36, pl. 13, figs 10-19.

Description: External. Shell small, subcircular, almost as long as wide, with greatest width at midlength, and unequally biconvex; hinge about two-thirds as wide as shell; median sinus on both valves shallow, giving a gently emarginate anterior, and a rectimarginate commissure; radial ornament not observed; concentric growth lines on one specimen with original shell material.

Pedicle valve highest at two-thirds its length, with greatest convexity in posterior half of valve; umbo incurved and poorly differentiated from lateral slopes; lateral slopes evenly convex, divided by a narrow rounded median sinus; sinus commencing at umbo, and usually remaining narrow along its entire length; interarea apsacline, gently concave, triangular, and well differentiated from umbonal shoulders by beak ridges; interarea usually moderately high (2 mm on a valve 8.5 mm wide), but sometimes lower; delthyrium partly closed by narrow plates along lateral margins, delthyrial angle 20-30°.

Brachial valve subrectangular to subovate, and lower and less convex than pedicle valve; valve highest and most convex a short distance in front of umbo, becoming flatter towards front and lateral margins; umbo bluntly rounded, projecting a short distance behind hinge; lateral slopes mainly convex, becoming flat on their posterolateral extremities; median sinus narrow as on pedicle valve, commencing in front of umbo, and extending to front of valve; interarea very short, possibly orthocline, with a triangular notothyrium.

Internal. Pedicle valve. Dental plates extending down inner margins of delthyrium, and supporting broad bluntly rounded teeth; muscle field not observed.

Brachial valve. Cardinal process small, peg-like, and situated in apex of notothyrial cavity; crural plates broad and ventrally concave at posterior margin, extending from either side of cardinal process, supporting dorsal margins of inner socket ridges with broad curving laterally directed extensions, and running parallel with one another along floor of valve; crural plates 1 mm long on a valve 5 mm long and 7 mm wide; one specimen 5 mm long and 7.5 mm wide having a low rounded median ridge arising at apex of valve, and extending anteriorly for more than 1 mm.

Measurements	Length (pedicle valve)	Length (brachial valve)	Width
CPC 8261	—	5	7
CPC 8262	6	4	6.5
CPC 8263	7.5	—	7.4
CPC 8265	6	—	6.5
CPC 8266	7	5	7.5
CPC 8267	7.5	5	7.5
CPC 8268	8.5 est.	—	9

Remarks: Crurithyris apena Veevers is the index species for the lower Frasnian *apena* Zone in the Canning Basin, Western Australia. The *apena* Zone is placed beneath Teichert's 'Upper *Manticoceras*' Zone, and may be equivalent to his 'Lower *Manticoceras*' Zone (Veevers, 1959).

Specimens of *C. apena* Veevers (1959, pp. 133-36, pl. 13, figs 10-19) from the Canning Basin have crural plates which are either parallel, or unite on the floor of the brachial valve to form a sessile cruralium (Veevers, 1959, figs 86E, 87C). This variation has not been observed in the material from the Kununurra Member of the Cockatoo Formation in which the crural plates are always parallel and well separated. When compared with specimens from the Canning Basin, those from the Kununurra Member generally have slightly higher ventral interareas. A single specimen from the Westwood Member of the Cockatoo Formation cannot be distinguished from Veevers' specimens from the Canning Basin.

Compared with the type species, *C. urei* (Fleming), specimens of *C. apena* from the Kununurra Member differ in having a smaller umbo on the pedicle valve, and usually a higher ventral interarea. The possession of a high ventral interarea also distinguishes this material from *C. inflata* (Schnur) figured by Vandercammen (1956, pl. 1, figs 10-27). According to Vandercammen, *C. inflata* ranges from the upper Couvinian to the upper Frasnian in Belgium, and is known also in the Devonian of Germany, the Carnic alps, and USSR.

Occurrence: Locality 28/10, Kununurra Member, Cockatoo Formation, Matheson Ridge; locality 12/3, Westwood Member, Cockatoo Formation, Westwood Creek.

Material: CPC 8261-8268. All from locality 28/10. About 30 specimens preserved as internal and external moulds in fine-grained sandstone in the Kununurra Member, and one specimen with shell material from the Westwood Member.

Age: Frasnian.

Superfamily SUESSIACEA Waagen, 1883

Family CYRTINIDAE Fredericks, 1912

Genus CYRTINA Davidson, 1858

Type species: Calceola heteroclita Defrance, 1828; by subsequent designation of Hall & Clarke, 1894.

CYRTINA sp.

(Pl. 58, figs 31-33)

Description: External. Pedicle valve hemipyramidal in outline, transverse, twice as wide as long, with obtusely pointed posterolateral extremities; greatest width at hinge; ventral interarea catacline to very slightly procline, mainly flat except for a slight concavity immediately beneath umbo and an anteriorly inclined dorsal margin; interarea 6.5 mm high on a valve 12 mm wide and 6.5 mm long; delthyrium narrowly triangular, delthyrial angle 30°, partially closed by a convex pseudodeltidium having an elongate-ovoid apical foramen 3.5 mm high; both interarea and pseudodeltidium ornamented with horizontal growth lines; umbo short, straight to slightly incurved, projecting a short distance behind interarea, and poorly differentiated from lateral slopes; lateral slopes extremely steep, and separated from interarea by rounded cardinal ridges; median sinus commencing at umbo, bordered by two low rounded plicae, and having a smooth concave floor; two plicae on each lateral slope consisting of a low plica outlining sinus, and one other very low plica on midpart of lateral slope; concentric lamellae imbricating, broad, approximately 1 mm long over most of valve, but much narrower and crowded at anterior margin.

Internal. Pedicle valve. Dental plates extending along inner margins of delthyrium, elongate, joining lower parts of median septum and forming a narrow V-shaped spondylium; teeth blunt, rounded, and projecting from median dorsal extremities of dental plates; spondylium bearing a septate elongate-oval tichorhinum, and having a medianly keeled posterior surface with blade-like posterior edge of median septum projecting above two elongate troughs; adductor muscles presumably located on median keel, and diductor muscles in troughs, but no muscle scars observed; median septum high, blade-like, extending in front of spondylium to approximately two-thirds length of valve, thickened at floor, having an abrupt anterior termination.

Remarks: Comparisons with other species have not been made because only one specimen of a ventral valve is available for study.

Occurrence: Locality 108/3, Utting Calcarenite, Utting Gap.

Material: CPC 11066. From locality 108/3. One silicified pedicle valve.

Age: Visean.

Superfamily SPIRIFERINACEA Davidson, 1884

Family SPIRIFERINIDAE Davidson, 1884

Genus PUNCTOSPIRIFER North, 1920

Type species: Punctospirifer scabricostus North, 1920, from the Visean, Ashfell Edge, Westmorland, England, by original designation of North, 1920.

PUNCTOSPIRIFER PAUCIPLICATUS sp. nov.

(Pl. 58, figs 17-30)

Diagnosis: Shell small for genus, unequally biconvex, and trigonal in the adult growth stage; pedicle valve high, with a short straight to slightly incurved umbo; brachial valve almost flat posteriorly, becoming higher anteriorly; sinus smooth, with a flat to slightly concave floor; fold smooth, low posteriorly, becoming progressively higher anteriorly; plicae rounded, few in number (four to five on each lateral slope on the pedicle valve), and separated by wide rounded furrows; micro-ornament of thread-like spines 0.25 mm long on the anterior of each concentric lamella; median septum extending to a point just short of half the length of valve, and high along its entire length; ventral adminicula short, divergent, and extending one-quarter the length of valve; cardinal process supported on a thick apical callus.

Description: External. Shell transverse, semi-elliptical to trigonal, approximately twice as wide as long, and unequally biconvex; juvenile growth stages with rounded posterolateral extremities, and adult stages with pointed extremities; hinge slightly less than greatest width in juvenile specimens, having greatest width in adult; commissure serrated and uniplicate; plicae few in number (four to five on each lateral slope of pedicle valve, and three to four on each lateral slope of brachial valve), rounded, moderately high, separated by wide rounded interplical furrows, and crossed by imbricating concentric lamellae; concentric lamellae usually poorly defined on posterior of shell, and at 5 mm from umbo numbering three per 1 mm; micro-ornament of prominent thread-like spines 0.25 mm long on each concentric lamella; spines narrow posteriorly, becoming wider and higher anteriorly, and producing a serrated anterior margin on each lamella; punctae sometimes unevenly distributed, and concentrated on inner surfaces of interplical furrows; in areas of uniform distribution on inner surface of shell, density of punctae fifteen to twenty per 1 sq mm.

Pedicle valve high posteriorly, and moderately to strongly convex; greatest convexity towards front of valve; umbo small, narrow, straight to slightly incurved, and projecting a short distance behind ventral interarea; umbonal shoulders slightly concave to nearly flat; lateral slopes steep, especially at front of valve, becoming almost flat at posterolateral extremities, and abruptly separated from ventral interarea by straight cardinal ridges; ventral interarea catacline to steeply apsacline, concave immediately beneath umbo, flat near hinge and on lateral extremities, and 5 mm high on a valve 9 mm long and 19 mm wide; delthyrium narrow, delthyrial angle 30° (one probably distorted specimen with an angle of 45°), plugged apically by a small callus at posterior of median septum, and bordered by wide dental flanges; no delthyrial cover observed; median sinus bordered by strong rounded plicae, having a smooth flat to gently concave floor, and steep sides; sinal angle 30-40°.

Brachial valve rounded-rectangular, much lower than pedicle valve, almost flat posteriorly, becoming higher and moderately convex anteriorly; umbo very low to completely obsolete, and in some specimens projecting a short distance behind hinge; fold narrow, commencing at umbo, low posteriorly, and becoming progressively higher anteriorly; dorsal interarea short, and orthocline.

Internal. Ventral median septum slender, high along almost its entire length, extending almost to half the length of valve; in larger specimens, posterior extremity of median septum covered by a saddle-shaped apical callus linking posterior portions of dental plates; dental plates extending down inner margins of delthyrium, flanged and curving laterally at their distal extremities, and bearing blunt rounded teeth; ventral adminicula supporting dental plates short, divergent, and extending approximately one-quarter length of valve; median septum and adminicula slightly thickened in large individuals; diductor muscle scars situated between adminicula and median septum posteriorly, becoming wider anteriorly, having an uneven rounded anterior termination between distal extremities of adminicula and median septum; adductor muscle scars not observed.

Brachial valve. Cardinal process supported on a thick apical callus, projecting ventrally, smooth and rounded anteriorly between inner socket ridges, and having a posteriorly or posteroventrally facing lamellose myophore bearing twelve delicate lamellae; sockets narrow posteriorly, deep and rounded anteriorly, bordered by high inner socket ridges; inner socket ridges originating from behind cardinal process, and on their inner dorsal margins giving off nearly flat flanges of crura; crura connected also with smooth anterior portion of cardinal process; median ridge low, narrow, commencing in front of apical callus, and extending to about two-thirds length of valve; adductor muscle scars elongate, located in depression of fold along median ridge, and bordered anterolaterally by narrow rounded ridges.

Measurements	Length (pedicle valve)	Length (brachial valve)	Width	Height (ventral interarea)
CPC 11059	6.5	—	11.5	3.5
CPC 11060 Paratype	7	—	12	3.5
CPC 11061 Holotype	9	—	19	5
CPC 11062 Paratype	5	—	10.5 est.	2.5
CPC 11063 Paratype	—	9	17 est.	—
CPC 11064 Paratype	—	9	15.5	—

Remarks: Punctospirifer pauciplicatus sp. nov. is distinguished from most species of the genus by the possession of a small number of rounded plicae on the lateral slopes. It differs from the type species, *P. scabricostus* North, recently redescribed by Campbell (1959), in having a more transverse outline, fewer plicae on the lateral slopes, a floor on the median sinus which is flat or slightly concave instead of V-shaped, and in the possession of a coarser spinose micro-ornament.

P. uttingi Thomas (1970, pp. 172-74, pl. 14, figs 5-9, 10b-13) from the Septimus Limestone in the Bonaparte Gulf Basin is distinguished from *P. pauciplicatus* by the possession of a median rib in the sinus and a corresponding furrow on the fold, a larger number of higher, more angular plicae on the lateral slopes (six to eight plicae on each lateral slope of the pedicle valve), more acute cardinal extremities which tend to be mucronate, a longer more prominent umbo, and a ventral median septum which extends farther anteriorly.

P. plicatosulcatus Glenister (Pl. 58, figs 1-16) is similarly distinguished from *P. pauciplicatus* by the possession of a median rib in the sinus and a furrow on the fold. *P. plicatosulcatus* is also larger, has a greater number of plicae on each lateral slope, a much lower ventral median septum, and thick deposits of apical callus in adult specimens.

The closest overseas species is *P. solidirostris* (White), from the Kinderhook of Iowa (Weller, 1914, pp. 292-94, pl. 36, figs 25-34), which resembles *P. pauciplicatus* in having relatively few plicae on the lateral slopes. *P. solidirostris* is distinguished by a deeper V-shaped sinus bearing a median rib, a fold with a deep median furrow, a less transverse outline, and a more prominent incurved ventral umbo.

The specific name, taken from the Latin *pauci* — few, and *plica* — a fold, refers to the small number of plicae on the lateral slopes.

Occurrence: Localities 108/0 (the type locality), 108/3, and possibly at 108/4A, Utting Calcarenite, Utting Gap.

Material: CPC 11059-11065. Holotype CPC 11061; paratypes CPC 11059, 11060, 11062, 11063, 11064. CPC 11059-11064 from locality 108/0; CPC 11065 from locality 108/3. Several hundred silicified specimens.

Age: Visean.

Punctospirifer plicatosulcatus Glenister

(Pl. 58, figs 1-16)

1955 *Punctospirifer plicatosulcatus* Glenister, p. 69, pl. 6, figs 26-29; pl. 8, figs 10-12.
1970 *Punctospirifer plicatosulcatus* Glenister, Thomas, pp. 168-72, pl. 25, fig. 7.

Description: External. Shell of average size for genus, unequally biconvex, wider than long; posterolateral extremities rounded in juvenile growth stages, becoming slightly mucronate in adult specimens; greatest width at hinge, highest at midpoint; commissure serrated and uniplicate; plicae moderately high, rounded, and separated by rounded interplical furrows; concentric lamellae regularly imbricating, having a density of approximately four per 1 mm near front of shell, but more widely spaced in sinus; micro-ornament of very small cone-shaped spine bases numbering ten to twelve along 1 mm near front of shell; density of punctae approximately twenty per 1 sq mm on lateral slopes.

Pedicle valve more strongly convex than brachial valve, most at umbo and at front of sinus; apical angle approximately 120°; umbo incurved, bluntly rounded, and projecting well over interarea; umbonal shoulders moderately convex, and poorly differentiated from umbo; cardinal ridges rounded; ventral interarea apsacline, concave beneath umbo, flat anteriorly and at margins, ornamented with parallel growth lines and poorly defined vertical striations, and 3 mm high at umbo on a valve 19 mm wide and 14 mm long; delthyrium extremely narrow, plugged apically by a small callus, and bordered laterally with prominent dental flanges; dental flanges below level of interarea; delthyrial angle 30°; median sinus commencing at umbo, moderately deep, having steep sides, a flat floor bearing a single rounded median rib, and produced into a short lingual extension; sinal angle 25°; lateral slopes gently convex, becoming flatter at posterolateral margins, each slope bearing eight to nine plicae.

Brachial valve subrectangular to alate, most convex at umbo; umbo broadly rounded, short, and nearly straight; umbonal shoulders small, and gently concave; lateral slopes moderately convex, becoming flat on anterolateral margins; fold narrow, rounded, low posteriorly, becoming higher towards front of valve, and bearing a shallow rounded median furrow on anterior two-thirds of its length; dorsal interarea low, flat, and orthocline.

Internal (Fig. 87). Pedicle valve. Dental plates extending along inner margins of delthyrium, curved and slightly thickened distally, and bearing small pointed teeth; ventral adminicula thickened towards apex of valve, especially in large specimens, becoming low and slender anteriorly, and 3 mm long on a valve 14 mm wide; median septum low and rounded posteriorly, tapering to a narrow ridge anteriorly, extending to midlength of valve, and in large individuals almost completely buried by apical callus; muscle field obscure in small individuals; in large specimens with a thick apical callus, diductor muscle scars form two triangular platforms enclosed laterally by distal ends of adminicula; adductor muscle scars probably in a narrow depression between diductor muscle scars.

Brachial valve. Socket plates arising from apex of valve, becoming thickened anteriorly; sockets shallow and rounded posteriorly, becoming wider and flaring laterally towards front, and bordered by high rounded inner socket ridges; crura commencing from beneath cardinal process, and forming wide flanges along inner dorsal margins of inner socket ridges; cardinal process having up to fourteen fine lamellar platelets, and supported by posterior extremity of median ridge; median ridges sharp, narrow, and extending one-third length of valve; adductor muscle scars narrow, elongate, having rounded extremities, impressed posteriorly, and situated along anterior two-thirds of median ridge in depression of fold.

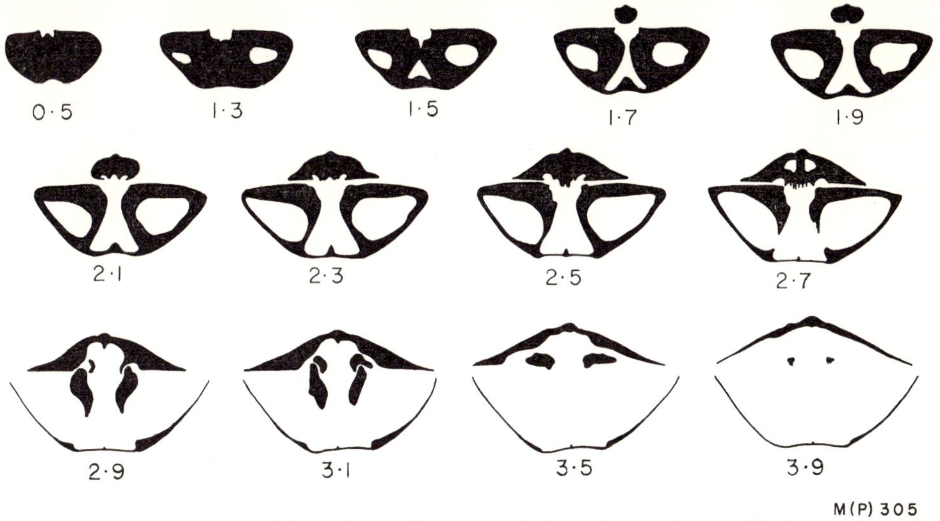

Figure 87. Transverse serial sections of *Punctospirifer plicatosulcatus* Glenister. Note the apical thickening, the low ventral median septum, and the dorsal median ridge supporting the cardinal process. CPC 11126; locality 109/8, Enga Sandstone. × 2.

Measurements	Length (pedicle valve)	Length (brachial valve)	Width	Height	Height (ventral interarea)
CPC 11052	9	7	13.5	6	2
CPC 11053	10	7.5	15.5	8	2.5
CPC 11054	14	11	19	11	3
CPC 11055	11.5	—	17.5	—	4.5
CPC 11056	7.5	—	13	—	2
CPC 11058	—	8.5	19	—	—

Remarks: The specimens of *Punctospirifer plicatosulcatus* Glenister from the Burt Range Formation and Enga Sandstone in the Bonaparte Gulf Basin differ from the holotype (Glenister, 1955, pl. 6, figs 26-29; pl. 8, fig. 10) from the Moogooree Limestone in the Carnarvon Basin in the morphology of the ventral interarea; the holotype, the only pedicle valve illustrated by Glenister, has a ventral interarea which is high, truncated laterally, and trapezoidal in outline, whereas the interarea in the former specimens is lower and triangular in outline. I consider this difference to be of minor taxonomic significance.

The type species, *P. scabricostus* North, has been studied by Sanders (1958), and redescribed by Campbell (1959). *P. plicatosulcatus* differs from *P. scabricostus* in the possession of a lower median septum, shorter adminicula, and a flat-floored sinus having a median rib in the pedicle valve; in the brachial valve there is a stronger median ridge which also supports the dorsal part of the cardinal process, and a median furrow on the fold. The micro-ornament of tiny cone-like spine bases contrasts with the hair-like spines (Sanders, 1958, p. 65) or thread-like lirae which are elevated into minute projections at the front of the lamellae (Campbell, 1959, p. 353) in *P. scabricostus*.

A spinose micro-ornament similar to that in *P. plicatosulcatus* has been described by Cvancara (1958, p. 880) on *P. amblys* Cvancara from the *Rhipidomella fortimuscula* Zone at Barrington, New South Wales. *P. amblys* is distinguished from *P. plicatosulcatus* by its higher median septum and completely smooth sinus in the pedicle valve, and lower but longer median ridge in the brachial valve.

P. uttingi Thomas (1970, pp. 172-74, pl. 14, figs 5-9, 10b-13) from the Septimus Limestone in the Bonaparte Gulf Basin differs from *P. plicatosulcatus* in having fewer, coarser, and more angular plicae, coarser concentric lamellae, and more mucronate posterolateral extremities. In the pedicle valve the median septum is higher and blade-like, the adminicula are more slender and shorter, the apical callus is smaller, and the interarea is less concave.

P. transversus (McChesney) described by Easton (1962, pp. 80-82, pl. 10, figs 6-17) from the Cameron Creek Formation, Montana, is characterized by the possession of a spinose micro-ornament also. It is distinguished from *P. plicatosulcatus* by its greater number of plications, and less convex lateral slopes and higher median septum in the pedicle valve.

P. sulcifer Sanders (1958, pp. 65-67, pl. 7, figs 25-30) from the Represo Limestone, northwestern Sonora, Mexico, has a lower ventral interarea, and a more obtuse apical angle on the pedicle valve.

Occurrence: Localities 109/7, 128/4, 128/6, middle part of Enga Ridge; locality 421/17, southern end of Enga Ridge; locality 113/7, Burt Range Amphitheatre; possibly locality 101/21, 4.5 miles west of Mount Septimus; Burt Range Formation. Locality 102/4, northern end of Enga Ridge, localities 103/17B and 109/8, middle part of Enga Ridge; Enga Sandstone. Locality 136/1, Spirit Hill, a possible equivalent of the Enga Sandstone.

Material: CPC 11052 - 11058, CPC 11126, CPC 11052 - 11054, 11126 from locality 109/8; CPC 11055 from locality 129/4; CPC 11056 - 11057 from locality 128/6; CPC 11058 from locality 102/4. Thirty-five specimens.

Age: Tournaisian.

Order TEREBRATULIDA Waagen, 1883
Suborder TEREBRATULIDINA Waagen, 1883
Superfamily DIELASMATACEA Schuchert, 1913
Family CRANAENIDAE Cloud, 1942
Subfamily CRANAENINAE Cloud, 1942
Genus CRANAENA Hall & Clarke, 1893

Type species: Terebratula romingeri Hall, 1863, by original designation of Hall & Clarke, 1893.

Remarks: The type species is redescribed, and the genus discussed in detail by Cloud (1942, pp. 132-39, pl. 24, figs 2-12). Stehli (1965, fig. 614-1b) gives a composite figure of the interior of the brachial valve of *Cranaena romingeri.*

CRANAENA MONTANA sp. nov.
(Pl. 59, figs 19-38)

Diagnosis: Shell equally biconvex, with a rectimarginate commissure; punctae with a density of approximately 250 per sq mm on the midportion of the shell; pedicle valve subelliptical to subovate, having a round to ovoid labiate foramen in a permesothyrid to mesothyrid position; dental plates robust, extending anteriorly to beneath the teeth; median ridge extending along the length of floor in pedicle valve; cardinal plate nearly flat, bordered by high inner socket ridges; crural bases prominent anteriorly; myophragm extending approximately one-third the length of brachial valve.

Description: External. Shell equally biconvex, longer than wide, greatest width at midlength; commissure rectimarginate; concentric growth lines irregularly spaced, and crowded anteriorly; punctae evenly spaced, having a density of approximately 250 per sq mm on midportion of shell.

Pedicle valve subelliptical to subovate, tapering anteriorly, usually evenly convex over entire valve, but in a few individuals having a faint median sinus at front; beak sub-erect; palintropes small, and gently concave; cardinal margin terebratulid; pedicle foramen permesothyrid to mesothyrid, round to subelliptical, and labiate, with a dorsally curved anterior lip extending from pedicle collar; upper part of delthyrium closed by two concave triangular conjunct deltidial plates.

Brachial valve round to ovoid, evenly convex, with no indication of a median fold; umbo small, pointed, and enclosed by delthyrium on pedicle valve.

Internal. Pedicle valve with a short cylindrical pedicle collar; dental plates straight, divergent, robust, and extending anteriorly to beneath teeth; teeth elongate, rounded, and tapering posteriorly; median ridge low, becoming slightly broader anteriorly, extending the length of valve, and bordered by two narrow furrows and two smaller lateral ridges probably bearing mantle canals; muscle scars poorly defined.

Brachial valve. Cardinal plate apically perforated, nearly flat, giving off crural bases from its lateral margins, and bordered by inner socket ridges; crural bases strong, and projecting slightly ventrally; loop not observed; sockets wide anteriorly, deep, grooved, and bordered by strong nearly vertical inner socket ridges; myophragm low, and extending approximately one-third length of valve; posterior adductor muscle scars narrow, elongate, and situated almost immediately above cardinal plate; anterior adductor muscle scars larger, triangular to subovate, and located on anterolateral flanks of posterior muscle scars; mantle canals extending anterolaterally from front of anterior scars.

Measurements	Length (pedicle valve)	Length (brachial valve)	Width	Height
CPC 11076	10	0.5	9	5
CPC 11077	13.5	—	13	—
CPC 11079	—	11.5	11.5	—
CPC 11080 Paratype	15	—	13	—
CPC 11082 Holotype	14	11.5	11.5	7 est.

Remarks: This species differs from the type species, *C. romingeri* (Hall), figured by Cloud (1942, pl. 24, figs 2-12), in being less globose, and in having a narrower beak, a more circular pedicle foramen, and a straight anterior commissure.

Several Mississippian species from USA appear to be morphologically related. *C. globosa* Weller (1914, pp. 249-51, pl. 34, figs 60-62) from the Burlington Limestone of Missouri differs only in being slightly longer, and in having weaker crural bases than those in *montana*. *C. occidentalis* (Miller) from the Chouteau Limestone of Missouri, figured by Moore (1928, pl. 12, figs 1-6), is not as wide, and has a less incurved ventral umbo bearing a more subcircular foramen when compared with *montana*. Specimens of the same species from the Bushberg Sandstone, Missouri, figured by Branson (1938, pl. 19, figs 7-10) may be closer externally to this species. The interior of *occidentalis* has not been described. Compared with *C. texana* Carter (1967, pp. 425-28, pl. 44, figs 1-10) from the Chappel Limestone of Texas, *montana* is distinguished by its slightly larger size, less prominent concentric growth halts, especially in the brachial valve, and shorter dental plates (Carter, 1967, fig. 48).

The only other *Cranaena* species so far described from Australia, *C. concava* (Campbell & Engel, 1963, pp. 77-79, pl. 2, figs 7-13) from the Tournaisian Tulcumba Sandstone at Rangari, New South Wales, is distinguished by its more concave cardinal plate.

The specific name is taken from the Latin *montis* — mountain, and refers to the type locality high on Enga Ridge in the Burt Range.

Occurrence: Locality 102/4, Enga Sandstone, at the northern end of Enga Ridge, the type locality; localities 150/4 and 150/6, Mount Septimus; locality 110/4, Mount Zimmermann; Septimus Limestone.

Material: CPC 11076 - 11085. Holotype CPC 11082; paratypes CPC 11080, 11081, 11083, 11084, 11085. CPC 11076 - 11079 from locality 150/4; CPC 11080 - 11085 from locality 102/4. More than 150 silicified specimens.

Age: Tournaisian.

Subfamily GIRTYELLINAE Stehli, 1965

Genus GIRTYELLA Weller, 1914

Type species: Harttina indianensis Girty, 1908, by original designation of Weller, 1914.

GIRTYELLA ACYMOSA sp. nov.

(Pl. 59, figs 1-18)

Diagnosis: Shell longitudinally subelliptical, dorsibiconvex, having a smoothly rounded anterior margin, and lacking sinuses on both valves; commissure nearly rectimarginate; pedicle valve umbo sub-erect, bearing a large round or ovoid foramen in a mesothyrid position; dental plates short; cardinal plate strongly concave; median septum extending approximately one-quarter the length of valve.

Description: External. Shell longitudinally subelliptical, slightly dorsibiconvex, possibly becoming globose in adult; anterior margin smoothly rounded; commissure rectimarginate or with a slight median dorsal flexure; greatest height near midlength, greatest width at or slightly in front of midlength; concentric growth lines crowded anteriorly; punctation and most of ornament obscured by coarse silicification.

Pedicle valve umbo sub-erect, bearing a large round or slightly ovate apical foramen in a mesothyrid position; some individuals with a slightly labiate foramen; palintropes wide, concave, and poorly differentiated from umbonal shoulders; delthyrium closed apically by two flat conjunct deltidial plates; valve evenly convex in anterior profile, most convex at umbo in lateral profile; median sinus absent.

Brachial valve subcircular to subelliptical, higher and more strongly convex than pedicle valve in adult specimens; umbo short, pointed, and well defined; fold or sinus absent.

Internal. Pedicle valve. Dental plates short, divergent along floor of valve, and supporting blunt robust teeth; pedicle collar short and cylindrical; muscle field subrectangular, impressed, and having narrow elongate diductor muscle scars separated by obscure adductor scars; mantle canals straight, slightly divergent, and arising from front of diductor scars.

Brachial valve. Cardinal plate entire, strongly concave, supported on a high median septum, and near lateral margins giving rise to strong ridge-like crural bases; median septum blade-like, extending approximately one-quarter the length of valve, and decreasing in height at front; inner socket ridges high, enclosing narrow elongate sockets; adductor muscle scars in elliptical depressions on floor of valve beneath front of hinge plate, separated by front of median septum; loop not observed.

Measurements	Length (pedicle valve)	Length (brachial valve)	Width	Height
CPC 11067 Holotype	9	8	7	5.5
CPC 11068 Paratype	7	6	5.5	3.5
CPC 11070 Paratype	7.5	—	—	—
CPC 11071 Paratype	8 est.	—	7	—

Remarks: Girtyella acymosa sp. nov. is characterized by a well-rounded anterior margin, and the lack of sinuses in both valves. In the holotype, one of the few whole specimens available for study, the brachial valve is swollen; it is not known whether this is representative of the species because of insufficient numbers of mature complete specimens. These characters distinguish *acymosa* from the species of *Girtyella* from the Mississippi Valley area, USA, described by Weller (1914).

Externally *G. acymosa* is close to a specimen of *G. wellsvillensis* Gunnell from the Brazer Limestone, Utah, illustrated by Easton (1962, p. 12, fig. 7). Easton (p. 87) places *wellsvillensis* in synonymy with *G. woodworthi* Clark, but illustrates it separately. His specimen is small, has a rounded anterior margin, and apparently lacks median sinuses at the front of both valves. Gunnell's original specimens (1932, pl. 27, figs 18-21) have narrow median furrows on the posterior of both valves, and hence are distinguished from this species.

G. injensis Besnosova (*in* Sarycheva et al., 1963, pl. 63, fig. 5) from the early Tournaisian Abishevsky Beds of the Kuznetsk Basin, Siberia, resembles this species in outline, and like the holotype has a swollen brachial valve.

The specific name is taken from the Greek *akymatos,* waveless, and refers to the smooth anterior region of both valves.

Occurrence: Localities 108/3 (the type locality) and 108/0, Utting Calcarenite, Utting Gap.

Material: CPC 11067-11075. Holotype 11067; paratypes CPC 11068-11072. CPC 11067-11072 from locality 108/3; CPC 11073-11075 from locality 108/0. Fifty-six silicified specimens.

Age: Visean.

REFERENCES

ALEKSEEVA, R. E., 1960 — A new sub-genus *Atrypa (Desquamatia),* family Atrypidae Gill. *Dokl. Akad. Nauk SSSR,* 131 (2), 421-424.

ALEKSEEVA, R. E., 1962 — DEVONIAN ATRYPIDA FROM THE KUZNETSK AND MINUSSINSK BASINS AND THE EASTERN PART OF THE NORTHERN URALS. *Acad. Sci. SSSR, Inst. Geol. Geophys.,* 196 pp., pls 1-12.

BANKS, M. R., ET AL., 1969 — Correlation charts for the Carboniferous, Permian, Triassic, and Jurassic Systems in Australia. *I Simposio internacional sobre Estratigrafia y Paleontologia de Gondwana, Mar del Plata, Argentina,* 1967.

BELL, W. A., 1929 — Horton-Windsor district, Nova Scotia. *Mem. geol. Surv. Brch Can.,* 155, 1-268, pls 1-36.

BENSON, W. N., 1922 — Materials for the study of the Devonian palaeontology of Australia. *Rec. geol. Surv. N.S.W.,* 10, 83-204, pls 13-14A.

BERRY, W. B. N., 1968 — GROWTH OF A PREHISTORIC TIME SCALE BASED ON ORGANIC EVOLUTION. *W. H. Freeman & Company, San Francisco.* 158 pp.

BESNOSOVA, G. A., 1959 — Lower Carboniferous brachiopods of the Kuznetsk Basin. *Trudy paleont. Inst. Akad. Nauk SSSR,* 75, 1-136, pls 1-11.

BESNOSOVA,, G. A. BENEDIKTOVA, R. N., SARYCHEVA, T. G., and SOKOLSKAYA, A. N., 1962 — Biostratigraphy of the Palaeozoic Sayano-Altai Mountain region. III. Upper Palaeozoic, description of index forms, Brachiopoda. *Trudy sib. nauchno-issled. Inst. Geol. Geofiz. miner. Syr.,* 21, 143-184, pls C12-C24.

BIERNAT, G., 1964 — Middle Devonian Atrypacea (Brachiopoda) from the Holy Cross Mountains, Poland. *Acta palaeont. pol.,* 9, 277-340, pls 1-15.

BION, H. S., 1928 — The fauna of the Agglomeratic Slate Series of Kashmir. *Mem. geol. Surv. India Palaeont. indica,* n.s., 12, 1-54, pls 1-8.

BOND, G., 1941 — Species and variation in British and Belgian Carboniferous Schizophoriidae. *Proc. Geol. Assoc.,* 52, 285-303, pls 21-22.

BOUCOT, A. J., and HARPER, C. W., 1968 — Silurian to Lower Middle Devonian Chonetacea. *J. Paleont.,* 42, 143-176, pls 27-30.

BOUCOT, A. J., JOHNSON, J. G., and STATON, R. D., 1965 — *In* MOORE, R. C. (Editor), TREATISE ON INVERTEBRATE PALAEONTOLOGY, PART H, BRACHIOPODA. *Univ. Kansas Press.*

BRADY, T. J., JAUNCEY, W., and STEIN, C., 1966 — The geology of the Bonaparte Gulf Basin. *J. Aust. Petrol. Expl. Ass.,* 1966, 7-11.

BRANSON, C. C., 1964 — *Neochonetes oklahomensis* (Snider). *Okla. Geol. Notes,* 24, 95-97.

BRANSON, E. B., 1938 — Stratigraphy and paleontology of the lower Mississippian of Missouri. Parts 1 & 2. *Univ. Mo. Stud.,* 13 (3), 1-205, pls 1-20; (4), 1-52, pls 21-25.

BRANSON, E. B., and GREGER, D. K., 1918— Amsden formation of the east slope of the Wind River Mountains of Wyoming and its faunas. *Bull. geol. Soc. Amer.,* 29, 309-326, pls 18-19.

BROWN, D. A., CAMPBELL, K. S. W., and CROOK, K. A. W., 1968 — THE GEOLOGICAL EVOLUTION OF AUSTRALIA AND NEW ZEALAND. *Pergamon Press, London.* 409 pp.

BROWN, R. A. C., 1952 — Carboniferous stratigraphy and palaeontology in the Mount Greenock area, Alberta. *Mem. geol. Surv. Brch Can.,* 264, 1-119, pls 1-5.

BROWNE, I. A., 1953a — Permian spirifers from Tasmania. *J. Proc. R. Soc. N.S.W.,* 86, 55-63, pls 5-6.

BROWNE, I. A., 1953b — *Martiniopsis* Waagen from the Salt Range, India. *J. Proc. R. Soc. N.S.W.,* 86, 100-107, pl. 9.

BRUNTON, C. H. C., 1966a — Silicified Productoids from the Visean of County Fermanagh. *Bull. Br. Mus. nat. Hist.,* Geology, 12, 175-243, pls 1-19.

BRUNTON, C. H. C., 1966b —The morphology of *Delepinea destinezi* (Vaughan) (Brachiopoda: Daviesiellidae). *Ann. Mag. nat. Hist.,* ser. 13, 9, 439-444, pl. 8.

BRUNTON, C. H. C., 1968 — Silicified brachiopods from the Visean of County Fermanagh (II). *Bull. Br. Mus. nat. Hist.*, Geology, 16, 1-70, pls 1-9.

BUCKMAN, S. S., 1908 — Brachiopod homeomorphy: *'Spirifer glaber'*. *Q. Jl geol. Soc. Lond.*, 64, 27-33.

CAMPBELL, K. S. W., 1955 — *Phricodothyris* in New South Wales. *Geol. Mag.*, 92, 374-384, pl. 18.

CAMPBELL, K. S. W., 1956 — Some Carboniferous productid brachiopods from New South Wales. *J. Paleont.*, 30, 463-480, pls 48-50.

CAMPBELL, K. S. W., 1957 — A Lower Carboniferous brachiopod-coral fauna from New South Wales. *J. Paleont.*, 31, 34-98, pls 11-17.

CAMPBELL, K. S. W., 1959 — The type species of three upper Palaeozoic punctate spiriferoids. *Palaeontology,* 1, 351-363, pls 58-60.

CAMPBELL, K. S. W., 1961 — Carboniferous fossils from the Kuttung rocks of New South Wales. *Palaeontology,* 4, 428-474, pls 53-63.

CAMPBELL, K. S. W., and ENGEL, B. A., 1963 — The faunas of the Tournaisian Tulcumba Sandstone and its members in the Werrie and Belvue Synclines, New South Wales. *J. geol. Soc. Aust.,* 10, 55-122, pls 1-9.

CAMPBELL, K. S. W., and MCKELLAR, R. G., 1969 — EASTERN AUSTRALIAN CARBONIFEROUS INVERTEBRATES: SEQUENCE AND AFFINITIES. Hill Memorial Vol., *A.N.U. Press, Canberra.*

CAMPBELL, K. S. W., and ROBERTS, J., 1964 — Two species of *Delepinea* from New South Wales. *Palaeontology,* 7, 514-524, pls 80-82.

CAMPBELL, K. S. W., and ROBERTS, J., 1969 — Carboniferous System: The faunal sequence and overseas correlation, *in* PACKHAM, G. H. (Editor), THE GEOLOGY OF NEW SOUTH WALES. *J. geol. Soc. Aust.,* 16, 261-264.

CARTER, J. L., 1967 — Mississippian brachiopods from the Chappel Limestone of central Texas. *Bull. Amer. Paleont.,* 53, no. 238, 253-488, pls 13-45.

CARTER, J. L., 1968 — New genera and species of early Mississippian brachiopods from the Burlington Limestone. *J. Paleont.,* 42, 1140-1152, pls 145-148.

CASTER, K. E., 1930 — Higher fossil faunas of the upper Allegheny. *Bull. Amer. Paleont.,* 15, no. 58, 1-174, pls 1-59.

CHAO, Y. T., 1927 — Productidae of China. Part I; Producti. *Palaeont. sin.,* ser. B, 5, fasc. 2, 1-244, pls 1-16.

CHAO, Y. T., 1929 — Carboniferous and Permian spiriferids of China. *Palaeont. sin.,* ser. B, 11, fasc. 1, 1-133, pls 1-11.

CHU, S., 1933 — Corals and Brachiopoda of the Kinling Limestone. *Monogr. natn. Res. Inst. Geol. Nanking,* ser. A, 2, 1-58, pls 1-5.

CLOUD, P. E., 1942 — Terebratuloid Brachiopoda of the Silurian and Devonian. *Spec. Pap. geol. Soc. Amer.,* 38, 1-182, pls 1-26.

COLEMAN, P. J., 1951 — *Atrypa* in Western Australia. *J. Paleont.,* 25, 677-690, pls 100-102.

CONDON, M. A., 1965 — The geology of the Carnarvon Basin, Western Australia. Part 1: Pre-Permian stratigraphy. *Bull. Bur. Miner. Resour. Aust.,* 77, 1-82.

COOPER, G. A., and MUIR-WOOD, H. M., 1951 — Brachiopod homonyms. *J. Wash. Acad. Sci.,* 41, 195-196.

COPE, F. W., 1935 — On *Daviesiella carinata* (Garwood). *Proc. Yorks. geol. Soc.,* 23, 79-90, pl. 3.

COPPER, P., 1965a — A new Middle Devonian atrypid brachiopod from the Eifel, Germany. *Senckenberg. leth.,* 46, 309-325, pl. 27.

COPPER, P., 1965b — Unusual structures in Devonian Atrypidae from England. *Palaeontology,* 8, 358-373, pls 46-47.

COPPER, P., 1966a — The *Atrypa zonata* brachiopod group in the Eifel, Germany. *Senckenberg. leth.,* 47, 1-55, pls 1-7.

COPPER, P., 1966b — Ecological distribution of Devonian atrypid brachiopods. *Palaeogeogr. Palaeoclimatol. Palaeoecol.*, 2, 245-266, pl. 1.

COPPER, P., 1967a — Adaptations and life habits of Devonian atrypid brachiopods. *Palaeogeogr. Palaeoclimatol. Palaeoecol.* 3, 363-379, pl. 1.

COPPER, P., 1967b — Brachidial structures of some Devonian atrypid brachiopods. *J. Paleont.*, 41, 1176-1183, pls 155-156.

COPPER, P., 1967c — *Spinatrypa* and *Spinatrypina* (Devonian Brachiopoda). *Palaeontology*, 10, 489-523, pls 76-83.

COPPER, P., 1967d — Frasian Atrypidae (Bergisches Land, Germany). *Palaeontographica* A, 126, 116-140, pls 19-27.

COPPER, P., 1967e — The shell of Devonian Atrypida (Brachiopoda). *Geol. Mag.*, 104, 123-131, pls 5-6.

CRICKMAY, C. H., 1952 — Discrimination of the late Upper Devonian. *J. Paleont.*, 26, 585-609, pls 70-78.

CRICKMAY, C. H., 1957 — ELUCIDATION OF SOME WESTERN CANADA DEVONIAN FORMATIONS. *Imperial Oil Limited, Calgary*. 15 pp., 1 pl.

CRICKMAY, C. H., 1967 — THE METHOD OF INDIVISIBLE AGGREGATES IN STUDIES OF THE DEVONIAN. *E. de Mille Books, Calgary*. 22 pp., pls 1-4.

CVANCARA, A. M., 1958 — Invertebrate fossils from the Lower Carboniferous of New South Wales. *J. Paleont.*, 32, 846-888, pls 109-113.

DAVIDSON, T., 1857-1863 — A monograph of the British Carboniferous Brachiopoda. *Palaeontogr. Soc. (Monogr.)*, part 5, 1-280, pls 1-55.

DAVIDSON, T., 1864-1865 — A monograph of the British Brachiopoda III, Pt VI. The Devonian Brachiopoda. *Palaeontogr. Soc. (Monogr.)*, 17, 1-131, 20 pls.

DAVIDSON, T., 1880 — A monograph of British Fossil Brachiopoda, Supplement to the Permian and Carboniferous species. *Palaeontogr. Soc. (Monogr.)*, Supplements, 4, (3), 243-316, pls 30-37.

DEAR, J., 1963 — Upper Palaeozoic biostratigraphy of the Yarrol Basin in the vicinity of Monto. *Unpubl. Ph.D. thesis, Univ. Queensland.*

DEHEE, R., 1929 — Description de la faune d'Etroeungt. Faune de passage du Devonien au Carbonifère. *Mem. Soc. geol. Fr.*, n.s., 11, 1-62, pls 1-8.

DEMANET, F., 1934 — Les brachiopodes du Dinantian de la Belgique. *Mem. Mus. Hist. nat. Belg.*, 61, 1-114, pls 1-10.

DEMANET, F., 1938 — La faune des couches de passage du Dinantian au Namurien dans le synclinorium de Dinant. *Mem. Mus. r. Hist. nat. Belg.*, 84, 1-201, pls 1-14.

DEMANET, F., 1958 — Contribution à l'étude du Dinantien de la Belgique. *Mem. Mus. r. Hist. nat. Belg.*, 141, 1-152.

DIENER, C., 1899 — Anthracolithic fossils of Kashmir and Spiti. *Mem. geol. Surv. India Palaeont. indica,* ser. 15, 1 (2), 1-95, pls 1-8.

DIENER, C., 1915 — The Anthracolithic faunae of Kashmir, Kanaur and Spiti. *Mem. geol. Surv. India Palaeont. indica,* n.s., 5 (2), 1-135, pls 1-11.

DRUCE, E. C., 1969 — Devonian and Carboniferous conodonts from the Bonaparte Gulf Basin, Western Australia, and their use in international correlation. *Bull. Bur. Miner. Resour. Aust.*, 98, 1-242, pls 1-43.

DUCHEMIN, A., and CREEVEY, K., 1966 — Well completion report, Aquitaine Kulshill No. 1. *Aust. Aquit. Petrol. Pty Ltd* (Unpubl.).

DUNBAR, C. O., and CONDRA, G. E., 1932 — Brachiopoda of the Pennsylvanian System in Nebraska. *Bull. Neb. geol. Surv.*, ser. 2, 5, 1-377, pls 1-44.

EASTON, W. H., 1962 — Carboniferous formations and faunas of central Montana. *Prof. Pap. U.S. geol. Surv.*, 348, 1-126, pls 1-13.

EINOR, O. L., VOYNOVSKY-KRIEGER, K. G., VASILUK, N. P., VDOVENSKO, M. V., GORAK, S. V., and DUNAYEVA, N. N., 1965 — Charactères généraux de la biogéographie de l' U.R.S.S. pendant que la période Carbonifère. *Bull. Soc. geol. Fr.*, ser. 7, 7, 110-123.

FENTON, C. L., and FENTON, M. A., 1924 — The stratigraphy and fauna of the Hackberry Stage of the Upper Devonian. *Contr. Mus. Geol. Univ. Mich.*, 1, 1-260, pls 1-45.

FENTON, C. L., and FENTON, M. A., 1935 — Atrypae described by Clement L. Webster and related forms (Devonian, Iowa). *J. Paleontology,* 9, 369-384, pls 37-43.

FERGUSON, J., 1966 — Variation in two species of the Carboniferous brachiopod *Pleuropugnoides. Proc. Yorks. geol. Soc.*, 35, 353-374, pl. 23.

FOTIEVA, N. N., 1961 — Tournaisian chonetids of Timan-Pechora province. *Paleont. Zh.*, 4, 100-108, pl. 9.

FRECH, F., 1891 — Ueber das Devon der Ostalpen. II. *Z. dt. geol. Ges.*, 43, 672-687, pls 44-47.

FREDERICKS, G., 1928 — Contributions to the classification of the genus *Productus. Bull. Comm. geol. Leningr.*, 46 (7), 1919, 773-792.

GAETANI, M., 1964 — *Rossirhynchus adamantinus* gen. n., sp. n. from the Tournaisian of central Elburz, Iran (Rhynchonellida) (1). *Riv. ital. Paleont. Stratigr.*, 70, 637-648, pl. 47.

GAETANI, M., 1965 — The geology of the upper Djadjerud and Lar Valleys (North Iran). II Palaeontology. Brachiopods and molluscs from Geirud Formation, Member A (Upper Devonian and Tournaisian). *Riv. ital. Paleont. Stratigr.*, 71, 679-770, pls 68-75.

GATINAUD, G., 1949 — Contributions à l'étude des brachiopodes Spiriferidae. 1. Exposé d'une nouvelle méthode d'étude de la morphologie externe des Spiriferidae a sinus plisse. *Bull. Mus. Hist. nat. Fr.*, ser. 2, 21 (1) 153-159; (2) 300-307; (3) 408-413; (4) 487-492.

GEORGE, T. N., 1927 — Studies in Avonian Brachiopoda. I. The genera *Brachythyris* and *Martinia. Geol. Mag.*, 64, 106-119.

GEORGE, T. N., 1932 — The British Carboniferous recticulate Spiriferidae. *Q. Jl geol. Soc. Lond.*, 88, 516-575, pls 31-35.

GEORGE, T. N., ET AL., 1967 — Geological Society of London. Report of the stratigraphical code sub-committee. *Proc. geol. Soc. Lond.*, 1967, no. 1638, 75-87.

GIRTY, G. H., 1899 — Geology of the Yellowstone National Park, part 2, section 2, Devonian and Carboniferous fossils. *Monogr. U.S. geol. Surv.*, 32, 479-599, pls 66-71.

GIRTY, G. H., 1911 — On the genus *Syringopleura* Schuchert. *J. Geol.*, 19, 548-554.

GIRTY, G. H., 1927 — The description of Carboniferous and Triassic fossils, *in* MANSFIELD, G. R., GEOGRAPHY, GEOLOGY, AND MINERAL RESOURCES OF PART OF SOUTH-EAST IDAHO. *Prof. Pap. U.S. geol. Surv.*, 152, 411-446, pls 22-25, 27-28.

GLENISTER, B. F., 1955 — Devonian and Carboniferous spiriferids from the North-West Basin, Western Australia. *J. Proc. R. Soc. W. Aust.*, 39, 46-71, pls 1-8.

GLENISTER, B. F., and KLAPPER, G., 1966 — Upper Devonian conodonts from the Canning Basin, Western Australia. *J. Paleont.*, 40, 776-842, pls 85-96.

GOLDRING, R., 1957 — The last toothed Productellinae in Europe (Brachiopoda, Upper Devonian). *Paläont. Z.*, 31, 207-228, pl. 24.

GRABAU, A. W., 1931a — Devonian Brachiopoda of China 1. Devonian Brachiopoda from Yunnan and other districts in south China. *Palaeont. sin.*, ser. B, 3 (3), 1-545, pls 1-54.

GRABAU, A. W., 1931b — Studies for students. 1. Studies of Brachiopoda II. *Sci. Q. natn. Univ. Peking,* 2, 397-422.

GRABAU, A. W., 1932 — Studies for students. Studies of Brachiopoda III. *Sci. Q. natn. Univ. Peking,* 3, 75-112.

GRANT, R. E., 1965 — The brachiopod superfamily Stenoscismatacea. *Smithson. misc. Collns,* 148 (2), 1-192, pls 1-24.

GRANT, R. E., 1968 — Structural adaptations in two Permian brachiopod genera, Salt Range, West Pakistan. *J. Paleont.*, 42, 1-32, pls 1-9.

GRECHISHNICOVA, I. A., 1966 — Stratigraphy and brachiopods of the lower Carboniferous of the Rudny Altai. *Trudy Mosk. Obschest. Ispytat. Prirod.*, 20, 1-184, pls 1-21.

GREGORY, J. W., 1930 — The geological history of the Pacific Ocean. *Q. Jl geol. Soc. Lond.*, 86, lxxii-cxxxvi.

GREINER, H., 1957 — *'Spirifer disjunctus'*: Its evolution and paleoecology in the Catskill Delta. *Bull. Peabody Mus. nat. Hist.*, 11, 1-75, pls 1-13.

GUILLAUME, R. E. F., 1966 — Petroleum geology in the Bonaparte Gulf Basin, N.T. *8th Commonw. Min. Metall. Congr.*, 5, 183-196.

GUNNELL, F., 1932 — The Brazer formation of northern Utah and its Telotremate brachiopods. *Am. Midl. Nat.*, 13, 282-300, pl. 27.

HALL, J., 1852 — Palaeontology of New York, II: containing descriptions of the organic remains of the lower middle division of the New York System. *Nat. Hist. N.Y.*, Albany, 1-362, pls 1-85.

HALL, J., 1858 — Palaeontology, *in* HALL, J., and WHITNEY, J. D., REPORT ON THE GEOLOGICAL SURVEY OF THE STATE OF IOWA: EMBRACING THE RESULTS OF INVESTIGATIONS MADE DURING PORTIONS OF THE YEARS 1855, 1856, AND 1857. Vol. 1, pt 2, 473-724, pls 1-29.

HALL, J., 1867 — Palaeontology of New York, 4, (1): containing descriptions and figures of the fossil brachiopods of the upper Helderberg, Hamilton, Portage and Chemung Groups. *Nat. Hist. N.Y.*, Albany, 1-428, pls 1-63.

HARRINGTON, H. J., 1939 — El aparato apical de *Spirifer verneuilli, Sp. leoncitensis* y *Sp. rugulatus. Notas Mus. La Plata Palaeontologia,* 4, 123-142, figs 1-3.

HARRINGTON, H. J., and LEANZA, A. F., 1952 — El aparada apical de *Spirifer striatus, S. crassus,* y *S. duplicostatus* del Carbonico de Gran Bretana. *Revta Assoc. geol. argent.*, 7 (4), 209-218.

HAVLICEK, V., 1959 — Spiriferidae of the Czech Silurian and Devonian. *Rozpr. ustred. Ust. geol.*, 25, 1-275, pls 1-28.

HAYNES, W. P., 1916 — The fauna of the Upper Devonian in Montana. Part 2. The stratigraphy and the Brachiopoda. *Ann. Carneg. Mus.*, 10, 13-54, pls 3-8.

HERNON, R. M., 1935 — The Paradise formation and its fauna. *J. Paleont.*, 9, 653-696, pls 80-82.

HOSKING, L. F. V., 1933 — Distribution of Devonian rocks in the Kimberley Division; and description of a recent collection of Devonian fossils from the Kimberley Division. *J. Proc. R. Soc. W. Aust.*, 19, 67-76, pl. 7.

HYDE, J. E., 1953 — Mississippian formations of central and southern Ohio. *Bull. geol. Surv. Ohio,* 51, 1-355, pls 1-54.

IRVING, E., 1964 — PALEOMAGNETISM AND ITS APPLICATION TO GEOLOGICAL AND GEOPHYSICAL PROBLEMS. *John Wiley, New York.* 399 pp.

JOHNSON, J. G., and BOUCOT, A. J., 1968 — External morphology of *Anatrypa* (Devonian, Brachiopoda). *J. Paleont.*, 42, 1205-1207, pl. 160.

JOHNSTONE, M. H., ET AL., 1967 — Devonian of Western and central Australia. *International Symposium on the Devonian System, Calgary, Alberta, Canada,* 1967, 1, 599-612.

JONES, P. J., 1968 — Upper Devonian Ostracoda and Eridostraca from the Bonaparte Gulf Basin, northwestern Australia. *Bull. Bur. Miner. Resour. Aust.*, 99, 1-108, pls 1-7.

JULIEN, A., 1896 — LE TERRAIN CARBONIFERE MARIN DE LA FRANCE CENTRALE. Paris. 304 pp, pls 1-17.

JUX, U., 1965 — 'Kragen' an Atrypiden aus Schillen des Unteren Plattenkalkes (Givet) von Bergisch Gladbach (Rheinisches Schiefergebirge). *Palaont. Z.*, 39, 147-164, pl. 20.

KALASHNIKOV, N. V., 1966 — Brakhiopody Nizhnego Karbona Verkhnei Pechory na Severnom Uraly. Stratigrafia i Paleontologii severn-vostoka europeiskoi chasti SSSR. *Akad. Nauk SSSR, Komi Filial Inst. Geol., Moscow, Leningrad,* 1966, 28-61, pls 1-12.

KEIDEL, J., and HARRINGTON, H. J., 1938 — On the discovery of Lower Carboniferous tillites in the Precordillera of San Juan, western Argentina. *Geol. Mag.*, 75, 103-129, pls 5-6.

KELUS, A. VON, 1939 — Devonische Brachiopoden und Korallen der Umgebung von Pelsza in Volhynien. *Biul. panst. Inst. geol.*, 8, 1-51, pls 1-3.

KINDLE, E. M., 1909 — The Devonian fauna of the Ouray limestone. *Bull. U.S. geol. Surv.*, 391, 1-60, pls 1-10.

KLAHN, G., 1912 — Die Brachiopoden der Frasne-Stufen bei Aachen. *Jb. preuss. geol. Landesanst.*, 33, 1-39, pls 1-2.

KONINCK, L. G. DE, 1887 — Faune du calcaire Carbonifère de la Belgique. *Ann. Mus. Hist. nat. Belg.*, 14, 1-154, pls 1-37.

KRESTOVNIKOV, V. N., and KARPYSHEV, V. S., 1948 — Fauna i stratigrafiya sloev Etroeungt reki Zigan (yuzhnyi Ural). *Trudy Inst. geol. Nauk, Mosk.*, 66, no. 21, 29-66, pls 1-3.

KRYLOVA, A. K., 1951 — Brachiopody devona Penzen skoi, Ul'ianovskoi i Stalingradskoi. oblastei. *Trudy vses. neft. nauchno-issled. geol.-razv. Inst.*, ns,, 45 (Geologia Povolzhia), 81-156, pls 1-7.

LANE, N. G., 1963 — A silicified Morrowan brachiopod faunule from the Bird Spring Formation, southern Nevada. *J. Paleont.*, 37, 379-392, pls 43-45.

LYASHENKO, A. I., 1959 — Atlas of the brachiopods and Devonian stratigraphy of the Russian Platform. *Vses nauchno-issled. geol.-razv. neft. Inst.*, 1-267, pls 1-87.

M'COY, F., 1844 — A SYNOPSIS OF THE CHARACTERS OF THE CARBONIFEROUS LIMESTONE FOSSILS OF IRELAND. *University Press, Dublin.* 207 pp., 29 pls.

MACKAY, R. M., 1964 — *Lepidophloios* and *Cyrtospirifer* from the Lambie Group at Mount Lambie, N.S.W. *J. Proc. R. Soc. N.S.W.*, 97, 83-89, pl. 1.

MCKELLAR, R. G., 1967 — The geology of the Cannindah Creek area, Monto district, Queensland. *Publs geol. Surv. Qld*, 331, 1-38.

MCLAREN, D. J., 1965 — *In* MOORE, R. C. (Editor), TREATISE ON INVERTEBRATE PALEONTOLOGY, PART H, BRACHIOPODA. *Univ. Kansas Press.*

MCLAREN, D. J., NORRIS, A. W., and MCGREGOR, D. C., 1962 — Illustrations of Canadian fossils, Devonian of Western Canada. *Pap. geol. Surv. Brch Can.*, 62-64, 1-35, pls 1-16.

MAMET, B. L., 1967 — The Devonian-Carboniferous boundary in Eurasia. *International Symposium on the Devonian System, Calgary, Alberta, Canada*, 1967, 2, 995-1007.

MAMET, B. L., and BELFORD, D. J., 1968 — Carboniferous foraminifera, Bonaparte Gulf Basin, northwestern Australia. *Micropaleontology*, 14, 339-347.

MANSUY, H., 1912 — Etude géologique du Yun-nan oriental. II Paleontologie. *Mem. Serv. geol. Indoch.*, 1, fasc. 2, 1-146, pls 1-25.

MARTINOVA, M. V., 1961 — Stratigraphy and brachiopods of the Famennian Stage of western central Kazakhstan. *Mater. Geol. Tsentral. Kazakhstana* (*Mosk. Gosudar. Univ. Tsentral. Kazakh. Geol. Uprav.*), 2, 1-210, pls 1-29.

MAXWELL, W. G. H., 1950 — An Upper Devonian brachiopod (*Cyrtospirifer reidi* sp. nov.) from the Mt Morgan district. *Pap. Dep. Geol. Univ. Qld*, 3 (12), 1-8, pl. 1.

MAXWELL, W. G. H., 1951 — Upper Devonian and middle Carboniferous brachiopods of Queensland. *Pap. Dep. Geol. Univ. Qld*, 3 (14), 1-27, pls 1-4.

MAXWELL, W. G. H., 1954 — Upper Palaeozoic formations in the Mt Morgan district — faunas. *Pap. Dep. Geol. Univ. Qld*, 4 (5), 1-69, pls 1-6.

MAXWELL, W. G. H., 1961 — Lower Carboniferous brachiopod faunas from Old Cannindah, Queensland. *J. Paleont.*, 35, 82-103, pls 19-20.

MINATO, M., 1951 — On the Lower Carboniferous fossils of the Kitakami Massif, northeast Honsyu, Japan. *J. Fac. Sc. Hokkaido Univ.*, ser. 4, 7, 355-382, pls 1-5.

MINATO, M., 1952 — A further note on the Lower Carboniferous fossils of the Kitakami Mountainland, northeast Japan. *J. Fac. Sci. Hokkaido Univ.* ser. 4, 8, 136-174, pls 1-11.

MINATO, M., 1953 — On some reticulate Spiriferidae. *Trans. Proc. palaeont. Soc. Japan*, n.s., 11, 65-73.

MONAKHOVA, L. P., 1959 — Lower Visean brachiopods from Uglia, central Kazakhstan. *Trudy Lab. Geol. Uglya*, 9, 68-152, pls 1-17.

MOORE, R. C., 1928 — Early Mississippian formations in Missouri. *Missouri Bur. Geol. Mines*, ser. 2, 21, 1-283, pls 1-13.

S

MUIR-WOOD, H. M., 1928 — The British Carboniferous Productii, II. *Productus* (sensu stricto); *semiretulatus* and *longispinus* groups. *Mem. geol. Surv. U.K. Palaeontology,* 3, 1-217, pls 1-12.

MUIR-WOOD, H. M., 1948 — MALAYAN LOWER CARBONIFEROUS FOSSILS AND THEIR BEARING ON THE VISEAN PALAEOGEOGRAPHY OF ASIA. *British Museum (Nat. Hist.), London.* 118 pp., 17 pls.

MUIR-WOOD, H. M., 1962 — ON THE MORPHOLOGY AND CLASSIFICATION OF THE BRACHIOPOD SUBORDER CHONETOIDEA. *British Museum (Nat. Hist.), London.* 132 pp., 16 pls.

MUIR-WOOD, H. M., and COOPER, G. A., 1960 — Morphology, classification and life habits of the Productoidea (Brachiopoda). *Mem. geol. Soc. Amer.,* 81, 1-447, pls 1-135.

NALIVKIN, D., 1930 — The Semiluki and the Voroneje beds of the Upper Devonian of Voroneje region. *Izv. glav. geol.-razved. Uprav.,* 49 (1), 53-93, pls 6-8.

NALIVKIN, D., 1937 — Brachiopoda of the Upper and Middle Devonian and Lower Carboniferous of north-eastern Kazakhstan. *Trudy tsent. nauchno-issled. geologo-razv. Inst.,* 99, 1-200, pls 1-39.

NALIVKIN, D., 1941 — Brachiopods of the Main Devonian Field, *in* BATALINA, M.A., ET AL., FAUNA OF THE MAIN DEVONIAN FIELD. *Akad. Nauk SSSR, paleont. Inst.,* 139-226, pls 1-8.

NOAKES, L. C., ÖPIK, A. A., and CRESPIN, I., 1952 — Bonaparte Gulf Basin, north-western Australia: a stratigraphical summary with special reference to the Gondwana System. *Int. geol. Congr. 19th Sess. Alger. Symposium sur les series de Gondwana,* 91-106.

NORTH, F. J., 1920 — On *Syringothyris* Winchell, and certain Carboniferous Brachiopoda referred to *Spiriferina* D'Orbigny. *Q. Jl geol. Soc. Lond.,* 76, 162-227, pls 11-13.

ÖPIK, A. A., 1934 — Uber Klitamboniten. *Acta Comment. Univ. tartu,* ser. A, 26, 1-239.

OZAKI, K., 1939 — On some Lower Carboniferous brachiopods from central Hunan, China. *J. Shanghai Sci. Inst.,* sec. 2, 2, 225-282, pls 1-9.

PAECKELMANN, W., 1913 — Das Oberdevon des Bergischen Landes. *Abh. preuss. geol. Landesanst.,* 70, 1-356, pls 1-7.

PAECKELMANN, W., 1930 — Die brachiopoden des deutschen Unterkarbons. 1. Die Orthiden, Strophomeniden und Chonetiden des mittleren und oberen Unterkarbons. *Abh. preuss. geol. Landesanst.,* n.f., 122, 143-326, pls 9-24.

PAECKELMANN, W., 1931 — Die brachiopoden des deutschen Unterkarbons. 2. Die Productinae und *Productus* — anlichen Chonetinae. *Abh. preuss. geol. Landesanst.,* n.f. 136, 1-440, pls 1-41.

PAECKELMANN, W., 1942 — Beitrage zur Kenntnis devonischer Spiriferen. *Abh. Reichstelle Bodenforsch.,* 197, 1-188, pls 1-8.

PARKINSON, D., 1969 — Relative growth, variation and evolutionary trends in a Carboniferous rhynchonellid brachiopod. *J. Paleont.,* 43, 95-110, pls 19-20.

PATTE, E., 1926 — Etudes paleontologiques relatives à la géologie de l'est du Tonkin. *Bull. Serv. geol. Indochine,* 15, fasc. 1, 1-240, pls 1-12.

PAUL, H., 1942 — *Thomasella* n. nom. = *Thomasina* Paeckelmann, 1931. (Brachiop., Productidae). *Zentbl. Miner. Geol. Palaont.,* abt. B, 191, 1942.

PHILLIPS, J., 1836 — ILLUSTRATIONS OF THE GEOLOGY OF YORKSHIRE; PART II. THE MOUNTAIN LIMESTONE DISTRICT. *John Murray, London.* 253 pp., 25 pls.

PHILLIPS, J., 1841 — FIGURES AND DESCRIPTIONS OF THE PALAEOZOIC FOSSILS OF CORNWALL, DEVON, AND WEST SOMERSET; OBSERVED IN THE COURSE OF THE ORDINANCE GEOLOGICAL SURVEY OF THAT DISTRICT. *Longman, Brown, Green & Longman, London.* 231 pp., 60 pls.

PITRAT, C. W., 1965 — Suborder Spiriferidina *in* MOORE, R. C. (Editor), TREATISE ON INVERTEBRATE PALEONTOLOGY, PART H, BRACHIOPODA. *Univ. Kansas Press.*

PLAYFORD, P. E., 1967 — Devonian reef complexes in the northern Canning Basin, Western Australia. *International Symposium on the Devonian System, Calgary, Canada,* 1967, 2, 351-364.

PLAYFORD, P. E., and LOWRY, D. C., 1966 — Devonian reef complexes of the Canning Basin, Western Australia. *Bull. geol. Surv. W. Aust.*, 118, 1-150.

POCOCK, Y., 1968 — Carboniferous Schizophoriid brachiopods from western Europe. *Palaeontology*, 11, 64-93, pl. 18.

PRENDERGAST, K. L., 1935 — Some Western Australian upper Palaeozoic fossils. *J. Proc. R. Soc. W. Aust.*, 21, 9-35, pls 2-4.

PROSSER, C. S., 1913 — *In* MARYLAND GEOLOGICAL SURVEY MIDDLE AND UPPER DEVONIAN. Maryland Geol. Surv. *The John Hopkins Press, Baltimore.* 720 pp., 171 pls.

RAMOVS, A., 1966 — Revision des *'Productus elegans'* (Brachiopoda) im ostalpinen Jungpalaozoikum. *N. Jb. Geol. Palaont. Abh.*, 125, 118-124, pl. 11.

REED, F. R. C., 1922 — Devonian fossils from Chitral and the Pamirs. *Mem. geol. Surv. India Palaeont. indica,* n.s., 6 (2), 1-167, pls 1-16.

REED, F. R. C., 1943 — Notes on certain Upper Devonian brachiopods figured by Whidborne. *Geol. Mag.*, 80, 69-78, 95-106, 132-138.

RHODES, F. H. T., AUSTIN, R. L., and DRUCE, E. C., 1969 — British Avonian (Carboniferous) conodont faunas, and their value in local and intercontinental correlation. *Bull. British Museum (Nat. Hist.)*, Geology, Supplement 5, 1-313, pls 1-31.

RIGAUX, E., 1894 — Notice geologique sur le Bas Boulonnais. *Mem. Soc. acad. Arrond. Boulogne Sur-Mer,* 16 (1891-1894), 1-108, pls 1-2.

RIGAUX, E., 1908 — LE DEVONIEN DE FERQUES ET SES BRACHIOPODES. *Mlle Deligny, Boulogne-Sur-Mer,* 5-34, pls 1-2.

ROBERTS, J., 1963 — A Lower Carboniferous fauna from Lewinsbrook, New South Wales. *J. Proc. R. Soc. N.S.W.*, 97, 1-31, **pls 1-6.**

ROBERTS, J., 1964 — Lower Carboniferous brachiopods from Greenhills, New South Wales. *J. geol. Soc. Aust.*, 11, 173-194, pls 1-6.

ROBERTS, J., 1965 — Lower Carboniferous zones and correlations based on faunas from the Gresford-Dungog district, New South Wales. *J. geol. Soc. Aust.*, 12, 105-122.

ROBERTS, J., 1968 — Mantle canal patterns in *Schizophoria* (Brachiopoda) from the Lower Carboniferous of New South Wales. *Palaeontology,* 11, 389-405, pls 74-75.

ROBERTS, J., JONES, P. J., and DRUCE, E. C., 1967 — Palaeontology and correlations of the Upper Devonian of the Bonaparte Gulf Basin, Western Australia and Northern Territory. *International Symposium on the Devonian System, Calgary, Alberta, Canada,* 1967, 2, 565-577.

ROBERTS, J., & VEEVERS, J. J., in press — Carboniferous geology of the Bonaparte Gulf Basin, northwestern Australia. *6th Congr. Avanc. Etud. Stratigr. carb. Sheffield, 1967* (in press).

ROTAI, A., 1931 — Brachiopods and stratigraphy of Lower Carboniferous of the Donetz Basin. *Trudy glav. geol.-razv. Uprav.*, 73, 35-144, pls 1-11.

ROWLEY, R. R., 1900 — Descriptions of new species of fossils from the Devonian and subcarboniferous rocks of Missouri. *Am. Geol.*, 25, 261-273, pl. 5.

ROZMAN, KH. S., 1962 — Stratigraphy and brachiopods of the Famennian Stage of the Mugodzhar and adjacent regions. *Trudy Inst. geol. Nauk, Mosk.*, 50, 1-196, pls 1-31.

RUDWICK, M. J. S., 1959 — The growth and form of brachiopod shells. *Geol. Mag.*, 96, 1-24.

RUDWICK, M. J. S., 1964 — The function of zigzag deflexions in the commissures of fossil brachiopods. *Palaeontology,* 7, 135-171, pls 21-29.

RZHONSNITSKAYA, M. A., 1964 — On Devonian atrypids from the Kuznetsk Basin. *Trudy vses. nauchno-issled. geol. Inst.*, 93, 91-112, pls 1-2.

SADLICK, W., 1963 — *Quadranetes,* a new Carboniferous chonetid. *J. Paleont.*, 37, 721-723.

SANDERS, J. H., 1958 — Brachiopoda and Pelecypoda *in* EASTON, W. H., ET AL., MISSISSIPPIAN FAUNA IN NORTH-WESTERN SONORA, MEXICO. *Smithson. misc. Collns,* 119, no. 3, 1-87, pls 1-9.

SARTENAER, P., 1961a — Late Upper Devonian (Famennian) rhynchonelloid brachiopods. *Bull. Inst. r. Sci. nat. Belg.*, 37, no. 24, 1-10, pls 1-2.

SARTENAER, P., 1961b —Etude nouvelle, en deux parties, du genre *Camarotoechia* Hall et Clarke, 1893. Deuxième partie: *Cupularostrum reticostatum* n. gen., n. sp. (1). *Bull. Inst. r. Sci. nat. Belg.*, 37, no. 25, 1-15, pls 1-2.

SARTENAER, P., 1961c — Note nomenclatoriale: *Yunnanella* Grabau, *Yunnanellina* Grabau, *Nayunnella* nom. nov. (rhynchonelles). *Bull. Inst. r. Sci. nat. Belg.*, 37, no, 2, 1-3,

SARTENAER, P., 1962 — A propos de l'éspece-type du génre *Yunnanella* Grabau, A. W., 1923. *Bull. Inst. r. Sci. nat. Belg.*, 38, no. 19, 1-3.

SARTENAER, P., 1964 — Refonte du genre *Pugnoides* Weller, S., 1910 (Rhynchonelloidea). *Bull. Inst. r. Sci. nat. Belg.*, 40, no. 12, 1-10, pls 1-2.

SARTENAER, P., 1966 — *Ripidorhynchus,* nouveau genre de brachiopode rhynchonellide du Frasnien. *Bull. Inst. r. Sci. nat. Belg.*, 42, no. 30, 1-15, pls 1-2.

SARTENAER, P., 1967 — Famennian rhynchonellid brachiopod genera as a tool for correlation. *International Symposium on the Devonian System, Calgary, Alberta, Canada,* 1967, 2, 1043-1060, pl. 1.

SARYCHEVA, T. G. (Editor), 1960 — 'OSNOVI PALEONTOLOGII' MANUAL FOR PALEONTOLOGISTS AND GEOLOGISTS OF U.S.S.R. BRYOZOA, BRACHIOPODA. *Akad. Nauk SSSR, Moscow.* 343 pp.

SARYCHEVA, T. G. (Editor), 1968 — Late Palaeozoic brachiopods from eastern Kazakhstan. *Trudy paleont. Inst. Akad. Nauk SSSR,* 121, 1-212, pls 1-33.

SARYCHEVA, T. G., and SOKOLSKAYA, A. N., 1952 — Index of Palaeozoic brachiopods from the Sub-Moscow Basin. *Trudy paleont. Inst. Akad. Nauk SSSR,* 38, 1-307, pls 1-71.

SARYCHEVA, T. G., SOKOLSKAYA, A. N., BESNOSOVA, G. A., and MAXIMOVA, S. V., 1963 — Brachiopods and palaeography of the Kuznetsk Basin. *Trudy paleont. Inst. Akad. Nauk SSSR,* 95, 1-547, pls 1-64.

SCHMIDT, H., 1964 — Neue Gattungen palaozoischer Rhynchonellacea (Brachiopoda). *Senckenberg. leth.*, 45, 505-506.

SCHMIDT, H., 1965 — Neue Befunde an palaozoischen Rhynchonellacea (Brachiopoda). *Senckenberg. leth.*, 46, 1-25, pl. 1.

SEMICHATOVA, S. V., 1941 — The group of *Spirifer trigonalis* Martin from the Lower Carboniferous supra-coal bearing beds of Moscow Basin. *Trudy paleont. Inst. Akad. Nauk SSSR,* 12 (3), 1-174, pls 1-17.

SEMICHATOVA, S. V., 1943 — Forms related to *Spirifer duplicostata* Phillips in Visean deposits of Moscow Basin. *Izv. Akad. Nauk SSSR,* ser. biol., 5, 249-264

SHAW, A. B., 1962 — Rhynchonellid brachiopods and a *Torynifer* from the Madison Group (Mississippian). *J. Paleont.*, 36, 630-637, pls 97-98.

SHIELLS, K. A. G., 1968 — *Kochiproductus coronus* sp. nov. from the Scottish Visean and a possible mechanical advantage of its flange structure. *Trans. R. Soc. Edinb.*, 67, 477-507, pl. 1.

SHIELLS, K. A. G., 1969 — Rare flange-bearing productacean brachiopods from the Carboniferous of the British Isles. *Scott. J. Geol.*, 5 (3), 224-247, pls 1-5.

SIDYACHENKO, A. I., 1962 — Spirifers and stratigraphy of the Famennian deposits of central and south-east Karatau. *Akad. Nauk SSSR, Siberian section, Inst. Geol. Geophys, Moscow,* 1-152, pls 1-26.

SIMPSON, G. B., 1889 — Descriptions of new species of fossils from the Clinton, lower Helderberg, Chemung, and Waverly Groups, found in the collections of the Geological Survey of Pennsylvania. *Trans. Am. phil. Soc.*, n.s., 16, 435-460.

SIMPSON, I. M., 1953 — *Daviesiella destinezi* (Vaughan), a Lower Carboniferous index fossil in northwest Ireland. *Geol. Mag.*, 90, 193-200, pl. 9.

SOKOLSKAYA, A. N., 1941 — Lower Carboniferous and Devonian-Carboniferous brachiopods of the Moscow Basin (Tschernyschino, Upa and Malevka-Murajevnia Beds). *Trudy paleont. Inst. Akad. Nauk SSSR,* 12 (2), 1-139, pls 1-12.

SOKOLSKAYA, A. N., 1948 — Evolution of the genus *Productella* Hall and related forms in the Paleozoic of the Moscow region. *Trudy paleont. Inst. Akad. Nauk SSSR,* 14 (3), 1-168, pls 1-10.

SOKOLSKAYA, A. N., 1950 — Chonetidae of the Russian platform. *Trudy paleont. Inst. Akad. Nauk SSSR,* 27, 1-108, pls 1-13.

SOKOLSKAYA, A. N., 1954 — Strophomenidae of the Russian platform. *Trud. Akad. Nauk Palaeont. Inst. SSSR,* 51, 1-187, pls 1-18.

SOWERBY, J., 1840-1846 — THE MINERAL CONCHOLOGY OF GREAT BRITAIN. *London.* 7, 1-80, pls 610-648.

STAINBROOK, M. A., 1945 — Brachiopoda of the Independence Shale of Iowa. *Mem. geol. Soc. Amer.,* 14, 1-74, pls 1-6.

STAINBROOK, M. A., 1947 — Brachiopods of the Percha Shale of New Mexico and Arizona. *J. Paleont.,* 21, 297-328, pls 44-47.

STAINBROOK, M. A., 1950 — Brachiopods and stratigraphy of the Aplington formation of northern Iowa. *J. Paleont.,* 24, 365-385, pls 53-54.

STEHLI, F. G., 1965 — Paleozoic Terebratulida, *in* MOORE, R. C. (Editor), TREATISE ON INVERTEBRATE PALEONTOLOGY, PART H, BRACHIOPODA. *Univ. Kansas Press.*

STRUVE, W., 1956 — *Spinatrypa kelusiana* n. sp., ein Zeitmarke im Rheinischen Mittel-Devon (Brachiopoda). *Senckenberg. leth.,* 37, 383-409, pls 1-3.

STRUVE, W., 1964 — Eroterung des Alters der Retrath-Schichten und Darstellung einiger devonischer Atrypinae. *Senckenberg, leth.,* 45, 523-532.

SUTTON, A. H., 1938 — Taxonomy of Mississippian Productidae. *J. Paleont.,* 12, 537-569, pls 62-66.

TEICHERT, C., 1943 — The Devonian of Western Australia. A preliminary review. *Am. J. Sci.,* 241, 69-94, 167-184.

TEICHERT, C., 1949 — Observations on stratigraphy and palaeontology of Devonian, western portion of Kimberley Division, Western Australia. *Rep. Bur. Miner. Resour. Aust.,* 2, 1-55, pls 1-6.

TERMIER, G., and TERMIER, H., 1950 — Paleontologie Marocaine. II Invertebres de l'era Primaire. Fascicule II Bryozoaires et Brachiopodes. *Notes Mem. Serv. geol. Protect. Republic. Franc. Maroc.,* 77, 1-253, pls 67-122.

THOMAS, G. A., 1959 — The Lower Carboniferous Laurel Formation of the Fitzroy Basin, Western Australia. *Rep. Bur. Miner. Resour. Aust.,* 38, 21-36.

THOMAS, G. A., 1962a — The Carboniferous stratigraphy of the Bonaparte Gulf Basin. *4th Congr. Advanc. Etud. Stratigr. carb., Heerlen, 1958,* 3, 727-732.

THOMAS, G. A., 1962b — The Carboniferous stratigraphy of Western Australia. *4th Congr. Advanc. Etud. Stratigr. carb., Heerlen, 1958,* 3, 733-740.

THOMAS, G. A., 1965 — *Delepinea* in the Lower Carboniferous of northwest Australia. *J. Paleont.,* 39, 97-102, pl. 18A.

THOMAS, G. A., 1970 — Carboniferous and Early Permian Brachiopoda from Western and Northern Australia. *Bull. Bur. Miner. Resour. Aust.,* 56.

TIEN, C. C., 1938 — Devonian Brachiopoda of Hunan. *Palaeont. sin.,* n.s. B, no. 4, whole ser. 113, 1-192, pls 1-22.

TOLMACHOFF, I. P., 1924 — Faune du calcaire Carbonifère du bassin houiller de Kousnetsk, 1. *Mater. Geol. Gen. Appl. Leningr.,* 25, 1-320, pls 1-12.

TRAVES, D. M., 1955 — The geology of the Ord-Victoria region, northern Australia. *Bull. Bur. Miner. Resour. Aust.,* 27, 1-133.

UTTING, E. P., 1957 — Report on exploration and geology within Permit 3 Northern Territory, during 1956. Private Report to Westralian Oil Ltd (Unpubl.).

VANDERCAMMEN, A., 1956 — Revision des Ambocoeliinae de la Belgique. *Bull. Inst. r. Sci. nat. Belg.,* 32, no. 43, 1-51, pls 1-2.

VANDERCAMMEN, A., 1959 — Essai d'étude statistique des *Cyrtospirifer* du Frasnian de la Belgique. *Mem. Inst. r. Sci. nat. Belg.*, 145, 1-175, pls 1-5.

VANDERCAMMEN, A., and LAMBIOTTE, M., 1962 — Observations sur les sarcoglyphes dans *Atrypa reticularis* (C. Linne, 1767). *Bull. Inst. r. Sci. nat. Belg.*, 38, no. 53, 1-15, pl. 1.

VAUGHAN, A., 1905 — The palaeontological sequence in the Carboniferous Limestone of the Bristol area. *Q. Jl geol. Soc. Lond.*, 61, 181-307, pls 22-29.

VEEVERS, J. J., 1959a — Devonian brachiopods from the Fitzroy Basin, Western Australia. *Bull. Bur. Miner. Resour. Aust.*, 45, 1-220, pls 1-18.

VEEVERS, J. J., 1959b — Devonian and Carboniferous brachiopods from north-western Australia. *Bull. Bur. Miner. Resour. Aust.*, 55, 1-42, pls 1-4.

VEEVERS, J. J., 1967 — The Phanerozoic geological history of northwest Australia. *J. geol. Soc. Aust.*, 14, 253-271.

VEEVERS, J. J., 1968 — Identification of reef facies by computer classification. *J. geol. Soc. Aust.*, 15, 209-215.

VEEVERS, J. J., and ROBERTS, J., 1966 — Littoral talus breccia and probable beach rock from the Visean of the Bonaparte Gulf Basin. *J. geol. Soc. Aust.*, 13, 387-403.

VEEVERS, J. J., and ROBERTS, J., 1967 — Upper Devonian geology of the Bonaparte Gulf Basin, Western Australia and Northern Territory. *International Symposium on the Devonian System, Calgary, Alberta, Canada,* 1967, 2, 89-92.

VEEVERS, J. J., and ROBERTS, J., 1968 — Upper Palaeozoic rocks, Bonaparte Gulf Basin of Northwestern Australia. *Bull. Bur. Miner. Resour. Aust.*, 97, 1-155.

VEEVERS, J. J., and WELLS, A. T., 1961 — The geology of the Canning Basin, Western Australia. *Bull. Bur. Miner. Resour. Aust.*, 60, 1-323.

VERNEUIL, E. DE, 1845 — *In* MURCHISON, R. I., VERNEUIL, E. DE, and KEYSERLING, A. DE, GEOLOGIE DE LA RUSSIE D'EUROPE ET DES MONTAGNES DE L'OURAL. 2 (3) PALEONTOLOGIE. *London & Paris*. 17-395, pls 1-43.

WARREN, P. S., 1944 — Index brachiopods of the Mackenzie River Devonian. *Trans. R. Soc. Can.*, 3rd ser., 38, sec. 4, 105-135.

WARREN, P. S., and STELCK, C. R., 1956 — Reference fossils of Canada Part 1. Devonian faunas of Western Canada. *Spec. Pap. geol. Ass. Can.*, 1, 1-15, pls 1-29.

WATERHOUSE, J. B., 1968 — The classification and description of Permian Spiriferida (Brachiopoda) from New Zealand. *Palaeontographica* A., 129 (1-3), 1-94, pls 1-18.

WELLER, J. M., ET AL., 1948 — Correlation of the Mississippian formations of North America. *Bull. geol. Soc. Amer.*, 59, 91-196.

WELLER, S., 1914 — The Mississippian brachiopods of the Mississippi Valley Basin. *Monogr. geol. Surv. Ill.*, 1, 1-508, pls 1-83.

WEYER, D., 1967 — *Kitakamithyris* Minato 1951 (Brachiopoda, Spiriferida) aus dem Etroeungt (Oberdevon) und Tournai (Unterkarbon) des Rheinischen Schiefergebirges. *Geologie,* 16 (4), 433-451, pls 1-2.

WHIDBORNE, G. F., 1897 — A monograph of the Devonian fauna of the south of England. Vol. III, part II. The fauna of the Marwood and Pilton Beds of North Devon and Somerset. *Palaeontogr. Soc. (Monogr.)*, 51, 113-178, pls 17-21.

WHITE, C. A., 1861 — Observations upon the geology and palaeontology of Burlington, Iowa, and its vicinity. *Boston. J. nat. Hist.*, 7, 209-235.

WILLIAMS, A., and ROWELL, A. J., 1965 — *In* MOORE, R. C. (Editor), TREATISE ON INVERTEBRATE PALEONTOLOGY, PART H, BRACHIOPODA. *Univ. Kansas Press.*

WINKLER PRINS, C. F., 1968 — Carboniferous Productidina and Chonetidina of the Cantabrian Mountains (NW Spain): systematics, stratigraphy and palaeoecology. *Leid. geol. Meded.*, 43, 41-126, pls 1-9.

YANG, S., 1964 — Brachiopods of the Tournaisian Stage of the Lower Carboniferous of southeast Gueizhou. *Acta palaeont. sin.*, 12 (1), 82-109, pls 1-3.

YANISHEVSKY, M., 1918 — Material for the study of the Lower Carboniferous faunas of Fergana. *Trudy geol. Kom.*, n.s., 162, 1-145, pls 1-8.

APPENDIX 1. LIST OF LOCALITIES, DETERMINATIONS, AND RANGE CHARTS FOR DEVONIAN AND CARBONIFEROUS BRACHIOPODA, BONAPARTE GULF BASIN

DEVONIAN

Cockatoo Formation

Kununurra Member

28/10 — Matheson Ridge (Appendix 2, Fig. 2)
Schuchertella sp. cf. *S. gratillica* Veevers
Globosochonetes? mathesonensis sp. nov.
Steinhagella sp.
Cyrtospirifer sp.
Crurithyris apena Veevers

406/1 — Abney Hill near Kununurra Township (Appendix 2, Fig. 3)
Globosochonetes? mathesonensis sp. nov.
Steinhagella sp.
Calvinaria? sp.
Cyrtospirifer sp.

438/2 — Abney Hill near Kununurra Township (Appendix 2, Fig. 3)
Globosochonetes? mathesonensis sp. nov.
Calvinaria? sp.

Hargreaves Member

64/1 — In high ridges of sandstone on the margins of Waggon Creek, 0.5 miles south-west of the outlier of Tournaisian breccia, northern edge of the Pretlove Hills.
Spinatrypa prideri larga subsp. nov.
Tenticospirifer columnaris sp. nov.

250/8 — Hargreaves Hills (Appendix 2, Fig. 4)
Tenticospirifer columnaris sp. nov.

427/6 — Hargreaves Hills (Appendix 2, Fig. 4)
Cyrtospirifer sp.
Tropidoleptus? sp.

428/5 — Hargreaves Hills (Appendix 2, Fig. 4)
Cyrtospirifer sp.
Rhynchonellid gen. et sp.
Productella westwoodensis sp. nov.

429/2 — Hargreaves Hills (Appendix 2, Fig. 4)
Cyrtospirifer sp.

Westwood Member

37/1 — Westwood Creek (Appendix 2, Fig. 5)
Spinatrypa prideri larga subsp. nov.
Desquamatia (Synatrypa) kimberleyensis (Coleman)
Tenticospirifer columnaris sp. nov.

37/2 — Westwood Creek (Appendix 2, Fig. 5)
Retichonetes arenarius sp. nov.
Spinatrypa prideri larga subsp. nov.
Desquamatia (Synatrypa) kimberleyensis (Coleman)
Tenticospirifer columnaris sp. nov.

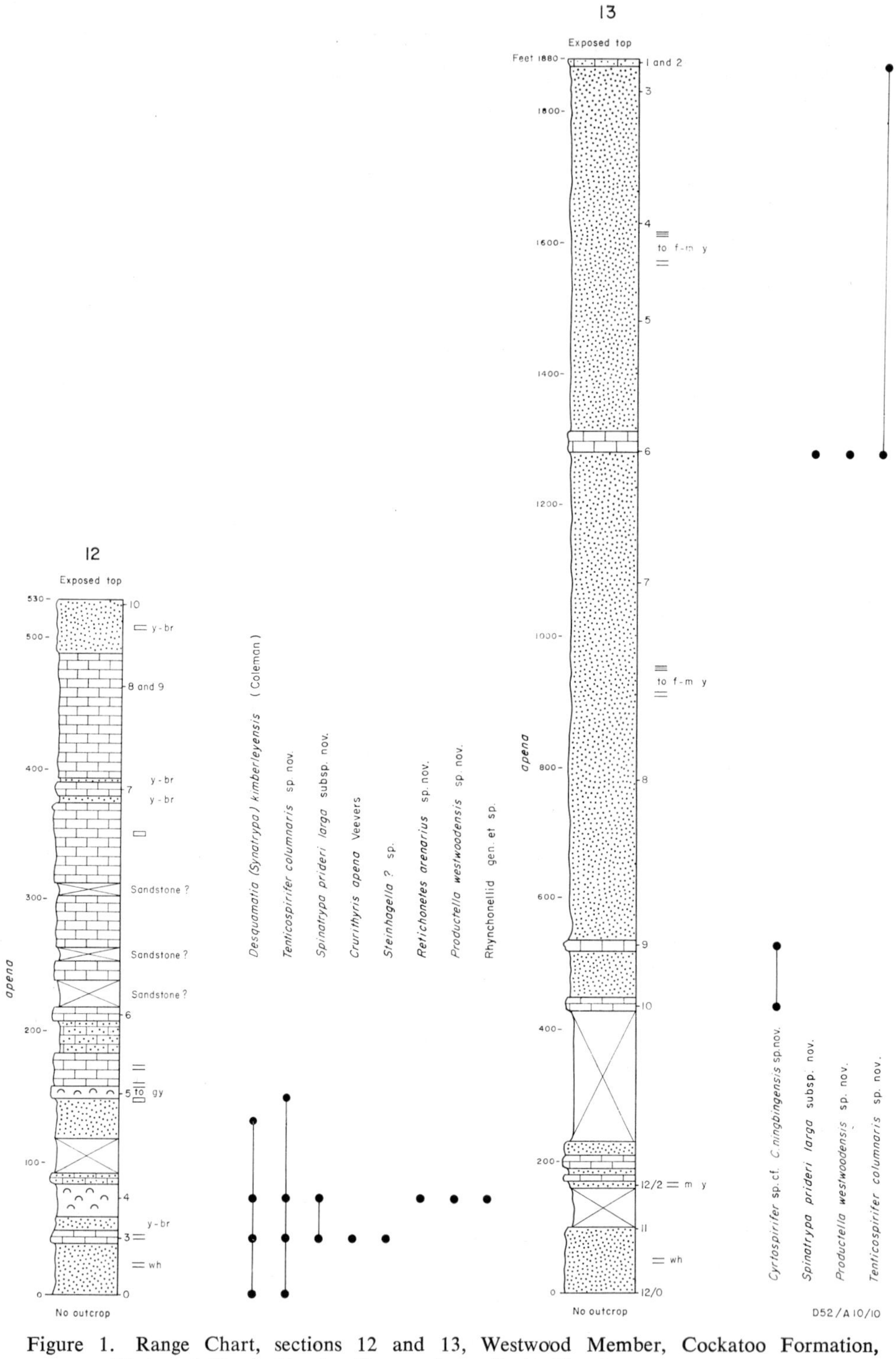

Figure 1. Range Chart, sections 12 and 13, Westwood Member, Cockatoo Formation, Westwood Creek. For locality see Appendix 2, Fig. 5.

Jeremiah Member

21/2B — 2 miles west of section 21 on the western side of the Ningbing Range.
Cyrtospirifer sp. cf. *C. ningbingensis* sp. nov.

Ningbing Limestone

4/1 — Southern margin of Ningbing Range (Appendix 2, Fig. 6). Back-reef.
Cyrtospirifer ningbingensis sp. nov.

8/1 — Surprise Creek, Ningbing Range (Appendix 2, Fig. 6). Fore-reef.
Nayunnella turgida sp. nov.

8/6 — Near Surprise Creek, Ningbing Range (Appendix 2, Fig. 6). Back-reef.
Nayunnella turgida sp. nov.
Cyrtospirifer ningbingensis sp. nov.

9/4 — Near Surprise Creek, Ningbing Range (Appendix 2, Fig. 6). Back-reef.
Rugaltarostrum australe sp. nov.

11/2 — Last outcrop of Ningbing Limestone 4 miles north of Utting Gap on the track to Knob Peak Bore (*see* Veevers & Roberts, 1968).
Sentosia subquadrata sp. nov.
Productella sp.

21/16 — Ningbing Range (Appendix 2, Fig. 6). Back-reef.
Cyrtospirifer ningbingensis sp. nov.

21/22 — Ningbing Range (Appendix 2, Fig. 6). Back-reef.
Cyrtospirifer ningbingensis sp. nov.

21/26 — Ningbing Range (Appendix 2, Fig. 6). Back-reef.
Mesoplica jeremiahensis sp. nov.

140/1 & 2 Southern margin of Ningbing Range (Appendix 2, Fig. 6). Back-reef.
Cyrtospirifer ningbingensis sp. nov.

140/3 — Southern margin of Ningbing Range (Appendix 2, Fig. 6). Back-reef.
Ptychomaletoechia lucida (Veevers)
Cyrtospirifer ningbingensis sp. nov.

140/4 — Southern margin of Ningbing Range (Appendix 2, Fig. 6). Back-reef.
Schuchertella sp. cf. *S. gratillica* Veevers

140/5 — Southern margin of Ningbing Range (Appendix 2, Fig. 6). Back-reef.
Cyrtospirifer ningbingensis sp. nov.

141/1 — Ningbing Range near Surprise Creek (Appendix 2, Fig. 6). Back-reef.
Schuchertella sp. cf. *S. gratillica* Veevers
Productacean gen. et sp. indet.
Cyrtospirifer ningbingensis sp. nov.

141/5 — Ningbing Range (Appendix 2, Fig. 7). Back-reef.
Cyrtospirifer ningbingensis sp. nov.

149/2 — Southeastern margin of Ningbing Range (Appendix 2, Fig. 6). Reef.
Cyrtospirifer ningbingensis sp. nov.

442/12— Jeremiah Hills (Appendix 2, Fig. 8). Back-reef.
Cyrtospirifer ningbingensis sp. nov.

447/11— 1040 feet above the base of section 447, southeastern part of the main outcrops of Ningbing Limestone (back-reef) in Jeremiah Hills.
Cyrtospirifer ningbingensis sp. nov.

455/2 — Eastern margin of Ningbing Range (Appendix 2, Fig. 7). Reef.
Cyrtospirifer ningbingensis sp. nov.

455/5 — Eastern margin of Ningbing Range (Appendix 2, Fig. 7). Inter-reef.
Cyrtospirifer ningbingensis sp. nov.

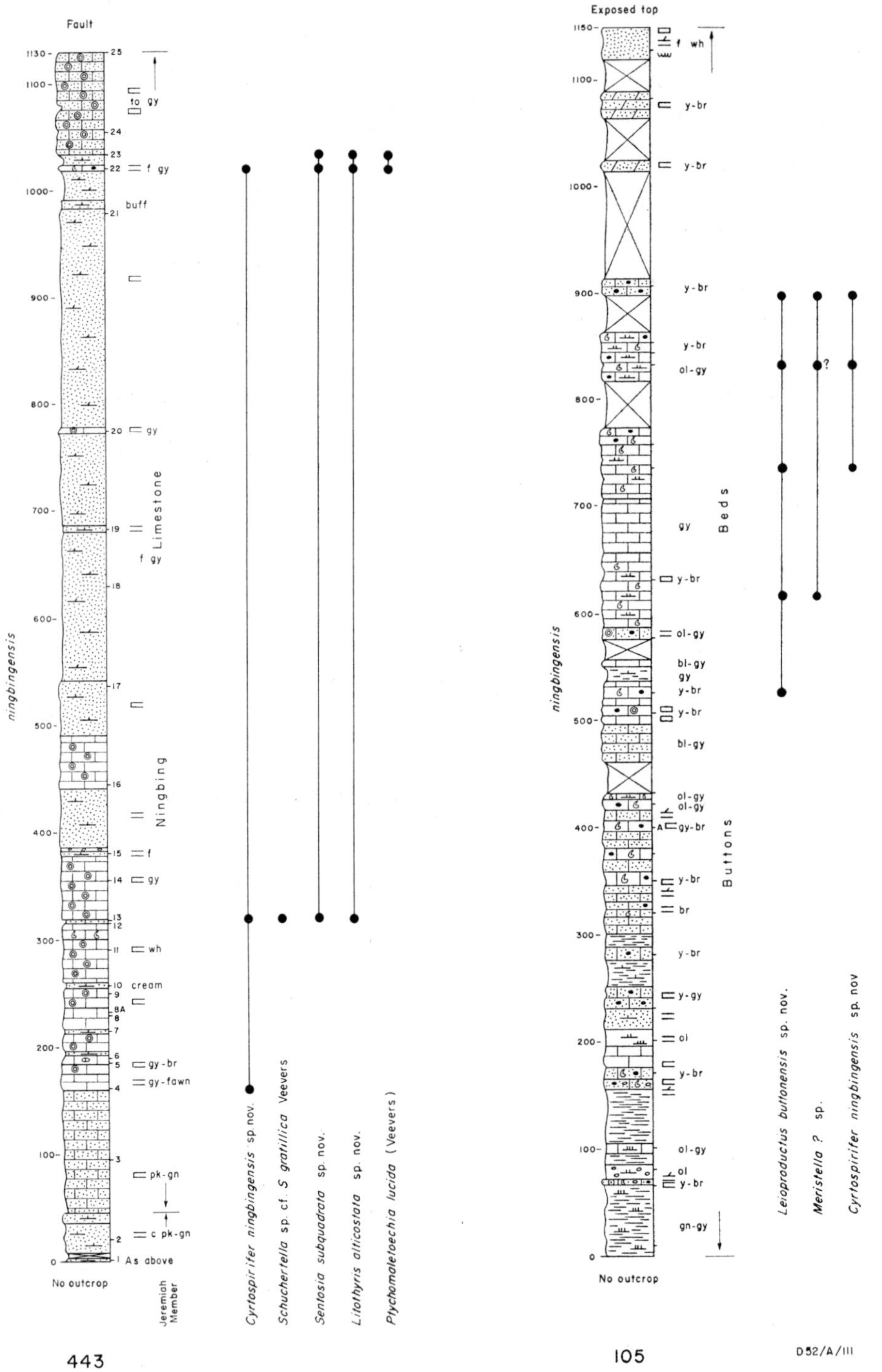

Figure 2. Range chart, sections 443, Ningbing Limestone (back-reef), Jeremiah Hills, and 105, Buttons Beds, Ord River at Buttons Crossing. For localities *see* Appendix 2, Figs 8 and 9.

Buttons Beds

145/5 — 7.5 miles northwest of Mount Septimus (Appendix 2, Fig. 10)
Cyrtospirifer depressus sp. nov.

146/11— 7 miles northwest of Mount Septimus (Appendix 2, Fig. 10)
Productacean gen. et sp. B

CARBONIFEROUS

Ningbing Limestone

7/1 — Southeastern part of Ningbing Range (Appendix 2, Fig. 6). Reef.
Schizophoria sp. cf. *S. resupinata* (Martin)
Leptagonia analoga (Phillips)
Schuchertella peltata sp. nov.
Acanthocosta teicherti gen. et sp. nov.
Rhytiophora sp. cf. *R. calhounensis* (Moore)
Ovatia sp. A
Productacean gen. et sp. A
Ningbingella flexuosa gen. et sp. nov.
Cardiothyris bisulcata gen. et sp. nov.
Spirifer otwayi sp. nov.
Brachythyris planulata sp. nov.
Crassumbo? jonesi sp. nov.
Kitakamithyris occidua sp. nov.

Burt Range Formation

113/5 — Southern margin of Burt Range Amphitheatre (Appendix 2, Fig. 10)
Schuchertella peltata sp. nov.
Spinauris cristata gen. et sp. nov.
Grammorhynchus eganensis (Veevers)
Cleiothyridina minilya Thomas

113/7 — Southern margin of Burt Range Amphitheatre (Appendix 2, Fig. 10)
Spinaurus cristata gen. et sp. nov.
Grammorhynchus eganensis (Veevers)
Cleiothyridina minilya Thomas
Punctospirifer plicatosulcatus Glenister

114/9 — Southeastern Spirit Hill (Appendix 2, Fig. 11)
Lomatiphora aquila gen. et sp. nov.
Cleiothyridina gloveri Thomas

115/3 — Southeastern Spirit Hill (Appendix 2, Fig. 11)
Cleiothyridina gloveri Thomas
Austrochoristites levisulcatus gen. et sp. nov.

115/7 — Southeastern Spirit Hill (Appendix 2, Fig. 11)
Lomatiphora aquila gen. et sp. nov.
Cleiothyridina gloveri Thomas

117/1 — Spirit Hill cave (Appendix 2, Fig. 11)
Lomatiphora aquila gen. et sp. nov.
Rugosochonetes obtectus sp. nov.
Cleiothyridina gloveri Thomas
Austrochoristites levisulcatus gen. et sp. nov.
Kitakamithyris occidua sp. nov.

122/2 — Sandy Creek (Appendix 2, Fig. 12)
Schuchertella dorsiplana Thomas
Rugosochonetes obtectus sp. nov.
Lomatiphora aquila gen. et sp. nov.
Rhytiophora sp. cf. *R. raricostata* (Herrick)
Cleiothyridina gloveri Thomas
Prospira incerta Thomas
Ectochoristites? arenatus Thomas

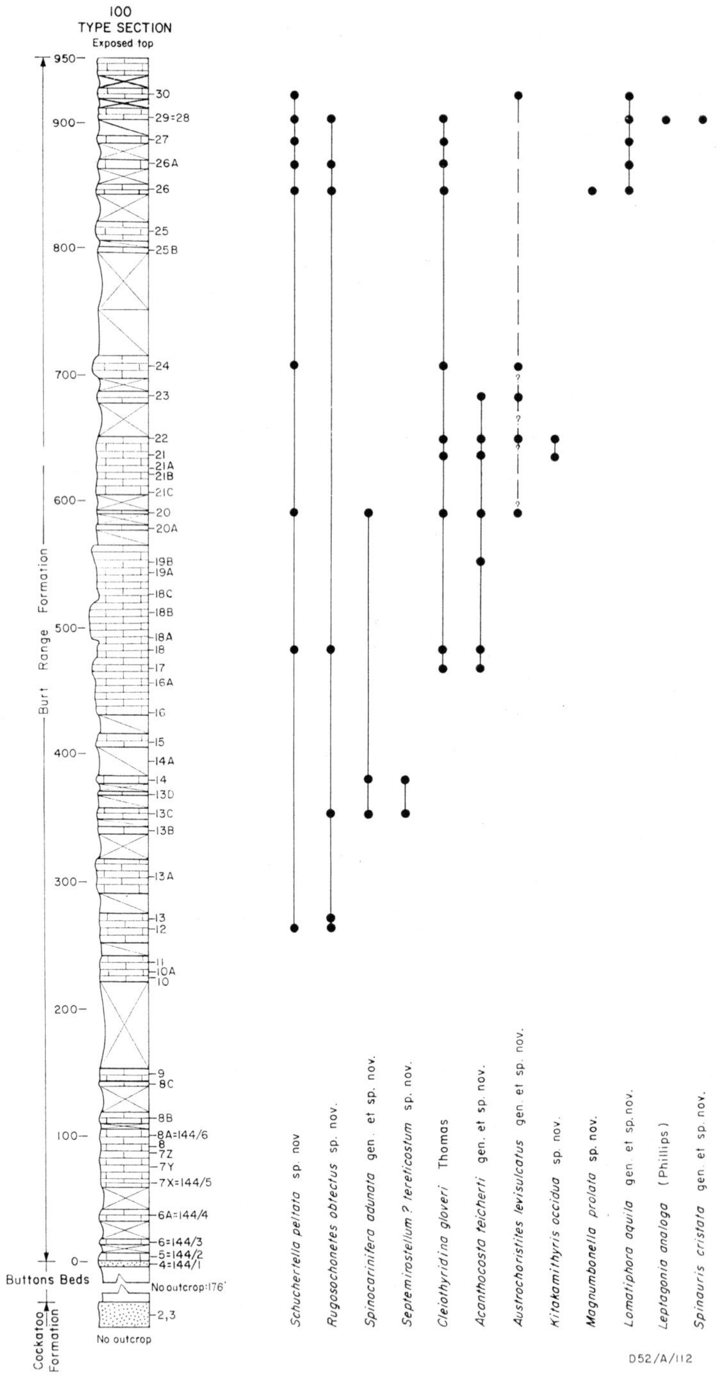

Figure 3. Range chart, section 100, Burt Range Formation. For locality *see* Appendix 2, Fig. 10.

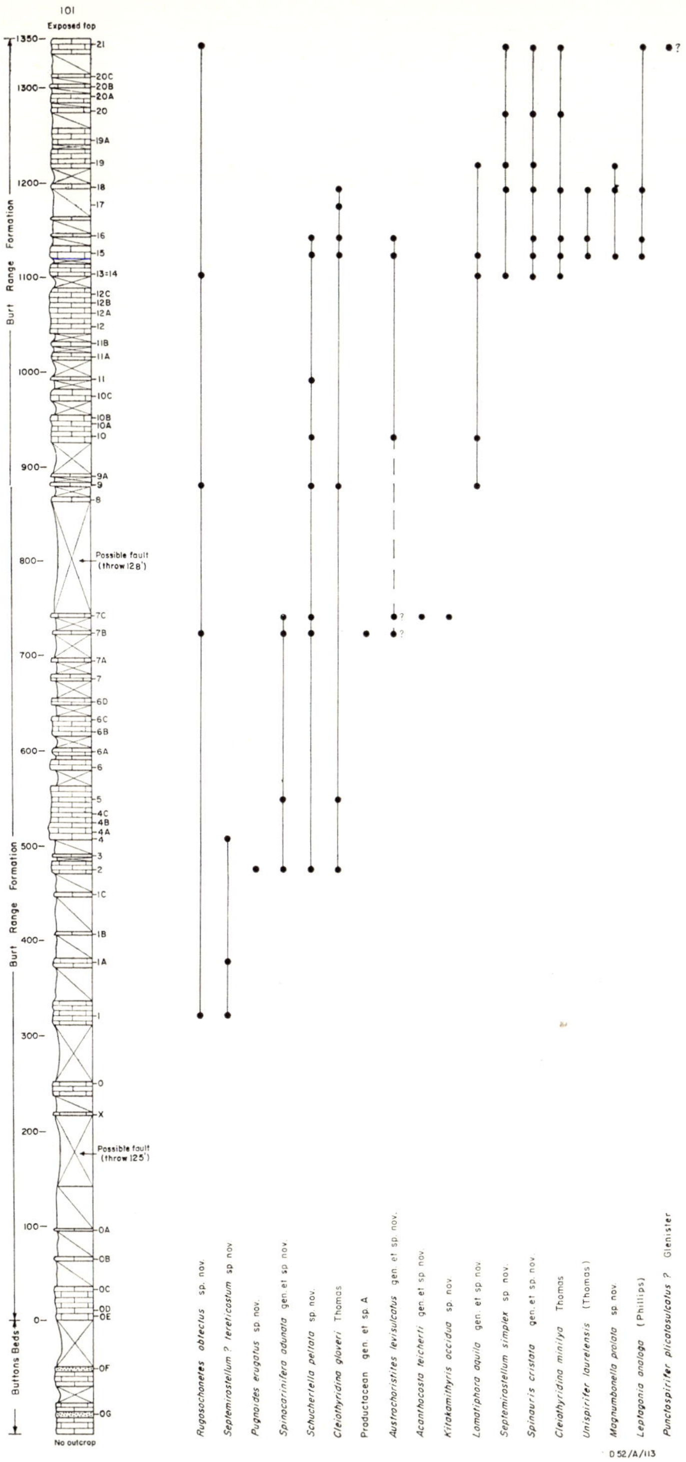

Figure 4. Range chart, section 101, Burt Range Formation. For locality *see* Appendix 2, Fig. 10.

122/3 — Sandy Creek (Appendix 2, Fig. 12)
Lomatiphora aquila gen. et sp. nov.
Cleiothyridina gloveri Thomas
Ectochoristites? arenatus Thomas

123/14— 11 miles southwest of Legune Homestead (Appendix 2, Fig. 12)
Schuchertella sp.
Rugosochonetes obtectus sp. nov.
Lomatiphora aquila gen. et sp. nov.
Prospira incerta Thomas

123/16— 10 miles southwest of Legune Homestead (Appendix 2, Fig. 12)
Leptagonia analoga (Phillips)
Prospira incerta Thomas

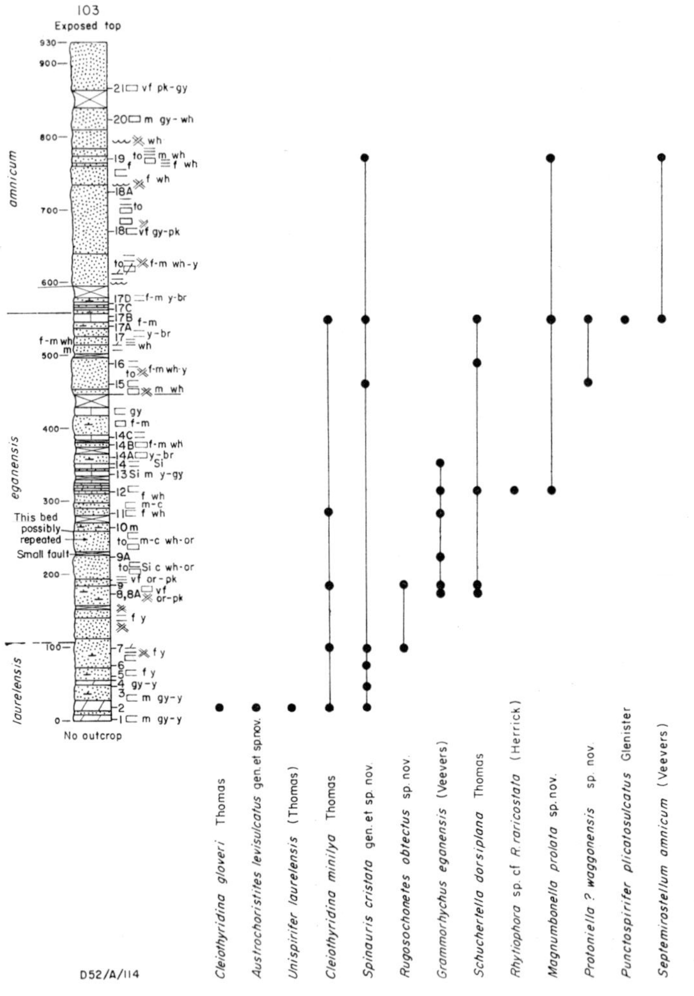

Figure 5. Range chart, section 103, Burt Range Formation — Enga Sandstone. For locality *see* Appendix 2, Fig. 10.

124/1 — Flapper Hill, 4 miles northwest of Legune Homestead
Schuchertella dorsiplana Thomas
Rugosochonetes obtectus sp. nov.
?Lomatiphora aquila gen. et sp. nov.
Septemirostellum sp.
Cleiothyridina minilya Thomas
Brachythyris sp.
Austrochoristites levisulcatus gen. et sp. nov.

126/5 — Sorby Hills (Veevers & Roberts, 1968, fig. 37)
Lomatiphora aquila gen. et sp. nov.
Septemirostellum sp.
Productacean gen. et sp. indet.
Cleiothyridina gloveri Thomas

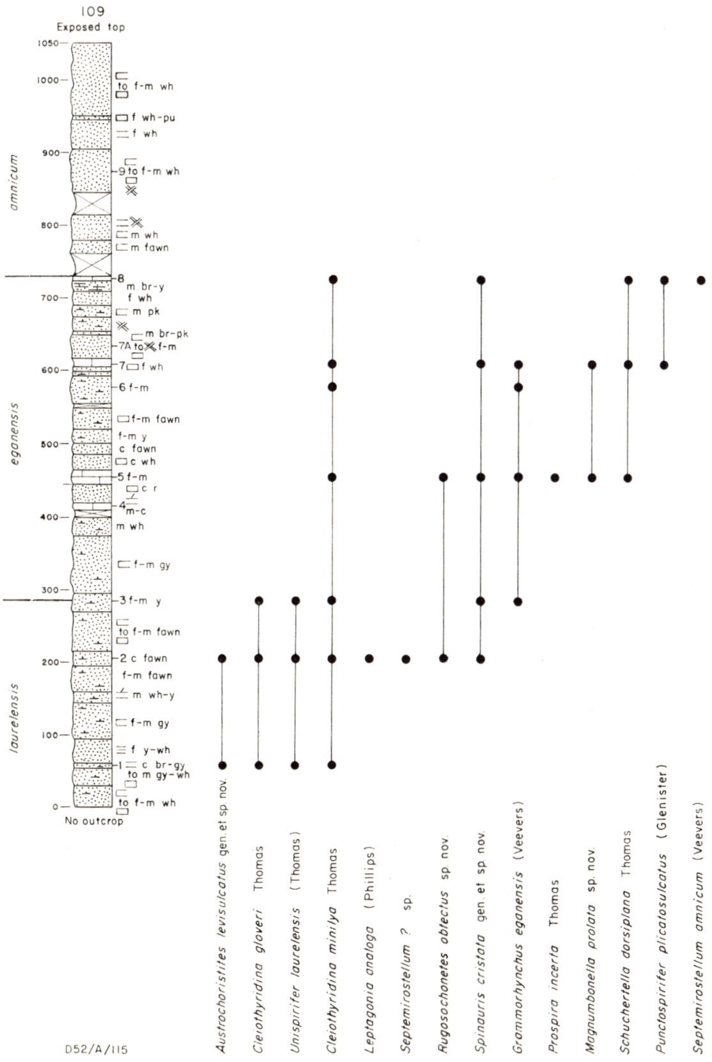

Figure 6. Range chart, section 109, Burt Range Formation — Enga Sandstone. For locality *see* Appendix 2, Fig. 10.

133/2 — Alpha Hill (Veevers & Roberts, 1968, fig. 18)
Schuchertella dorsiplana Thomas
Septemirostellum simplex sp. nov.
Cleiothyridina gloveri Thomas
Cleiothyridina minilya Thomas
Prospira incerta Thomas

Unnamed Tournaisian breccia

210/6 — Eastern outlier in Waggon Creek valley (Veevers & Roberts, 1966, fig. 1)
Schizophoria sp. cf. *S. resupinata* (Martin)
Rhipidomella michelini? (Léveillé)
Leptagonia analoga (Phillips)
Rugosochonetes obtectus sp. nov.
Magnumbonella prolata sp. nov.
Protoniella? waggonensis sp. nov.
Spinauris cristata? gen. et sp. nov.

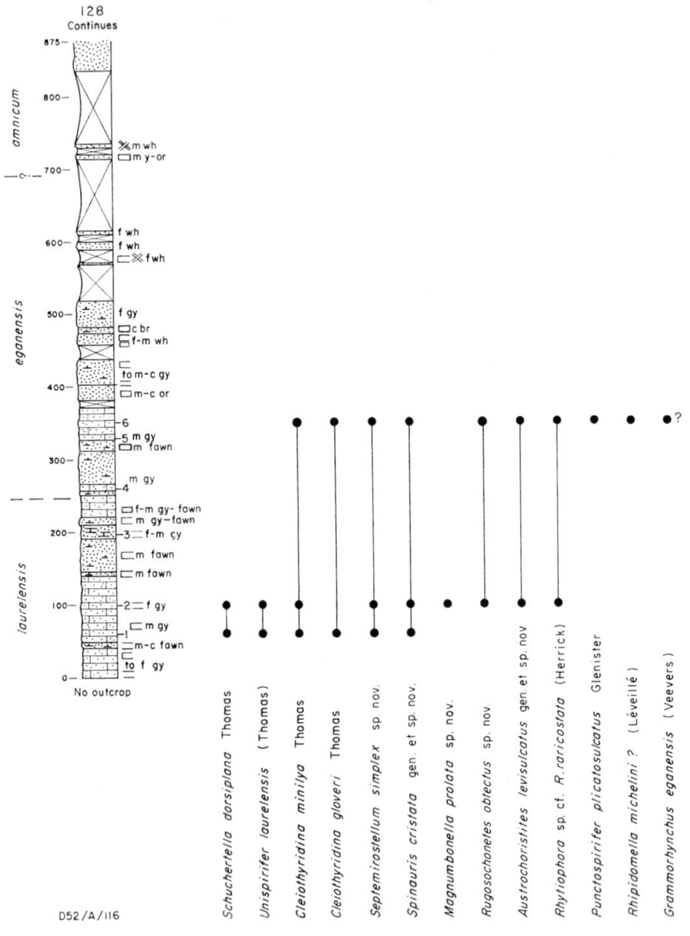

Figure 7. Range chart, section 128, Burt Range Formation. For locality *see* Appendix 2, Fig. 10.

Septemirostellum amnicum (Veevers)
Eomartiniopsis costata sp. nov.
Prospira incerta Thomas
Cyrtina sp.
Composita sp.
Cleiothridina sp.

Enga Sandstone

136/1 — Possible equivalent of the Enga Sandstone at Spirit Hill (Appendix 2, Fig. 11)
Rhipidomella michelini? (Léveillé)
Rugosochonetes obtectus sp. nov.
Septemirostellum sp.
Cleiothyridina sp.
Austrochoristites levisulcatus gen. et sp. nov.
Punctospirifer plicatosulcatus Glenister
Eomartiniopsis costata sp. nov.

139/1 &
139/2 — Eastern slope of Enga Ridge (Appendix 2, Fig. 10)
Septemirostellum amnicum (Veevers)

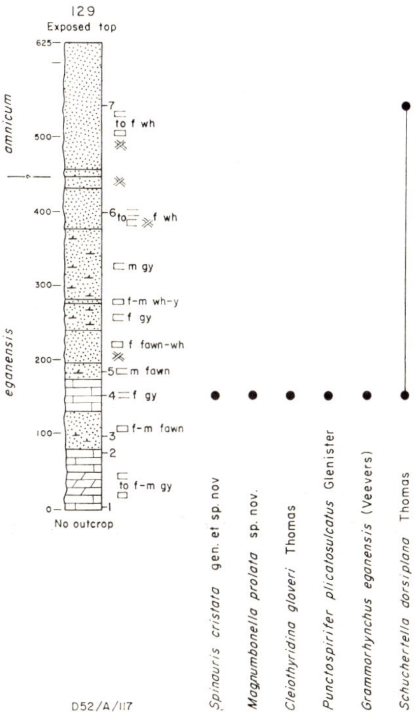

Figure 8. Range chart, section 129, Burt Range Formation — Enga Sandstone. For locality *see* Appendix 2, Fig. 10.

T

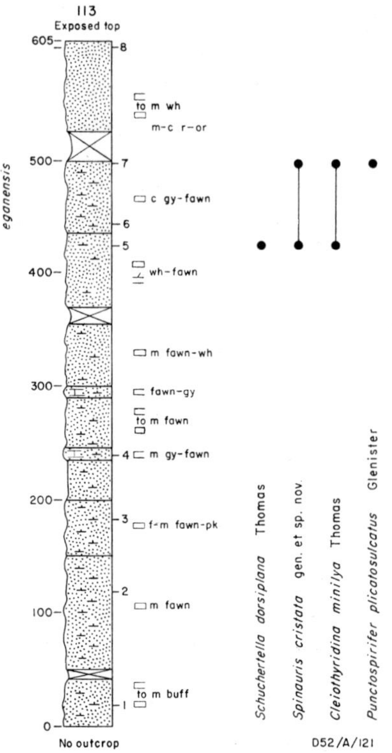

Figure 9. Range chart, section 113, Burt Range Formation — Enga Sandstone. For locality *see* Appendix 2, Fig. 10.

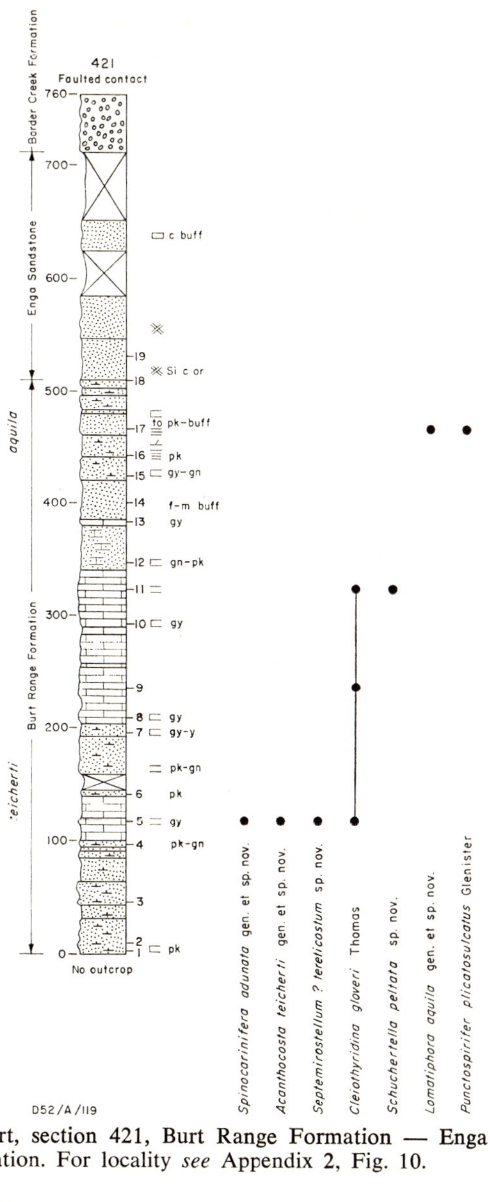

Figure 10. Range chart, section 421, Burt Range Formation — Enga Sandstone — Border Creek Formation. For locality *see* Appendix 2, Fig. 10.

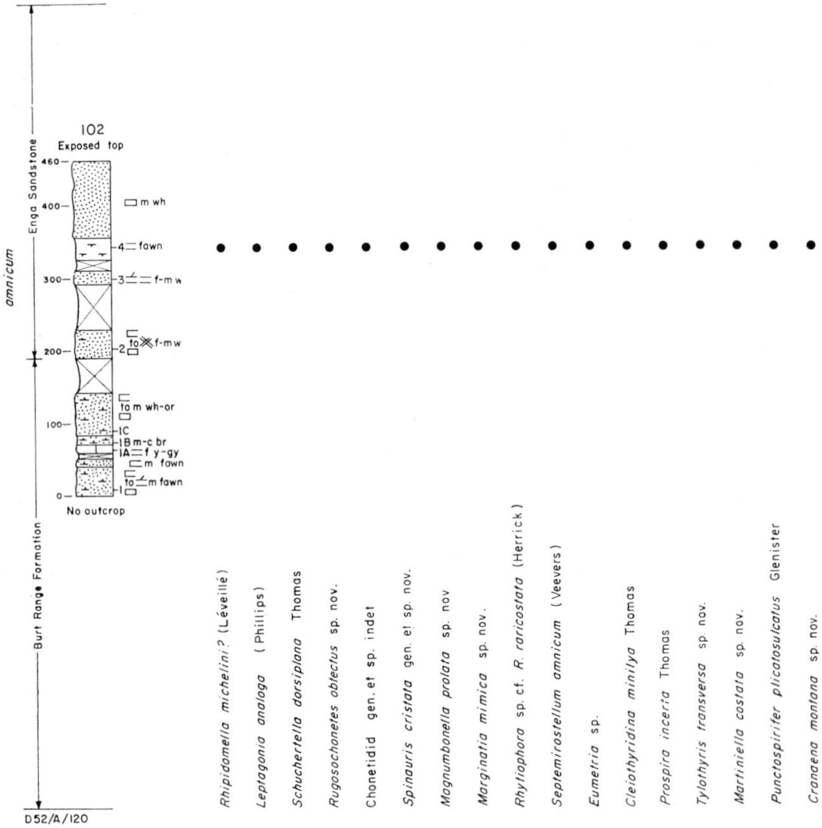

Figure 11. Range chart, section 102, Burt Range Formation — Enga Sandstone. For locality *see* Appendix 2, Fig. 10.

Septimus Limestone

138/1 — Southern margin of Burt Range Amphitheatre (Appendix 2, Fig. 10)
Schizophoria sp. cf. *S. resupinata* (Martin)
Schellwienella australis Thomas
Septemirostellum septimum (Veevers)
Composita bonapartensis Thomas
Torynifer dorsiseptatus Thomas
Punctospirifer uttingi Thomas

138/2 — Southern margin of Burt Range Amphitheatre (Appendix 2, Fig. 10)
Schellwienella australis Thomas
Septemirostellum septimum (Veevers)
Composita bonapartensis Thomas
Unispirifer septimus Thomas
Syringothyris langfieldensis sp. nov.
Torynifer? dorsiseptatus Thomas
Punctospirifer uttingi Thomas

416/4 — 2 miles east of Mount Septimus (Appendix 2, Fig. 10)
Schizophoria sp. cf. *S. resupinata* (Martin)
Rhipidomella michelini? (Léveillé)
Productacean gen. et sp. indet.
Spirifer spiritus Thomas
Unispirifer septimus Thomas

416/6 — 2 miles east of Mount Septimus (Appendix 2, Fig. 10)
Rhipidomella michelini? (Léveillé)
Leptagonia analoga (Phillips)
Schellwienella australis Thomas
Productacean gen. et sp. indet.
Septemirostellum septimum (Veevers)
Composita bonapartensis Thomas
Cleiothyridina minilya Thomas
Punctospirifer uttingi Thomas

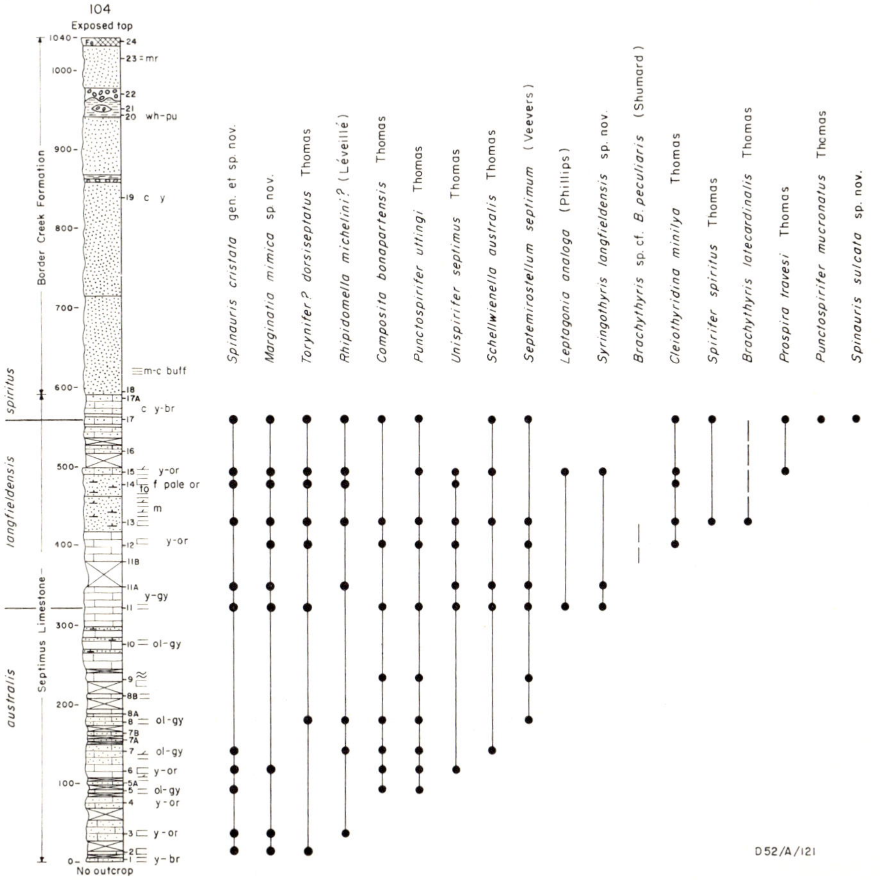

Figure 12. Range chart, section 104, Septimus Limestone — Border Creek Formation at Mount Septimus. For locality *see* Appendix 2, Fig. 10.

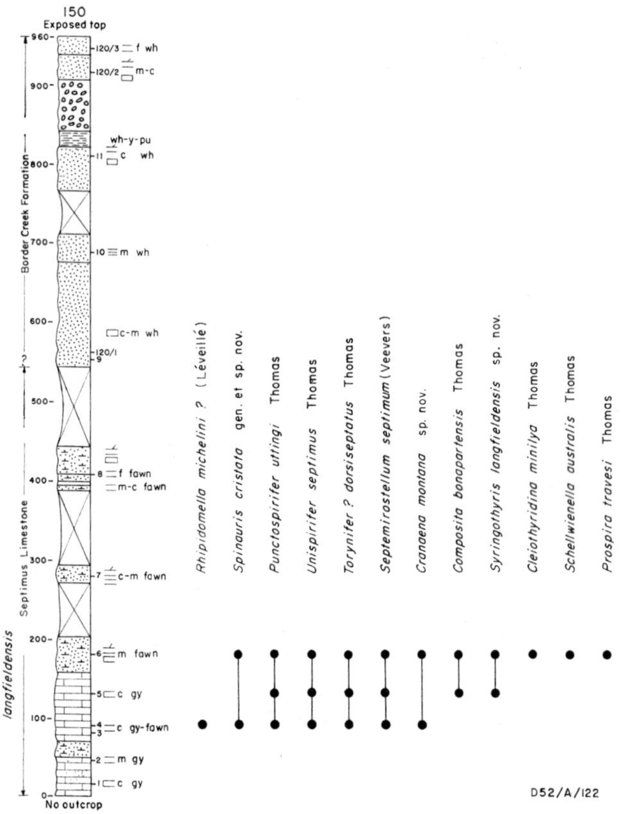

Figure 13. Range chart, section 150, Septimus Limestone — Border Creek Formation at Mount Septimus. For locality *see* Appendix 2, Fig. 10.

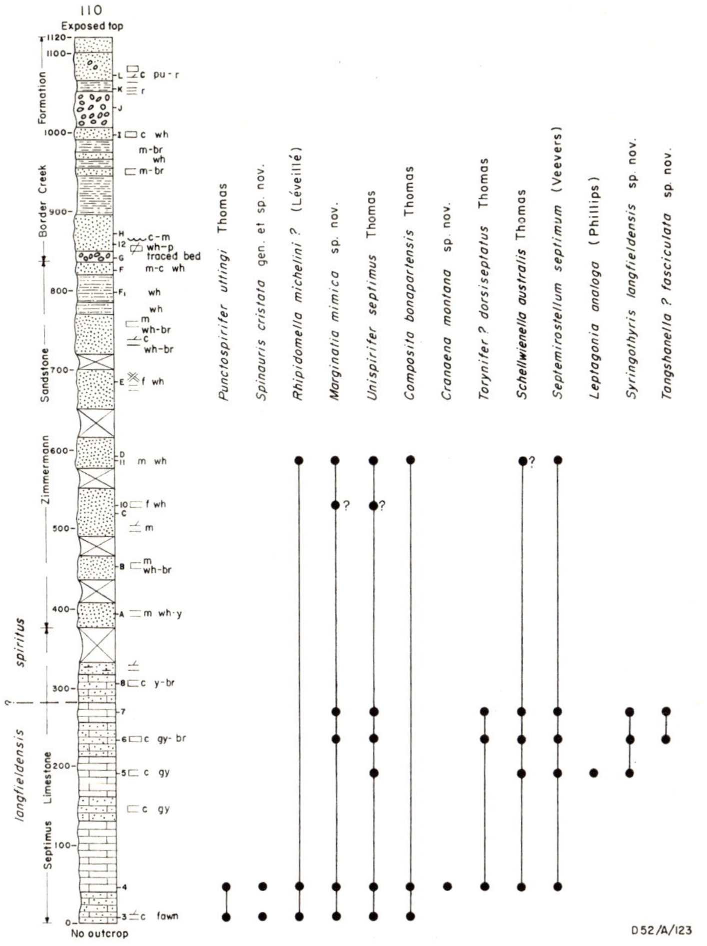

Figure 14. Range chart, section 110, Septimus Limestone — Zimmermann Sandstone — Border Creek Formation at Mount Zimmermann. For locality *see* Appendix 2, Fig. 10.

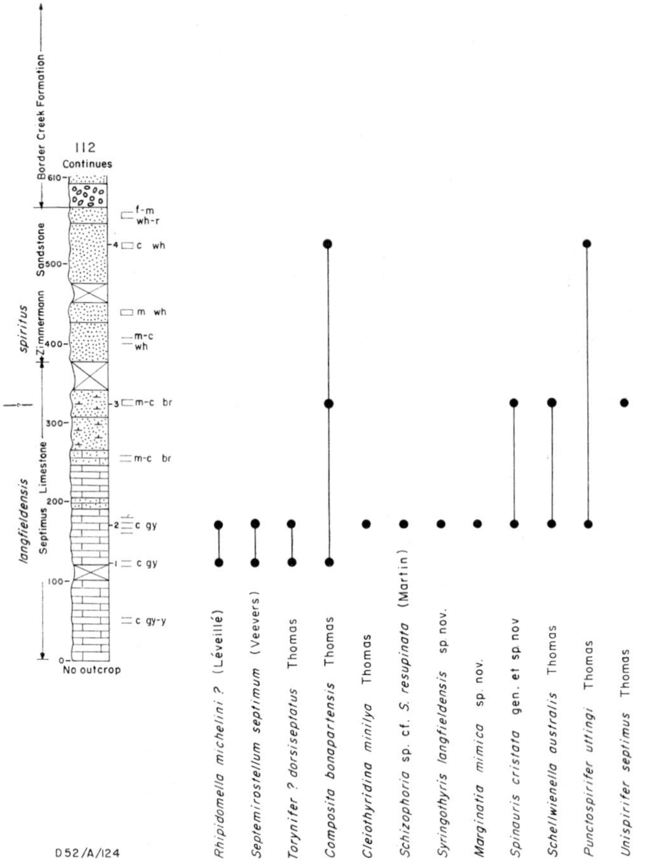

Figure 15. Range chart, section 112, Septimus Limestone — Zimmermann Sandstone — Border Creek Formation at Langfield Point. For locality *see* Appendix 2, Fig. 10.

Milligans Beds

73/2 — Southern margin of Spirit Hill (Appendix 2, Fig. 11)
Schuchertella sp.
Rugosochonetes sp.
Spinauris cristata gen. et sp. nov.
Dictyoclostus? funiferus sp. nov.
Elythidid gen. et sp.

Utting Calcarenite

143/1 — Utting Gap (Appendix 2, Fig. 13)
Schizophoria sp. cf. *S. resupinata* (Martin)
Rhipidomella michelini? (Léveillé)
Leptagonia analoga (Phillips)
Rugosochonetes macgregori sp. nov.
Delepinea uttingi Thomas
Dictyoclostus? funiferus sp. nov.
Cleiothyridina minilya Thomas
Podtsheremia? humilicostata sp. nov.
Podtsheremia? thomasi sp. nov.

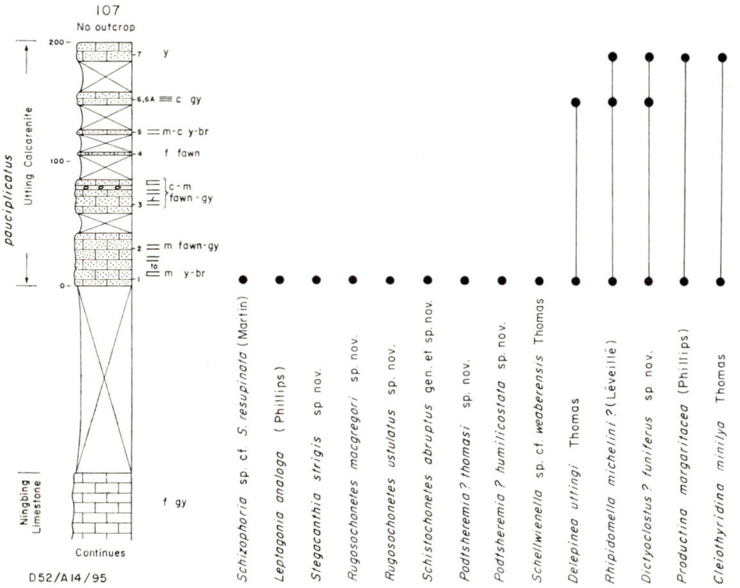

Figure 16. Range chart, section 107, Utting Calcarenite, Utting Gap. For locality *see* Appendix 2, Fig. 13.

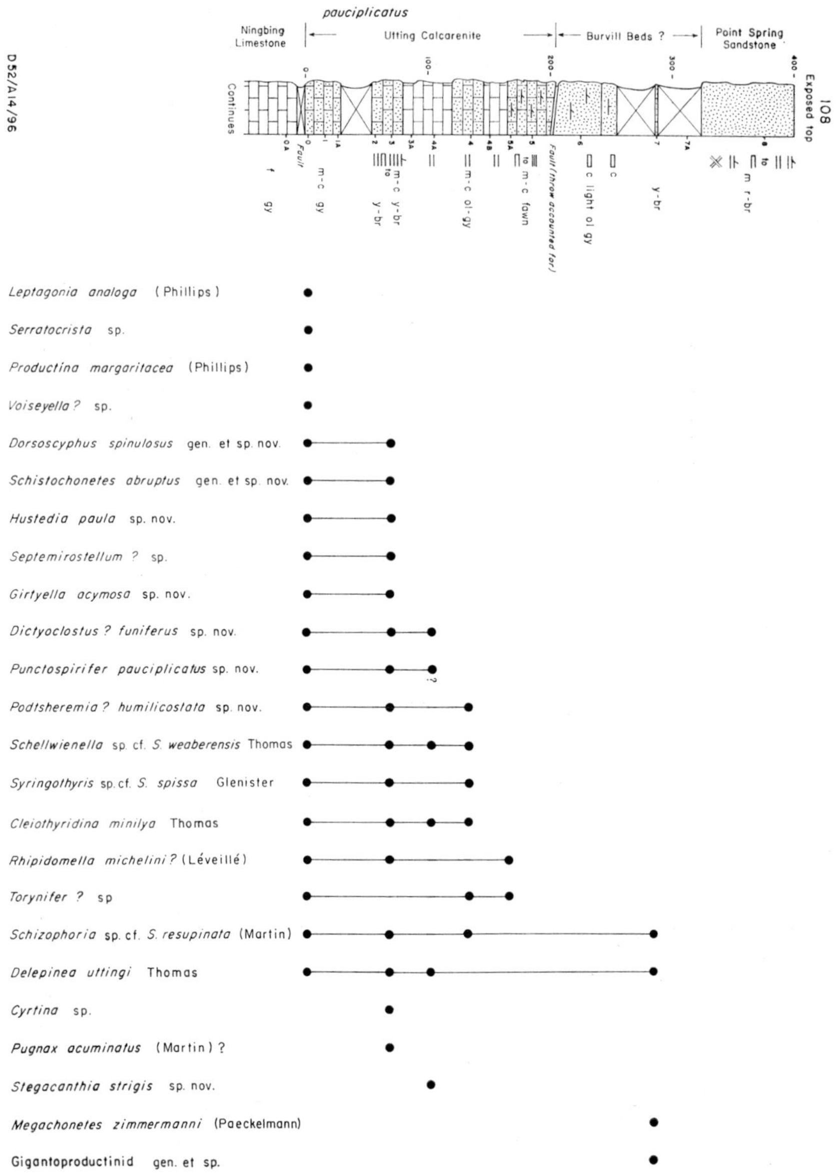

Figure 17. Range chart, section 108, Utting Calcarenite — Burvill Beds — Point Spring Sandstone at Utting Gap. For locality *see* Appendix 2, Fig. 13.

TABLE 1. FAUNA OF THE BURVILL BEDS AND WAGGON CREEK BRECCIA

Localities of the collections are given in Appendix 2; 108/7 — Fig. 13; 5/2, 305/1, 306, 453/1, 453/5, and 453/7 — Fig. 6; 25/0 and 435/0 — Fig. 14; 301 — Fig. 10. Locality 446/1 is in the Waggon Creek Breccia in Waggon Creek valley (Veevers & Roberts, 1966, fig. 1).

	108/7	5/2	305/1	306	453/1	453/5	453/7	25/0	435/0	301	446/1
Rhipidomella michelini? (Léveillé)	—	×	—	×	—	×	—	×	—	—	×
Schizophoria sp. cf. *S. resupinata* (Martin)	×	—	×	—	—	—	—	—	×	—	—
Leptagonia analoga (Phillips)	—	—	—	×	—	—	—	—	—	—	—
Schellwienella weaberensis Thomas	—	×	×	×	×	×	×	×	×	×	—
Orthotetes sp.	—	—	—	×	—	—	—	—	—	—	—
Rugosochonetes sp. cf. *R. kennedyensis* Maxwell	—	—	—	×	×	×	—	×	×	—	—
Globosochonetes burvillensis sp. nov.	—	—	×	—	—	—	—	—	—	—	—
Delepinea uttingi Thomas	×	—	—	—	—	—	—	—	—	—	—
Megachonetes zimmermanni (Paeckelmann)	×	—	—	—	—	—	—	—	—	—	—
Stegacanthia strigis sp. nov.	—	×	×	×	—	—	—	—	×	—	—
Dictyoclostus? funiferus sp. nov.	—	×	×	×	—	×	—	×	×	×	×
Echinoconchus gradatus Campbell	—	—	—	×	—	×	—	—	—	—	—
Ovatia sp. B	—	—	—	—	—	—	—	—	—	×	—
Gigantoproductinid gen. et sp.	—	—	×	—	—	—	—	—	×	—	—
Pleuropugnoides sp.	—	—	—	×	—	—	—	—	—	—	—
Spirifer sp.	—	—	—	×	—	—	—	—	—	—	—
Podtsheremia? humilicostata sp. nov.	—	—	×	×	—	—	—	—	—	—	—
Podtsheremia? thomasi sp. nov.	—	×	—	×	—	—	—	—	—	×	×
Anthracospirifer milliganensis Thomas	—	—	—	—	—	—	—	—	—	×	—
Syringothyris fontanalis sp. nov.	—	—	—	—	—	×	—	×	×	—	—
Brachythyris sp. cf. *B. peculiaris* (Shumard)	—	—	—	—	—	—	—	—	—	—	×
Composita sp.	—	×	—	—	—	—	—	—	—	—	—
Composita bonapartensis Thomas	—	—	—	—	—	—	—	—	—	×	×
Cleiothyridina? sp.	—	—	×	—	—	—	—	×	×	—	×
Punctospirifer sp.	—	—	—	—	—	—	—	—	—	—	×
Torynifer? sp.	—	—	—	—	—	—	—	—	—	—	×

Point Spring Sandstone

435/400— 5 miles northeast of Point Spring near the top of Weaber Range (Appendix 2, Fig. 14)
Gigantoproductinid gen. et sp.

436 — 250 feet above the base of the Point Spring Sandstone, 2.5 miles east-northeast of Point Spring, Weaber Range (Appendix 2, Fig. 14)
Schellwienella weaberensis Thomas
Dictyoclostus? funiferus sp. nov.

458/12— 4.25 miles northeast of Point Spring near the top of Weaber Range (Appendix 2, Fig. 14)
Echinoconchus gradatus Campbell
Ovatia sp. B
Brachythyris sp. cf. *B. peculiaris* (Shumard)
Punctospirifer sp.

APPENDIX 2. LOCALITY MAPS, BONAPARTE GULF BASIN

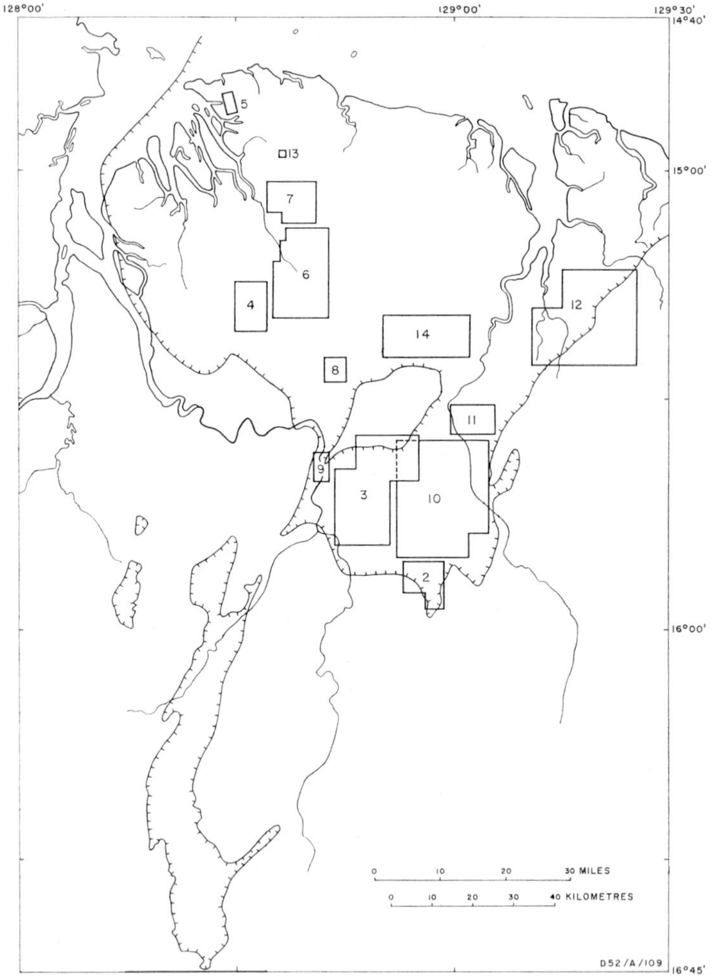

Figure 1. The southwestern part of the Bonaparte Gulf Basin showing the location of detailed geological maps.

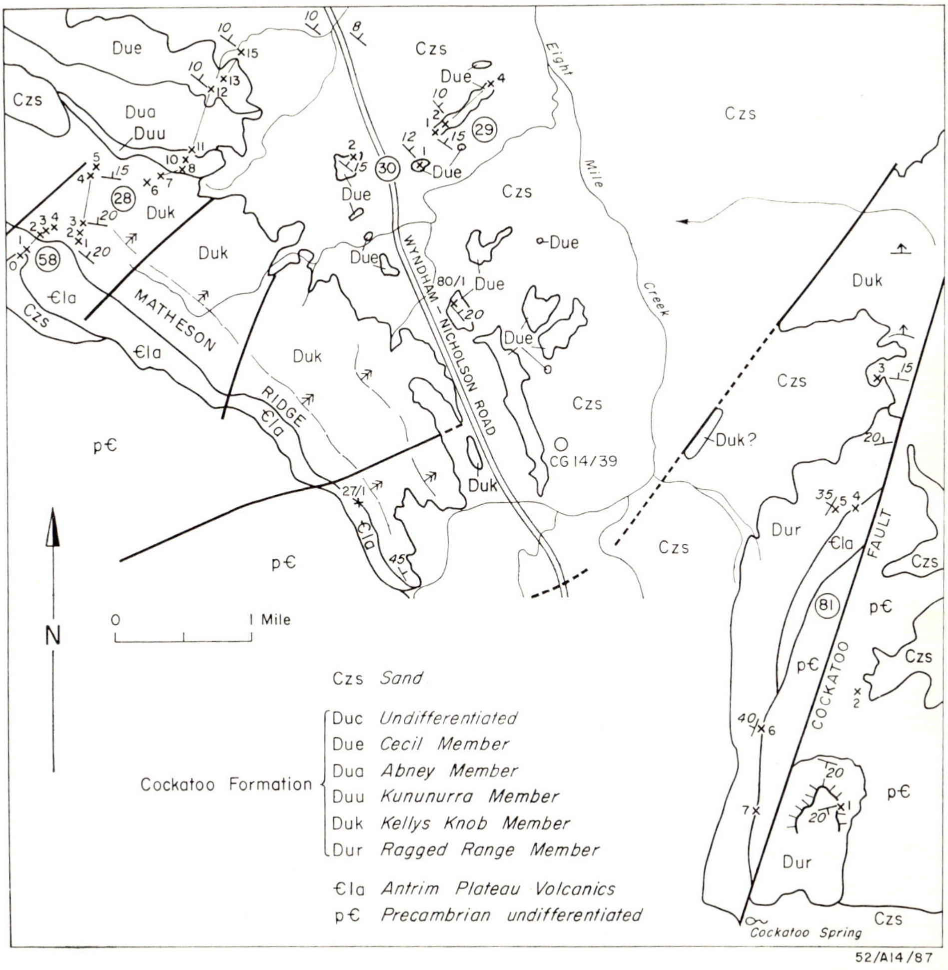

Figure 2. Geological map of the Matheson Ridge area.

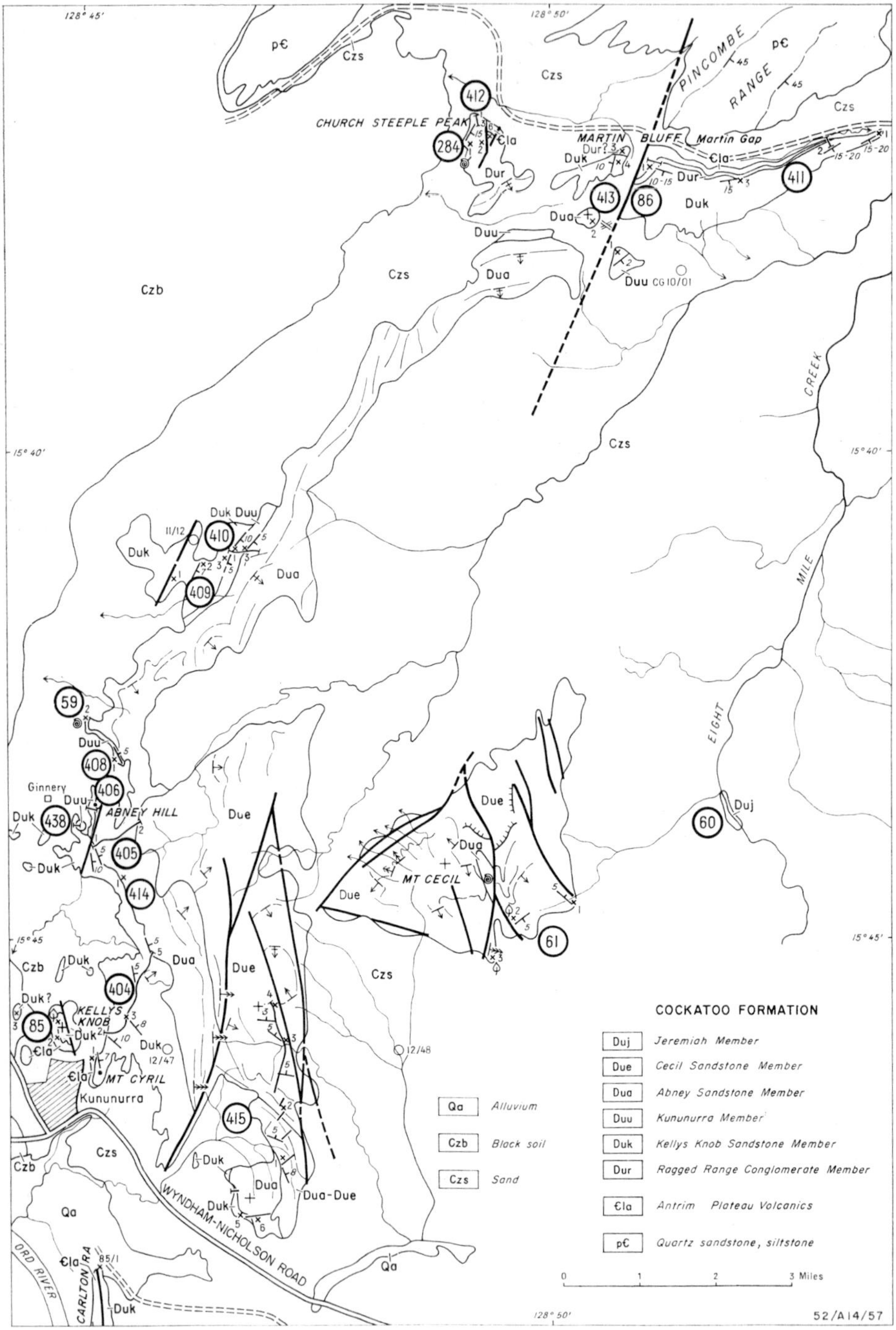

Figure 3. Geological map of the Kununurra-Mount Cecil-Martin Bluff area.

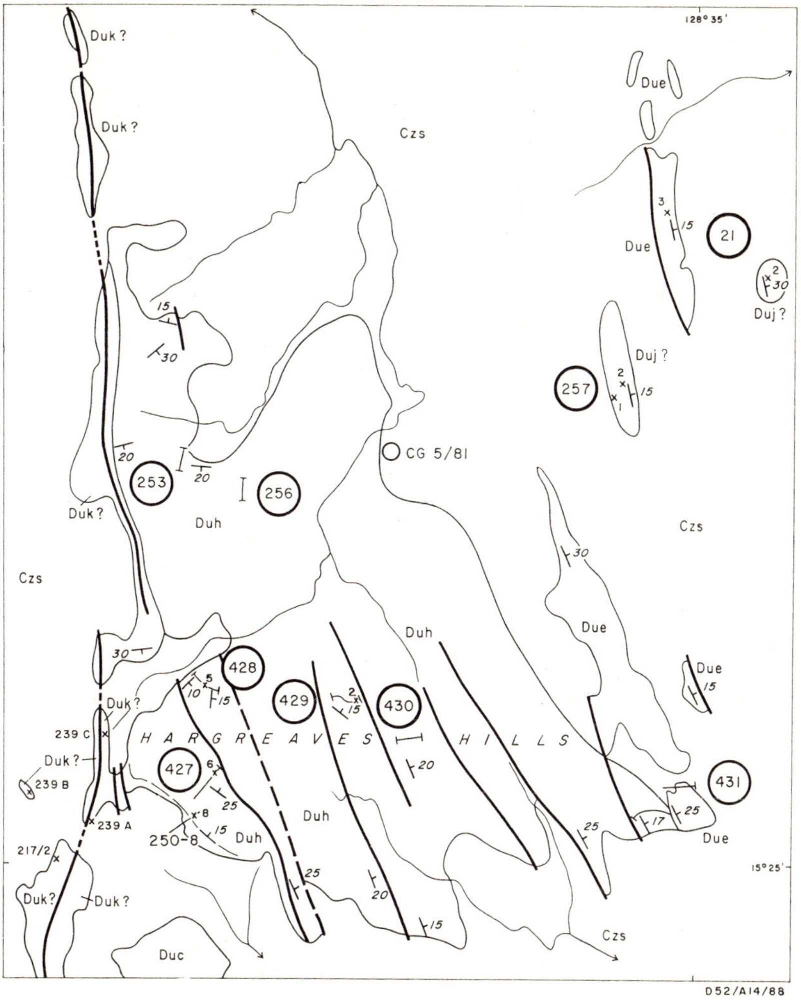

Cockatoo Formation

- Duj Jeremiah Member
- Due Cecil Sandstone Member
- Duh Hargreaves Member
- Duk Kellys Knob Sandstone Member
- Duc *Undifferentiated*

0 1 MILE

Figure 4. Geological map of the Hargreaves Hills area.

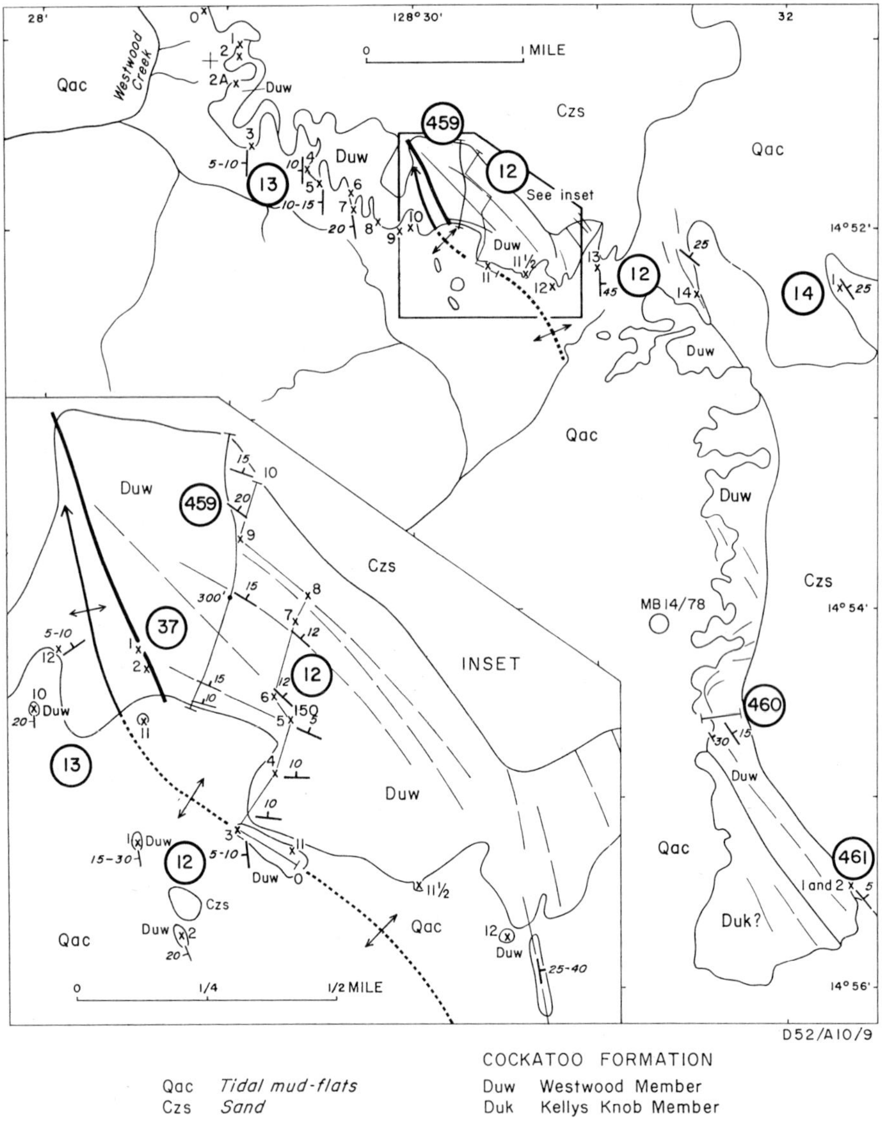

Figure 5. Geological map of the Westwood Creek area.

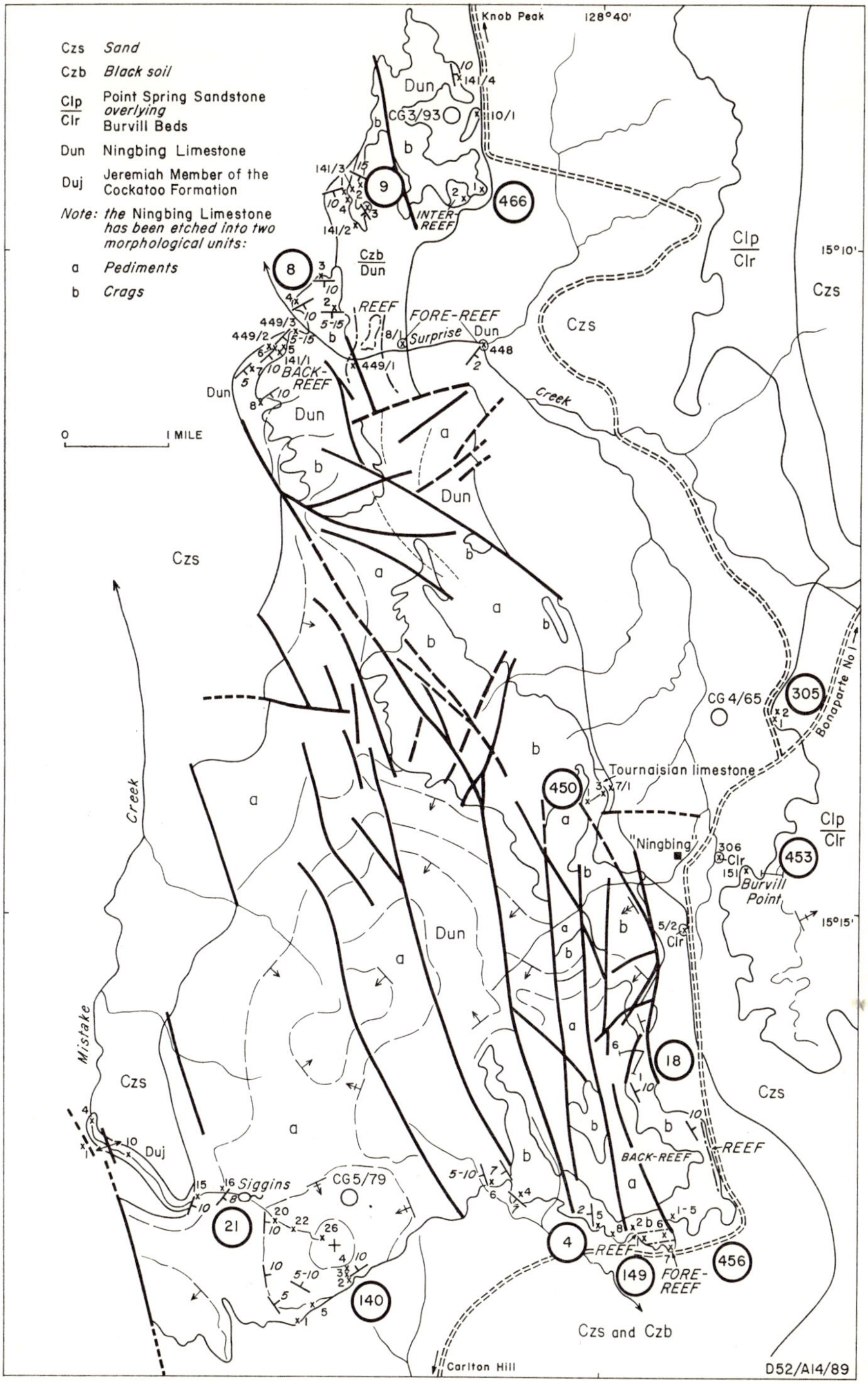

Figure 6. Geological map of the southern part of the Ningbing Range.

U

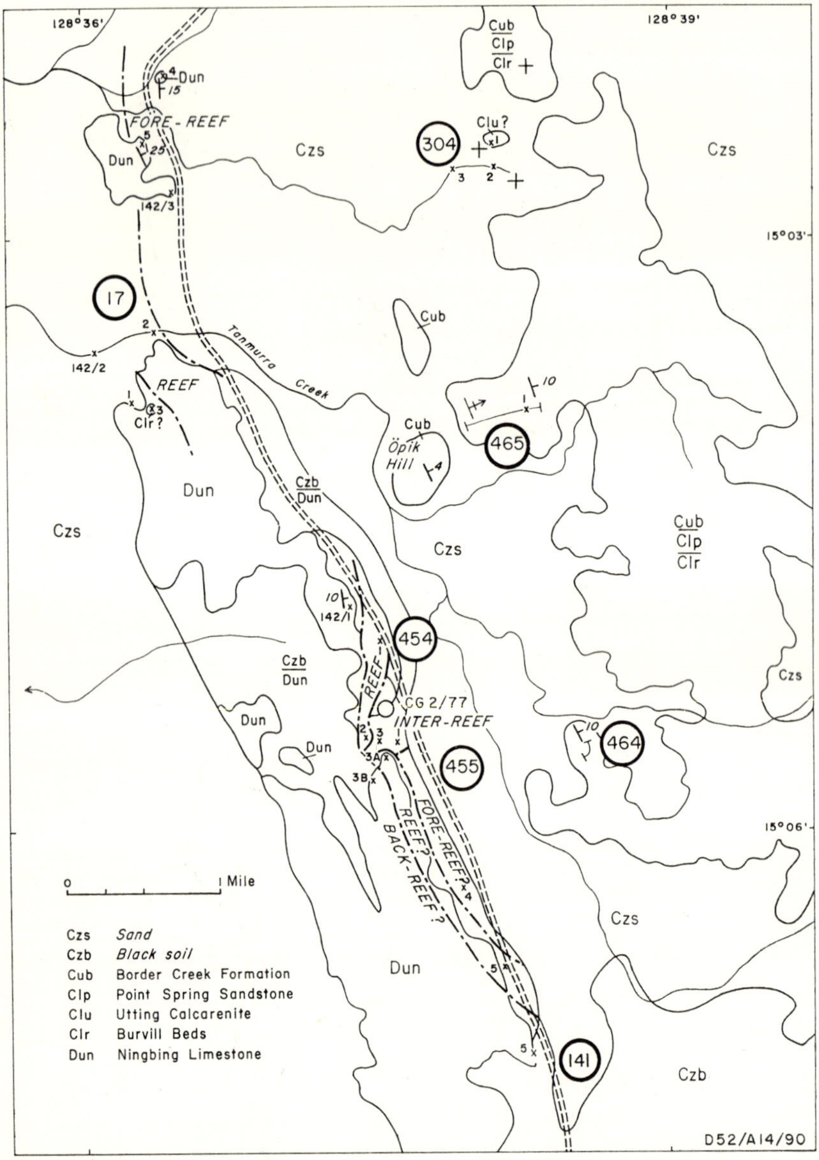

Figure 7. Geological map of the central part of the Ningbing Range.

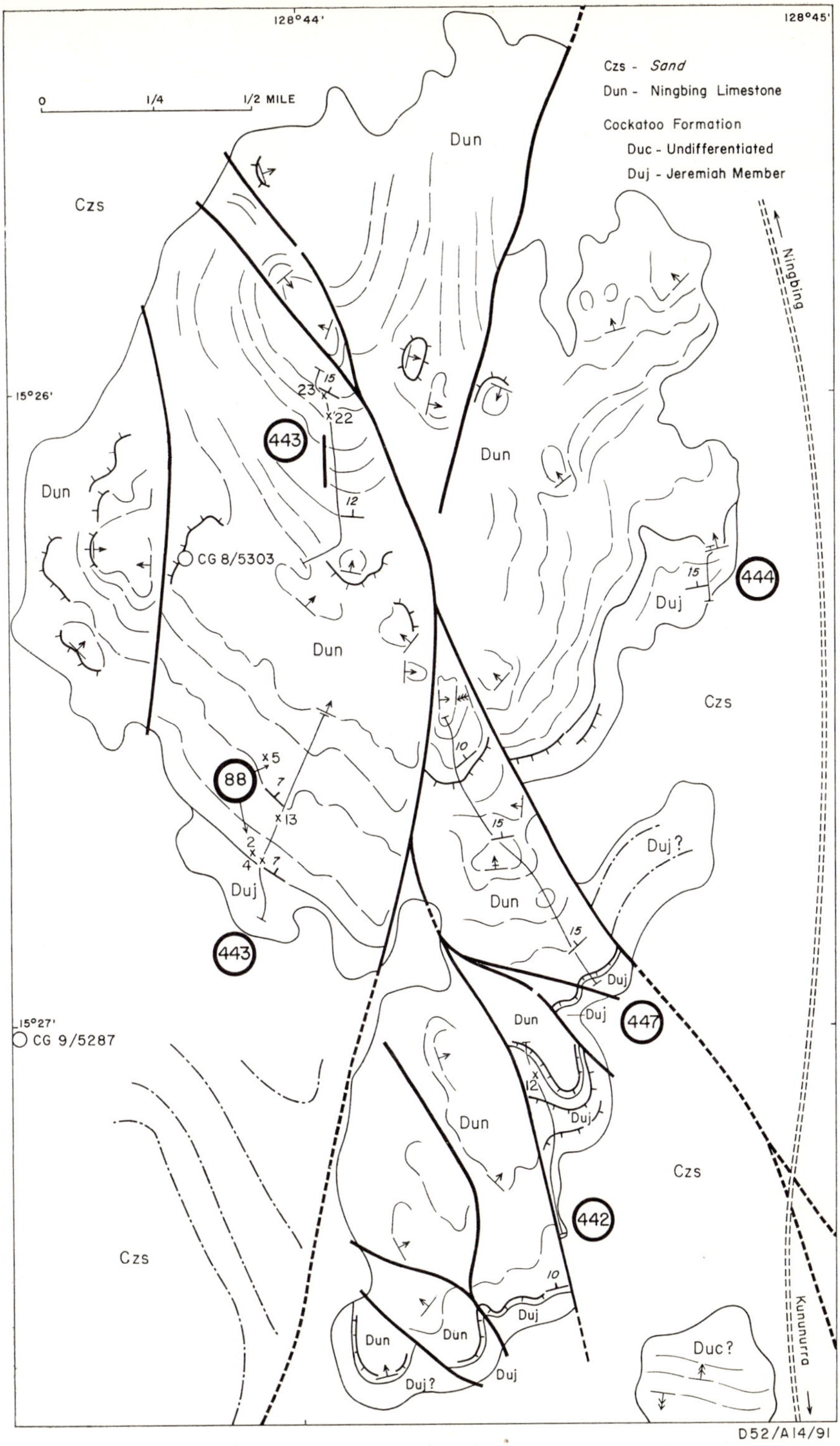

Figure 8. Geological map of the Jeremiah Hills.

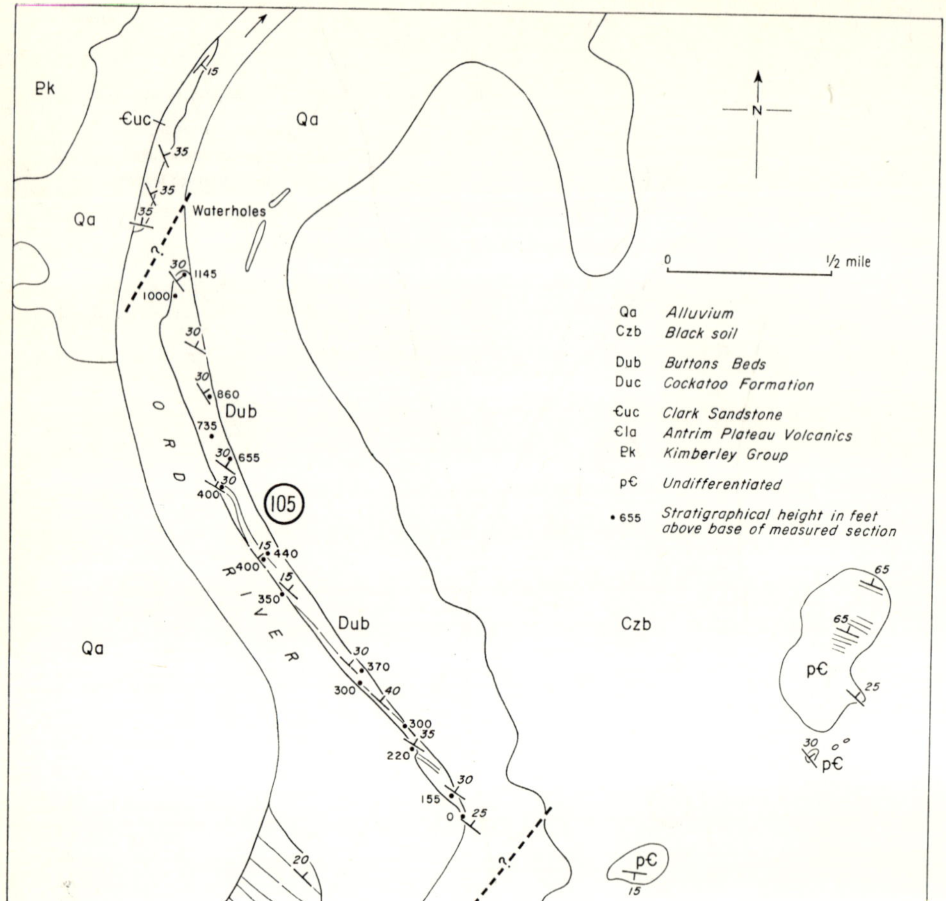

D52/A14/36

Figure 9. Geological map of the Buttons Crossing area on the Ord River, downstream from Kimberley Research Station.

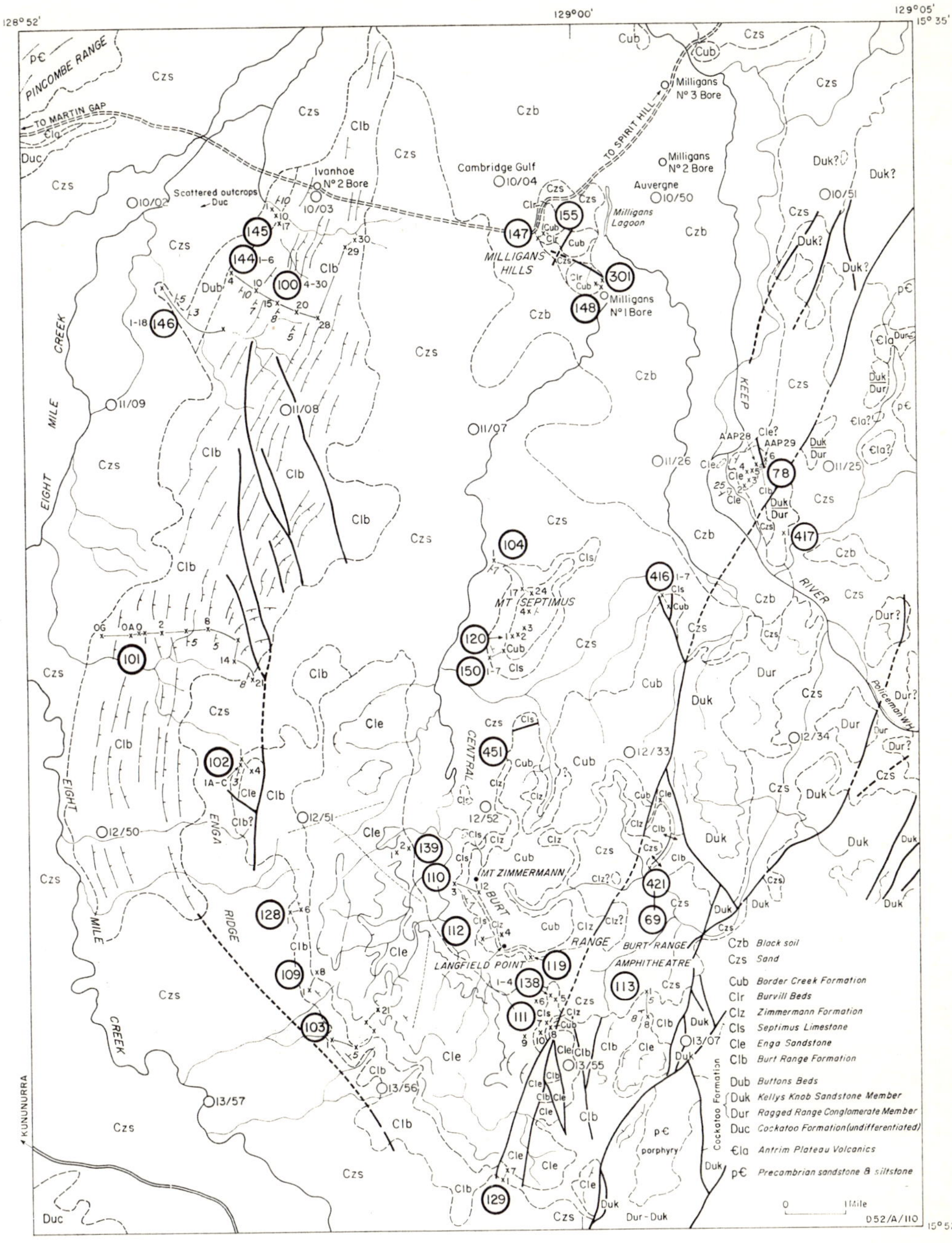

Figure 10. Geological map of the Burt Range area.

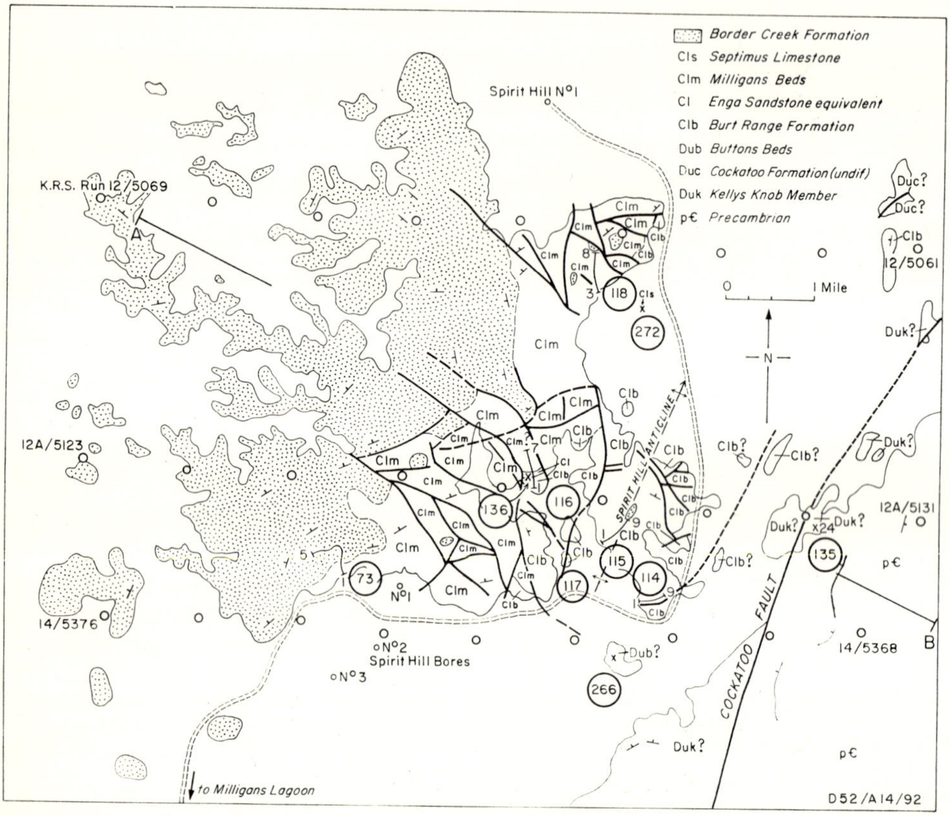

Figure 11. Geological map of the Spirit Hill area.

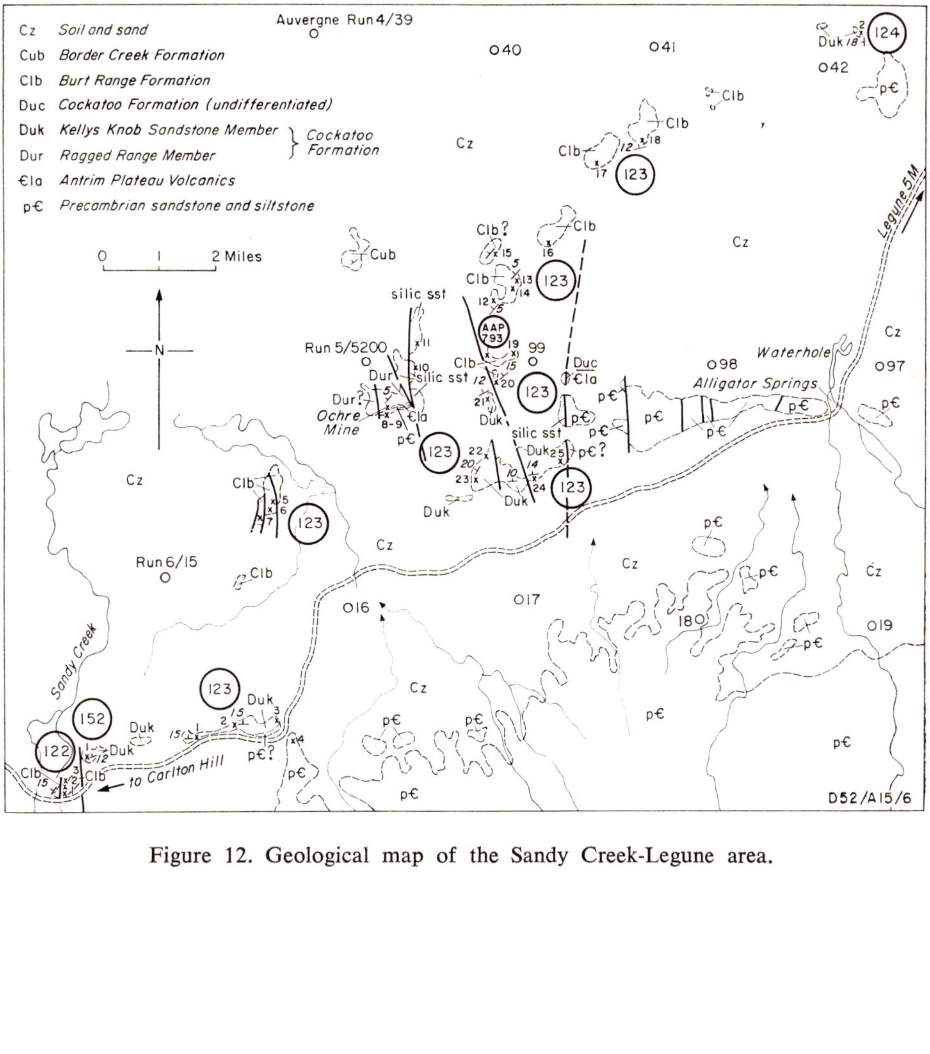

Figure 12. Geological map of the Sandy Creek-Legune area.

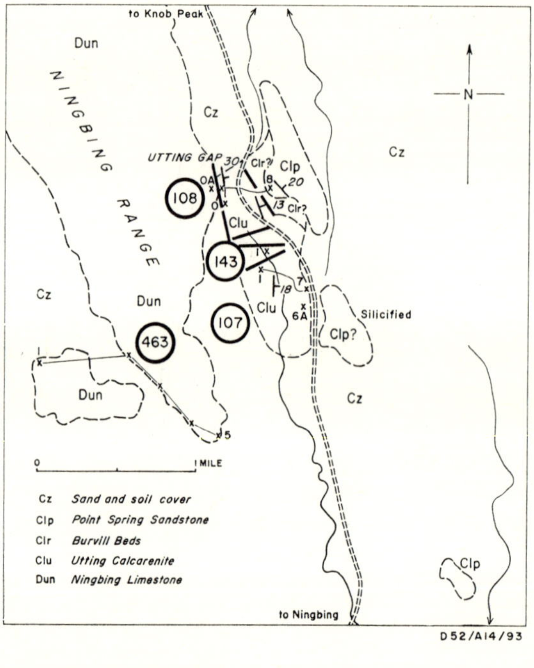

Figure 13. Geological map of the Utting Gap area.

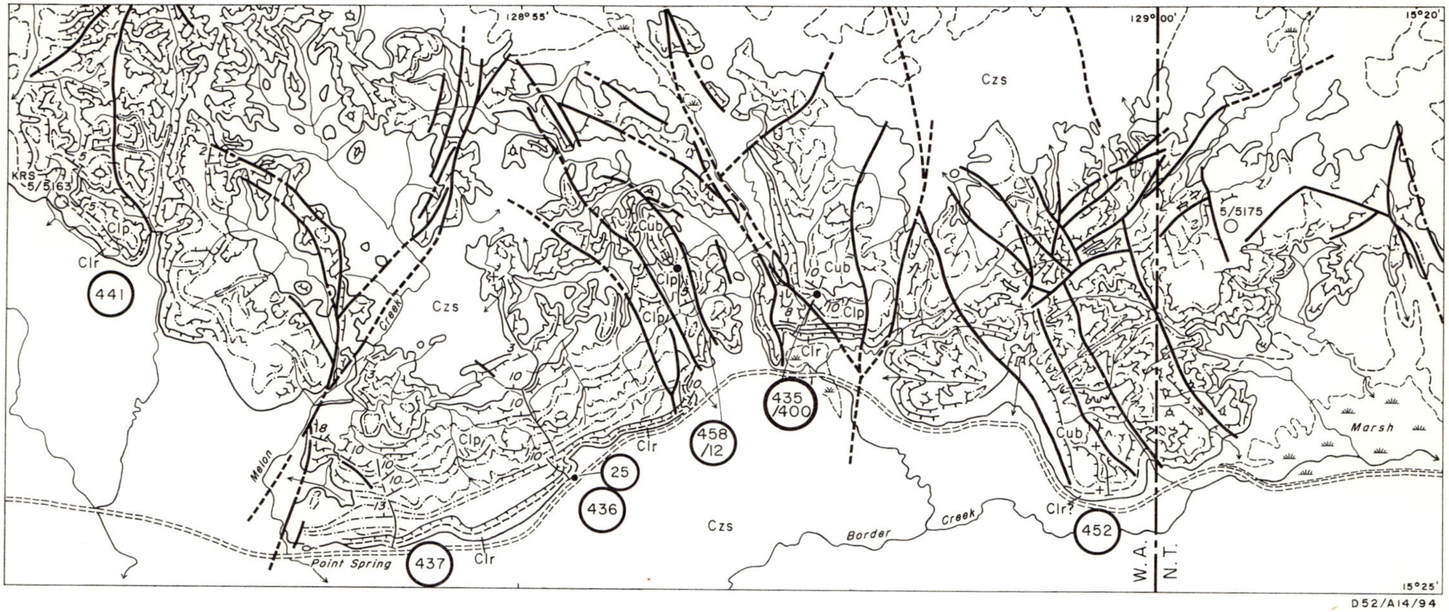

Czs *Sand*
Cub Border Creek Formation
Clp Point Spring Sandstone
Clr Burvill Beds

Note: Unlabelled outcrops either
Clp *or* Cub

Figure 14. Geological map of the eastern part of the Weaber Range.

V